Adaptive Evolution *of* Genes *and* Genomes

Adaptive Evolution *of* Genes *and* Genomes

AUSTIN L. HUGHES

New York Oxford

Oxford University Press

1999

Oxford University Press

Oxford New York
Athens Auckland Bangkok Bogotá Buenos Aires Calcutta
Cape Town Chennai Dar es Salaam Delhi Florence Hong Kong Istanbul
Karachi Kuala Lumpur Madrid Melbourne Mexico City Mumbai
Nairobi Paris São Paulo Singapore Taipei Tokyo Toronto Warsaw

and associated companies in
Berlin Ibadan

Published by Oxford University Press, Inc.
198 Madison Avenue, New York, New York 10016

Library of Congress Cataloging-in-Publication Data
Hughes, Austin L., 1949–
Adaptive evolution of genes and genomes / by Austin L. Hughes.
p. cm.
Includes bibliographical references and index.
ISBN 0-19-511626-7
1. Evolutionary genetics. 2. Proteins—Evolution.
3. Adaptation (Physiology) I. Title.
QH390.H84 1999
572.8'38—dc21 98-46584

9 8 7 6 5 4 3 2 1

Printed in the United States of America
on acid-free paper

For Austin Jr. and Helen

Trysor hynod yw gwybodeath—a roes
Drwy'r oesoedd gynhalieath
Gyson i feibion a faeth,
A'r olew i awr alaeth.

PREFACE

Biology is among the most fractious of sciences. Historically, as Ernst Mayr has often reminded us, there has been a rift between functional biologists interested in working out how organisms work and evolutionary biologists interested in explaining how organisms came to be the way they are. For the past half century or more, the majority of biologists have acknowledged that evolutionary theory provides, at least potentially, a unifying conceptual framework for all of biology. But in practice, evolution has had little day-to-day relevance for functional biologists.

The arrival of an era of abundant DNA sequence data has changed this situation to some extent. Functional biologists working at the DNA level now routinely use evolutionary reasoning, if only as a way of identifying potential functionally important sequences. At the same time, a new rift has opened up within evolutionary biology itself, a rift between those who work at the DNA sequence level and traditional evolutionary biologists working at the phenotypic level. At the phenotypic level, evolutionists have been accustomed to devising adaptive explanations for many, perhaps even most, of the observable traits of organisms. But at the DNA sequence level a very different type of evolution seems to predominate. Although initially Kimura's "neutral theory of molecular evolution" met fierce resistance from evolutionary biologists, by now virtually all have conceded that the bulk of evolution at the DNA level is probably selectively neutral. How to accommodate the existence of adaptive phenotypic traits with the predominance of neutral evolution at the DNA level poses a major challenge for contemporary biology.

For those interested in the evolution of adaptive traits, neutral molecular evolution is noise through which we must try to pick up the signal of adaptive evolution. The purpose of this book is to give examples of cases where it has been possible to detect such signals. I emphasize cases in which our knowledge of the biological function of proteins has led us to expect adaptive evolution to have occurred in certain gene regions and thus to predict certain patterns of nucleotide substitution. There are other ways of searching for evidence of adaptive evolution, in cases where such biological knowledge is lacking, but these are not my primary interest. Rather, I feel that using our knowledge of molecular function to make testable predictions regarding adaptive evolution is a particularly powerful strategy for testing the Darwinian hypothesis of natural selection. Indeed, it is my belief that such studies can play an important role in bridging the current conceptual rift between molecular and phenotypic evolutionary studies.

The research strategy I advocate here is reductionist in that it focuses on particular protein molecules and, indeed, on particular domains of these proteins. This is in contrast to approaches that involve surveying patterns of genetic variation at numerous loci—whether through protein electrophoresis or through examination of DNA polymorphism—and comparing the observed pattern with that expected under various models, such as the models of selective neutrality. I do not mean to denigrate survey methods, which can yield important insights. My own preference, however, is for cases where our previous biological knowledge enables us to make specific, testable predictions about the focus of adaptive evolution.

A number of simple statistical techniques are available for such analyses. In coding regions, one can compare the frequency of nucleotide substitutions that cause amino acid changes with that of those that leave the amino acid unchanged. Although it is only a little over a decade since Hill and Hastie first introduced this research strategy, it has already proved to be a remarkably fruitful one, and a major goal of the present book is to make this type of analysis known to a wider audience of both evolutionary and functional biologists.

The text is aimed at a general biological audience, including upper-level undergraduates in the biological sciences, graduate students, and research professionals. Because currently available texts in molecular evolution do not emphasize adaptive evolution, it can serve as supplementary reading for courses in molecular evolution. I assume that readers will have a diversity of backgrounds and will include both evolutionary biologists and molecular biologists. Readers with training in evolutionary biology may include both molecular evolutionists and more traditional evolutionary biologists working at the phenotypic level. Readers with a background in molecular biology, by contrast, may have had little formal exposure to evolutionary biology. Their interest in adaptive evolution may be driven by the desire to extract the maximum evolutionary information from molecular sequence data that they have obtained in their research. I have tried to structure the introductory material (Chapters 1 through 3) so that the remainder of the text will be understandable to readers with any of these backgrounds.

The statistical methods available for evolutionary analysis of molecular sequence data have been extensively treated elsewhere; here, I provide only an elementary introduction in one chapter (Chapter 2) to the methods particularly relevant to the study of adaptive evolution. No doubt this chapter will be elementary to advanced students of molecular evolution. However, I have included it in the hope that it will clarify to others—particularly those who are new to the field of molecular evolutionary genetics—some of the basic ideas behind statistical methods useful in detecting adaptive evolution. The bulk of the book is devoted to application of these methods to cases of biological interest.

I am grateful to the Pennsylvania State University for providing the facilities that permitted me to conduct many of the analyses described here. The National Institutes of Health (NIH) provided support for my research on the evolution of the immune system and the immune-evasion strategies of parasites, which has yielded some interesting examples of adaptive evolution at the molecular level. In particular, a Research Career Development Award from NIH provided me the freedom from other responsibilities that enabled me to complete the manuscript. I am grateful to Jack da Silva, Robert Friedman, and Federica Verra for detailed comments on an earlier version of the manuscript. Marianne Hughes has provided encouragement and support in all phases of this project.

University Park, Pennsylvania A. L. H.
July 1999

CONTENTS

Adaptive
Evolution
of Genes
and Genomes

1

What Is Adaptive Evolution?

Few modern biologists would question that the following is a reasonable hypothesis: most adaptive traits of organisms have evolved as a result of natural selection. But the precise meanings of terms like "adaptation," "evolution," and "natural selection" have often been controversial. Such controversies have been particularly intense since 1979, when Gould and Lewontin published a famous critique of what they called "the adaptationist programme." This paper seems to have precipitated something of a crisis in the study of adaptive evolution. In this chapter I analyze some of the probable reasons for this conceptual confusion. The basic aim of this book is to introduce a set of approaches that provide a way forward for the study of adaptive evolution, free from many encumbering misconceptions of the past.

I argue that the widespread availability of molecular sequence data heralds the beginning of a new era in the study of adaptive evolution and that many past confusions and controversies have now become moot. In the past, when biologists tried to reconstruct the evolution of traits whose genetic basis was unknown, they were forced into complex paths of reasoning, where there were many pitfalls. A failure to recognize these pitfalls led to the circular reasoning and kindred absurdities for which "adaptationists" were pilloried by Gould and Lewontin (1979). Here, I briefly review the historical development and current meaning of the concepts of *adaptation, natural selection,* and *evolution,* with the goal of showing how the availability of molecular data makes it possible to separate these concepts conceptually and study each in an appropriate fashion.

Adaptation

Adaptation is clearly an important concept in modern biology, but it is one that is rarely explicitly defined. Some landmark attempts at definition—by no means all in agreement with one another—are those of Muller (1949), Medawar (1951),

Dobzhansky (1956), Williams (1966), Stern (1970), Brandon (1978), Bock (1980), and Mayr (1982a). Mayr summarizes the situation as follows:

> The ambiguous meaning of the term adaptation is particularly disturbing. For instance, it is used both for the process of becoming adapted and for the stage of having achieved adaptedness. It is, on the one side, used for the somatic or nongenetic adaptation of an individual, as for instance for the adaptation of a person to low oxygen pressure at high altitudes and, in the case of bacteria, to the presence or absence of certain nutrients in the substrate, or else it is used in a strictly genetic sense to denote the reconstruction of the genotype owing to continued selection pressure over many generations. Some authors (Bock, 1980) would restrict the term adaptation to single components of the phenotype, while most others speak of the adaptation of an individual in an overall sense, considering the term as synonymous with fitness. (1982a, pp. 161–162)

Bypassing for the time being the problems arising from multiple meanings of the same word, we can start with a commonsense definition of adaptation that is independent of genetics or evolutionary biology. At the most basic level, the concept of adaptation is closely related to that of function. Some trait of an organism is said to be adaptive if it performs a function that is in some way beneficial to the organism. This trait could exist at any level from the biochemical (e.g., the production of some enzyme like alcohol dehydrogenase) to the morphological (e.g., the opposable thumb of primates) or behavioral (e.g., the tendency to flee when a potential predator approaches).

Viewed in this way, the use of the term "adaptation" implies no theory about how adaptive traits have originated but only summarizes an observation about the beneficial consequences the trait has for the organism. And in this naive sense, there is little doubt that human beings have known about adaptation for many centuries, and probably for many millennia. One defect, of course, in the naive concept of adaptation is that it leaves unspecified exactly in what sense the adaptive trait is beneficial to the organism possessing it. But, being animals themselves, humans have long been able to empathize with their fellow animals to the extent of appreciating the benefit of any trait that helped its possessor find food or protection from enemies or harsh weather.

In the prescientific era, people seem to have seldom felt any need to explain the origin of adaptive traits. If pressed for an explanation, many, no doubt, would have tended to fall back on ideas of religious origin and attribute the marvels of biological adaptation to divine providence. This sort of piety found a more formal philosophical expression in the school of thought known as "natural theology," which reached its peak in the eighteenth century (Mayr 1982b).

Arguably, natural theology was as bad theology as it was science. The English Fabian socialist writer Robert Blatchford once received a letter from a reader containing the following comments:

> There is a little animal called an aye-aye. This animal has two hands. Each hand has five fingers. The peculiar thing about these hands is that the middle finger is elongated a great deal . . . to enable it to scoop a special sort of insect out of special cracks in the special tree it frequents. . . . In this, as in scores of other instances is shown the infinite goodness of God.

To this Blatchford replied:

> The infinite goodness of God to whom? To the animal whose special finger enables him to catch the insect? Then what about the insect? Where does he come in? Does not the long finger of the animal show the infinite *badness* of God to the insect? (quoted in Mortimer 1972), p. 201).

The novelty of the theory of natural selection proposed by Darwin and Wallace lay in the fact that it provided an intellectually satisfying, thoroughly materialistic hypothesis to explain the existence of adaptations. Despite the fact that Darwin entitled his great work *The Origin of Species*, he actually was able to provide little insight into the process of speciation but a great deal of insight into the origin of adaptations. As noted by Brandon (1978, p. 181), "[What] Darwin did was to offer a radically new type of explanation of adaptations and in so doing he altered the conception."

The basic idea of natural selection, of course, is that the evolution of a new adaptive trait can be explained if we assume (1) some process by which a new trait can appear in a population; (2) some biological mechanism by which the trait is passed from one generation to the next; (3) a "struggle for existence" arising from the fact that the population's reproductive capacity greatly exceeds the number its environment can support; and (4) a benefit in terms of survival and/or reproduction in this environment conferred by the trait in question. This process was described in Mendelian terms by the architects of the Neo-Darwinian synthesis that gave rise to modern evolutionary biology in the 1920s and 1930s. In the case of simple genic selection, an adaptive trait first appears in a population as the phenotypic effect of a new mutant allele. The effect of the allele on the phenotype is such that an individual will, on average, leave more offspring in the next generation than one lacking it. Thus, over a number of generations, the frequency of the mutant allele will increase. The process can be complicated when dominance effects are taken into account or when the phenotypic trait of interest is polygenic. But in any case, "evolution" can be defined as a change in gene frequencies; and natural selection as the differential contribution of different genotypes to the next generation. As Dobzhansky (1956, p. 340) memorably summarized the process: "Selection favors genotypes the carriers of which transmit their genes to succeeding generations more efficiently than do the carriers of other genotypes."

As a corollary to the hypothesis of natural selection, we now have a more precise notion of the way in which an adaptive trait is expected to benefit its bearer: namely, it benefits the individual in terms of representation of its genes in the next generation, which is what the term *fitness* means in a Darwinian context. The transmission of the individual's genes may be either through offspring or through the offspring of close, nondescendant relatives (Hamilton 1963, 1964). To emphasize that nondescendant relatives may also contribute to fitness, particularly in the case of social animals, the term *inclusive fitness* is sometimes used. Traits conferring benefits to the individual organism (in terms of longevity, for example) that are not translated into inclusive fitness are not predicted to arise as a result of natural selection. Thus, the Darwinian hypothesis provides an answer to the question sidestepped by the earlier, naive understanding of adaptations beneficial to their possessors: "Beneficial in what way?" In what follows, I will speak of a trait as adaptive "in the Darwinian sense" if I need to distinguish the Darwinian from the naive understanding of adaptation.

It is important to realize, as pointed out by Gould and Lewontin (1979), that the hypothesis of natural selection does not imply that all traits that are adaptive in the Darwinian sense have arisen as a result of natural selection. At least in theory, it is possible to imagine counterexamples. Consider a case in which a phenotypic trait arises in a species as the result of chance fixation of a selectively neutral allele ("genetic drift"). Later, the species' environment changes, and in the new environment the previously neutral trait turns out to be adaptive.

Likewise, in some cases, an adaptive phenotype can arise as the result of some short-term phenotypic adjustment occurring in the life of an individual in response to its environment, rather than arising as a result of gene frequency change occurring over evolutionary time. The ability to make such adjustments is sometimes referred to as "phenotypic plasticity." In the passage quoted here, Mayr refers to such "adaptation" with the examples of a person's response to high altitude or a bacterium's response to changes in its nutrient supply.

In the case of humans, of course, the most dramatic form of phenotypically plastic adaptation occurs in the phenomenon of culture. There is evidence that many human cultural traits are adaptive in the Darwinian sense, in that they yield greater fitness than would alternative behaviors in response to various aspects of the physical and biotic environment (Forde 1934) or of the social environment (Alexander 1979; Hughes 1988). However, such cultural differences are certainly not caused by any genetic difference between two populations adopting different cultural strategies but rather by the great behavioral plasticity of humans that allows them to assess their environment and, at least in many cases, design an adaptive response (Hughes 1988).

In many cases of phenotypic plasticity, it is a reasonable hypothesis that the underlying mechanism making adjustment to the environment itself has a genetic basis and has arisen as a result of natural selection. This remains merely a hypothesis, however, in most cases. The genetic basis of the inducible or repressible operon systems of bacteria, which lead to expression of metabolic enzymes in response to the presence of their substrates in the environment, is, of course, well understood; but the evolutionary origin of such systems remains to be elucidated. The genetic basis of phenotypic plasticity in multicellular organisms is rarely understood, but it has often been proposed that such capacities have a genetic basis. It has even been speculated that human beings have an innate "capacity for culture" that has arisen as a result of natural selection (Cavalli-Sforza and Feldman 1981; Lumsden and Wilson 1981).

However, it is also true that a species is sometimes able to adjust phenotypically to an extreme or unusual environment that surely played no role in the species' evolutionary history. Consider the ability humans have to adjust (with the help of an oxygen tank) to the extraordinary water pressures encountered in deep-sea diving. The ability to survive such conditions is surely not the result of any innate mechanism but simply a fortuitous consequence of various aspects of human physiology. As emphasized by Gould and Lewontin (1979), it is important to realize that adaptive traits may, as a result of developmental constraints, have consequences that are not themselves in any meaningful sense adaptive.

It is often possible to demonstrate empirically that a given trait is adaptive in the Darwinian sense. Suppose that a phenotypic trait x is found in most members of species A. However, suppose that some members of species A fail to express x for some reason, whether genetic or nongenetic. If we can show that those possessing trait x have greater fitness than those lacking x, this supports the hypothesis that x is adaptive. If natural variation with respect to the expression of x is not found, we may be able to prevent expression of x in certain individuals by some experimental manipulation. Again, if we show that individuals lacking x have reduced fitness, that is evidence that x is adaptive.

In practice, there are no doubt cases where it is not really necessary to conduct such an experiment. For example, it hardly seems necessary to pinion a cohort of phoebes to demonstrate that the ability to fly is adaptive for these birds in enabling them to capture insects on the wing. For a larva of the fruitfly genus *Drosophila*, living in an environment consisting of rotting fruit, the production of an enzyme that detoxifies alcohol (alcohol dehydrogenase) seems obviously beneficial. Molecular biologists could no doubt create "knockout" flies lacking expression of the alcohol dehydrogenase gene; but it seems safe to predict that their survival would be poor in this species' natural environment. Where the adaptive advantage conferred by a trait is obvious, it may be sufficient to conduct a "thought experiment" regarding the disadvantages an individual lacking the trait is likely to encounter.

In contrast, the adaptive aspects of some traits, such as features of protein structure, are generally uncertain without experimental evidence. For example, even when the three-dimensional structure of an enzyme is known, it may still not be obvious which amino acid residues are critical for the enzyme's function. In cases like this, site-directed mutagenesis and other techniques of "protein engineering" may be useful in identifying adaptive features (Golding and Dean 1998).

Often biologists have emphasized the importance of the "comparative method" in the study of adaptation, but comparative studies can be risky if they are merely correlative and not experimental. Suppose there is one species that has adapted to an environment different from the environments occupied by related species. Suppose, also, that this species has a particular trait that the related species lack. It might be tempting to infer that this unique trait is an adaptation to the species' unique environment. Actually, however, the species' unique trait might have no effect on its fitness but rather has arisen by chance fixation of a selectively neutral mutation. Experimental data linking the trait to its supposed benefit can help resolve the issue.

There are traits, however, that are resistant to experimental manipulation; for example, general properties of the genome. The genomes of birds and mammals have higher proportions of the bases G and C (as opposed to A and

T) in their DNA than do those of fishes, reptiles, and amphibians (Bernardi 1993). It has been proposed that, because G-C base pairs are more heat stable than A-T base pairs, the higher GC content of bird and mammal genomes is an adaptation to the fact that these vertebrates are endothermic ("warm blooded") (Bernardi 1993); that is, able to maintain a relatively constant high body temperature by internal heat production (Schmidt-Nielsen 1979). Although this hypothesis may be true, it has been very difficult to test. Comparisons with other species turn up observations that are somewhat inconsistent with this theory. For example, average body temperature of a desert lizard may actually be higher than that of a bird or mammal, while some fishes such as tuna are, to all intents and purposes, endothermic (Schmidt-Nielsen 1979). Yet, as far as we know, the genomes of these species do not have high GC contents. In addition, as noted by Bernardi (1993), some bacteria living in very high-temperature environments have AT-rich genomes.

Yet these counterexamples do not really disprove Bernardi's hypothesis. There may be other factors, including their phylogenetic history, that have prevented other species with high body temperatures from adopting the GC-rich genome characteristic of birds and mammals. The fact that one species living in a particular environment has not evolved some specific adaptation to that environment does not prevent other species, with a different genetic heritage, from evolving that adaptation. On the other hand, such inconsistencies in the comparative data make it very difficult, if not impossible, to prove Bernardi's hypothesis at present. To genetically engineer an AT-rich bird or mammal remains beyond our capabilities. (For further discussion of this hypothesis, see Chapter 8, Codon Usage in Multicellular Organisms).

In any event, because adaptive traits may arise by processes other than evolution as a result of natural selection, showing that a trait is adaptive is not equivalent to showing that it has evolved by natural selection. This point has not always been appreciated by evolutionary biologists. Indeed, a recent series of articles by Orzack and Sober (1994a,b) explicitly assert that a way to test the hypothesis that "natural selection is a sufficient explanation of the evolution" of a given trait is to conduct a test of the trait's adaptiveness.

Definitions of "adaptation" that implied a process of natural selection contributed to such confusion. For example, Mayr stated: "Whenever the words adaptive or adaptation occur in the following discussion, they are used in a descriptive sense to indicate the result of a selective process" (1942, p. 85). If we adopt such a definition, the statement, "most adaptations of organisms have arisen as a result of natural selection," becomes a tautology rather than, as it was in Darwin's time, a bold hypothesis to explain the origin of adaptations. The hypothesis of natural selection can be tested only when we clearly distinguish the following: (1) evolution (the observable phenomenon of change in gene frequency over time); (2) natural selection (the mechanism by which evolution is hypothesized to have taken place in certain cases); and (3) adaptation (the observable phenotypic result, which may in some cases be explained as a result of evolution by natural selection). If cause (natural selection) and effect (adaptation) are confounded, evolutionary biology loses scientific rigor.

Before concluding this section, I should like to comment briefly on some of the issues of terminology raised by Mayr (1982a) in the passage quoted here. First, Mayr alludes to a distinction, which is due to Dobzhansky (1968a,b), between *adaptedness* (the degree to which an organism is able to live and reproduce in a series of environments) and *adaptation* (the process of becoming adapted or more adapted). Certainly, this is a valid distinction, but I see no reason why the word "adaptation" cannot cover both of these concepts. It is commonplace in English for abstract nouns to refer to both a state of being and the process of attaining that state (e.g., "education" or "intoxication").

Mayr (1982a) also refers to the use of "adaptedness" and "adaptation" as a property of an organism as a whole rather than of particular phenotypic trait. However, I think it is preferable to avoid this usage. As Mayr notes, if we view adaptation as the property of an individual, it becomes equivalent to fitness. This either makes one of the terms redundant or it confuses adaptation in the Darwinian sense with the way that it is measured. Some authors (e.g., Dobzhansky 1968a,b; Dunbar 1982; Endler 1986) have struggled to establish a meaningful distinction between the adaptedness of an individual and its fitness, but I find their efforts unconvincing. In any event, if we restrict the application of adaptedness and adaptation to traits of individuals, the need for such a distinction is removed.

The Adaptationist's Dilemma

The triumph of the Neo-Darwinian synthesis in the 1920s and 1930s ushered in a golden age of comparative biochemistry and physiology, when the mechanistic basis of numerous traits of organisms—from the citric acid cycle to the echolocation of bats—was worked out by elegant experiments. It seems likely that biologists were inspired in these endeavors by the confidence that many, if not most, traits of organisms were adaptive, and that their adaptive significance could be understood through physical and chemical dissection of their functioning.

However, many traits of organisms—particularly behavioral traits of animals—did not yield themselves so easily to such dissection. Given the widespread assumption that virtually all traits of organisms must be adaptive, biologists used their ingenuity to formulate conjectural accounts of the adaptive significance of traits whose function was obscure. Such speculation might sometimes give rise to fruitful, testable hypotheses (Williams 1985), but in some cases the mere existence of a plausible account of possible adaptive significance was taken as proof that the trait was adaptive (and that it had arisen as a result of natural selection).

A common type of research program in evolutionary biology at this time worked roughly as follows. The researcher would make predictions about how, for example, some animal "should" behave if its behavior evolved as a result of natural selection. The logic behind predictions of this sort is that natural selection would only favor certain types of behaviors (namely, those that are "selfish" in a Darwinian sense—that is, from the point of view of maximizing the individual's inclusive fitness) and not others. Then the researcher would examine the animal's behavior, observe that it had indeed behaved as expected, and congratulate himself or herself on having tested and verified evolutionary theory. This approach has been extended beyond behavioral characteristics to all sorts of other traits, including aspects of genome structure. But, in fact, this approach provided a very weak test of the theory of natural selection because the researcher's hypothesis was never in danger of being falsified. If the prediction failed to be supported by observation, there was no null hypothesis to fall back on, as is true in ordinary scientific research programs. Furthermore, it is a legitimate question what was learned from such research. How much have we advanced human knowledge if we demonstrate that we have the ability to predict how a certain animal will behave?

It was this approach to evolutionary biology that was branded "adaptive storytelling" by Gould and Lewontin (1979) in their colorful polemic against what they called the "adaptationist programme." This research program had, they claimed, "dominated evolutionary thought in England and the United States for the past 40 years." Gould and Lewontin emphasized the possibility that nonadaptive traits might exist, and the fact that adaptation does not in itself demonstrate the action of natural selection. They pointed out that certain traits for which biologists had felt compelled to devise adaptive explanations might actually be merely by-products of developmental processes.

In serving to discredit untested speculation, Gould and Lewontin's critique undoubtedly had a salutary effect on evolutionary biology. But it must be admitted that those authors completely failed to provide any satisfactory positive alternative to the adaptationist program that they rejected. The best they could come up with was to urge a holistic appreciation for the *Bauplan* or "body plan," according to which an organism is constructed. From our perspective at the present, when evolutionary biology has fully entered into the era of molecular biology, this seems like a retreat into mysticism and an abandonment of the reductionist approach that has characterized all successful scientific advances.

Gould and Lewontin criticized evolutionary biologists for breaking down organisms into a series of phenotypic "traits," the adaptive significance of each of which was studied separately. However, from my point of view, the problem was not studying traits separately—after all, in any scientific research program one has to focus on a system that is tractable to study—but the fact that in the past the genetic basis of few if any such traits was known. Gould and Lewontin advance pleiotropy (a case where one genetic locus influences many phenotypic characters) as a plausible alternative to the hypothesis that each of a series of traits is adaptive in itself. But this hypothesis itself cannot be tested unless the genetic basis of the trait is understood in detail. For example, Gould and Lewontin (1979) remark that the remarkable reduction of the forelimbs in dinosaurs of the genus *Tyrannosaurus*, rather than being an adaptation, may have been "a developmental correlate of allometric fields for relative increase in head and hindlimb size." That may be true, but in what sense does Gould and Lewontin's exercise in nonadaptive storytelling represent an improvement over the numerous adaptive stories that biologists have told over the years to account for forelimb reduction in *Tyrannosaurus?*

Gould and Lewontin (1979) claimed that their nonadaptive hypothesis about *Tyrannosaurus* "can be tested by conventional allometric methods." These methods involve plotting logarithms of size measures of various body parts against one another. At best, they can be used to establish correlation, not causation. To state, as do Gould and Lewontin (1979), that there are "laws" of allometry is to overstate the case. These methods may lead to empirical generalizations, but the underlying causes for these generalizations (if they are not simply statistical artifacts) lie at the genetic level and will not be exposed by any pencil-and-paper exercise.

Another paper published in the late 1970s offered a far more realistic hope of resolving the dilemmas posed by Gould and Lewontin, although few evolutionary biologists may have realized it at the time. This was Maxam and Gilbert's (1977) announcement of a rapid method of DNA sequencing, which has been followed by numerous technological advances in the isolation and sequencing of DNA. Evolution is nothing more or less than change of nucleotide sequences over time; thus, the availability of DNA sequence data makes it possible to study evolution much more directly than was ever possible before. In many cases, as we will see, DNA sequence data can be used to study both adaptive and nonadaptive evolution—and to distinguish between them. However, questions about the evolution of long-extinct organisms like *Tyrannosaurus* probably will never be addressed at the DNA level, given the absence of surviving material. Thus, such questions will probably never be answered.

A further important development, in addition to the advent of DNA sequencing, was the development by Motoo Kimura and others of the neutral theory of molecular evolution. Briefly, this theory states that, at the molecular level, most genetic polymorphisms are selectively neutral, and that, over evolutionary time, most changes occur as a result of random fixation of neutral variation (genetic drift) (Kimura 1968, 1983). Although Kimura had worked on the population genetics of selectively neutral traits since the 1950s and had first proposed the neutral theory in 1969, the impact of his theory was greatest after molecular sequence data became available in the 1980s. Evolutionary biology in the 1970s was dominated by often acrimonious but sterile debates between "selectionists" and "neutralists." But the availability of abundant sequence data largely resolved these debates by providing a substantial body of evidence in support of the neutral theory (Kimura 1983; Nei 1987).

The triumph of the neutral theory also provided a golden opportunity for the study of adaptive evolution because for the first time evolutionary biologists had a clear-cut alternative to the hypothesis of adaptive evolution. Many of the absurdities criticized by Gould and Lewontin had arisen because of the lack of a clearly stated null hypothesis in evolutionary studies. Indeed, sometimes adaptive evolution by natural selection was itself treated as a null hypothesis—which would imply that Darwinists had nothing left to prove.

In 1951, when the Neo-Darwinian synthesis was in its most triumphant phase, Cain published a short paper in *Nature* in which he argued that nonadaptive or selectivity neutral characters were unlikely to occur in organisms. According to Cain, in cases where an investigator had been unable to find evidence that a given trait was adaptively beneficial, he would often conclude that the trait was neutral—for lack of a better explanation, it would seem. Instead, said Cain, all that a researcher could really conclude was that he himself had been unable to find evidence of the trait's adaptive significance. This amounted to an attempt to treat adaptiveness of traits as the null hypothesis in evolutionary studies. This was an extraordinary proposal—surely unique in the history of science. For it is the usual practice in science that the null hypothesis is the hypothesis of no effect, which is accepted unless we are able to show that the data deviate significantly from what would be expected under the null hypothesis. By arguments like that of Cain, evolutionists were put in the impossible position of being required to prove the hypothesis of no effect—that is, of selective neutrality.

It is not surprising that those educated in such a baroque intellectual climate found Kimura's neutral theory puzzling and even infuriating. Yet the lack of a null hypothesis had a demonstrably detrimental effect on attempts to study adaptive evolution. Mitton (1997) describes studies of protein polymorphisms of marine invertebrates from tropical, temperate, and polar environments. The selectionist prediction was that heterozygosity should increase as one moves away from the tropics, because environmental variability is least in the tropics and greatest in the polar regions, and because environmental variability was thought to maintain genetic polymorphisms. In fact, exactly the opposite pattern was found, with heterozygosity being greatest in the tropics (Ayala and Valentine 1979). The obvious neutralist explanation for this is that tropical species generally have larger effective population sizes because their environments have been stable for a long time, whereas temperate and polar species have lower effective population sizes because of past extinction and recolonization episodes due to glaciation. The amount of neutral polymorphism that can be maintained in a species is a function of its effective population size (see Chapter 3, The Dynamics of Neutral Evolution). However, the neutralist explanation was not given serious consideration by these researchers. Rather, according to Mitton (1997), it was proposed that polar environments "forced species into a strategy of generalism," which was somehow assumed to entail homozygosity.

As in this example, if the data were found to contradict one selectionist hypothesis, the tendency was merely to concoct another "adaptive story" (Gould and Lewontin 1979). There was no null hypothesis available, because the obvious null hypothesis—that of selective neutrality—had been rejected a priori. Needless to say, such a stance made it difficult for selectionists to develop a conventional scientific research program.

From our present perspective, we can see that the neutral theory was tremendously liberating for the study of adaptive evolution because it provided a null hypothesis. Students of adaptive evolution were freed from an insoluble dilemma that they had imposed upon themselves. Indeed, it is not too much to say that, even if the neutral theory of molecular evolution turns out to be false as a generalization about the process of evolution at the molecular level, it will still be necessary that this theory be retained as a null hypothesis for evolutionary studies. For it really is the only conceivable null hypothesis for a Darwinian research program.

Natural Selection and Evolution

In his 1986 book *Natural Selection in the Wild*, Endler reviews empirical studies of natural selection in natural populations in the pre-DNA era. It is clear that many of the studies that he reviewed demonstrated natural selection, but they did not actually show any evidence of evolution. Natural selection is a phenomenon that occurs initially at the phenotypic level, and anyone who has spent time observing nature is likely to have observed natural selection at work at this level. For example, slower or weaker animals are caught by predators more easily than are fast and healthy conspecifics. But it is much harder to demonstrate that these observable differences have a genetic basis, and that selection acting on them thus gives rise to an evolutionary change.

Consider the widely cited study of Grant and Grant (1989a) on the Large Cactus Finch *Geospiza conirostris* on Isla Genovesa in the Galapagos islands. After an excessively wet year followed by two dry years, food availability on the island change in such a way that it placed longer-billed members of this species at a competitive disadvantage in obtaining food in the dry season. As a result, longer-billed individuals showed disproportionately high mortality. However, the authors had no direct evidence that bill length is under genetic control. They used methods of quantitative genetics (correlation of offspring values with midparent values) to estimate heritabilities of bill size and other metric traits, but because their calculations were based on field observations in the absence of controls, they could not rule out the possibility that these correlations resulted merely from a correlation of the rearing environments of parents and offspring (Grant and Grant 1989b). Even if we accept that bill-length variation in this species probably does have a genetic basis, this type of study tells us nothing at all about how evolution of bill length might take place, because the actual genes are completely unknown. Therefore, to call theirs a study of the "evolutionary dynamics of a natural population," as do Grant and Grant (1989b), is something of an overstatement.

Of the studies reviewed by Endler (1986) that did involve actual genetic data, the data were usually allelic frequencies for enzyme loci determined by protein electrophoresis. The test for natural selection usually involved a comparison of the distribution of allelic frequencies with that expected under selective neutrality. Although such studies can be interesting—particularly when the authors know of some environmental selective factor that might explain a deviation from the expectation of neutrality—they have a rather shallow depth in time. The great power of DNA sequence data is that it can be used, at least in some cases, to reconstruct natural selection that took place many millions of years ago.

Ideally, in this type of study, we need to know the genetic basis underlying an adaptive phenotype. Although in the case of morphological phenotypes there are still very few cases where the genetic basis is well understood, there are thousands of examples of biochemical phenotypes whose genetic basis is known. By comparing the sequence encoding a given phenotype with that encoding an alternative phenotype (e.g., in another species), we can reconstruct, at least to some extent, the pattern of change that has occurred over evolutionary time. The observed pattern can then be compared with what would be expected in the absence of natural selection; that is, with the expectations of the neutral theory of molecular evolution.

The purpose of this book is to review examples of this type of study, including both those that I myself have been involved in and those from other labs. The emphasis is on studies that involve comparative analyses of DNA sequence data in the study of adaptive evolution. Studies involving analysis of gene frequency or heterozygosity data are not emphasized, both because these have been reviewed extensively elsewhere (e.g., Endler 1986; Nei 1987; Gillespie 1991; Mitton 1997), and because I wish to emphasize both the power and the novelty of sequence-based approaches in evolutionary studies.

In some cases, authors have surveyed heterogeneous sets of sequence data to test hypotheses such as the clock-like nature of molecular evolution (as summarized by Gillespie 1991), which may have implications for the role of natural selection. In this book, however, I am not interested in such broad surveys but in studies with a very specific focus. Ideally, in such a study the following elements are available: (1) a knowledge of the function of a specific group or family of related proteins, providing, in turn, evidence of the adaptive advantage conferred by expression of proteins of this type; and (2) availability of DNA sequences for a number of members of this group of proteins that differ functionally in some way. These might be alleles at a polymorphic locus like the major histocompatibility complex loci of vertebrates or the self-incompatibility loci of plants (Chapters 4 and 5). They might be alleles at the same locus in a population exposed to an environmental factor that favors one allele (Chapter 6). Or, they might be members of a multigene family (Chapter 7). Given such a data set, it may be possible to correlate functional differences at the protein level with specific patterns of change at the DNA level, and thereby reconstruct the evolutionary mechanism responsible for maintaining or fixing such differences.

In addition to such sequence-based studies, I consider genome-wide traits of organisms for which adaptive explanations have been proposed (Chapter 8). Finally (Chapter 9), I address the question of what molecular data, particularly data regarding natural selection at the molecular level, imply for evolutionary theory in general. "I find it challenging," Roger Milkman wrote in 1972 (p. 224), "to realize that Darwinian evolution has not been proven quantitatively." The purpose of the present book is to show how molecular data have at last made it possible for biologists to rise to this challenge.

2

Methods of Sequence Analysis

To study adaptive evolution at the molecular level, it is useful to examine how the pattern of nucleotide changes in protein-coding genes affects the phenotype—that is, the protein the gene encodes. We can think of two major ways that changes at the DNA level can change protein function adaptively: (1) by altering the pattern of expression of the gene (e.g., so that the protein is expressed at a different time in development or in a new cell type); (2) by altering the amino acid sequence of the protein so that aspects of its structure, and thus its potential interactions with other molecules, are altered. So far, there have been relatively few evolutionary studies of the former type in comparison with the latter type. In either type of study, it is necessary to use statistical methods for extracting evolutionary information from DNA sequence data.

These methods have been treated in detail by Nei (1987) and Li (1997), and Li and Graur (1991) provide an elementary introduction. Several computer programs with excellent manuals also are available and provide a basic introduction to widely used methods (e.g., Maddison and Maddison 1992; Kumar et al. 1993; Swofford 1993). Because of the ready availability of these and other summaries, I will only briefly discuss the general principles involved in analyses of this type, with emphasis on the methods useful for study of adaptive evolution.

Genetic Information

There are 20 amino acids that commonly form the building blocks of proteins (Table 2.1). Amino acids form linear chains (polypeptides) in which amino acid residues are linked together by peptide bonds. These, in turn, fold first in relatively simple patterns such as α helices and β pleated sheets and then in more complex ways. Furthermore, a mature protein may be made up of two or more polypeptide chains. A protein composed of one polypeptide chain is called a monomer; one composed of two chains is a dimer; and so forth. If the two

Table 2.1. The 20 commonly occurring amino acids and the DNA codons that encode them in the universal genetic code

	Amino acid	Three- and one-letter code	Codons
Nonpolar	Phenylalanine	Phe, F	TTT, TTC
	Leucine	Leu, L	TTA, TTG, CTT, CTC, CTA, CTG
	Isoleucine	Ile, I	ATT, ATC, ATA
	Methionine	Met, M	ATG
	Valine	Val, V	GTT, GTC, GTA, GTG
	Proline	Pro, P	CCT, CCC, CCA, CCG
	Alanine	Ala, A	GCT, GCC, GCA, GCG
	Tryptophan	Trp, W	TGG
Polar, uncharged	Serine	Ser, S	TCT, TCC, TCA, TCG, AGT, AGC
	Threonine	Thr, T	ACT, ACC, ACA, ACG
	Tyrosine	Tyr, Y	TAT, TAC
	Asparagine	Asn, N	AAT, AAC
	Glutamine	Gln, Q	CAA, CAG
	Cysteine	Cys, C	TGT, TGC
	Glycine	Gly, G	GGT, GGC, GGA, GGG
Positively charged[a]	Histidine	His, H	CAT, CAC
	Lysine	Lys, K	AAA, AAG
	Arginine	Arg, R	CGT, CGC, CGA, CGG, AGA, AGG
Negatively charged[a]	Aspartic acid	Asp, D	GAT, GAC
	Glutamic acid	Glu, E	GAA, GAG
None	STOP (termination)	Ter, *	TAA, TAG, TGA

a. At pH 6.0.

chains composing a dimer are identical, then the protein is a homodimer; if they are different, it is a heterodimer.

The structure and function of proteins are, to a large extent, determined by the chemical properties of their constituent amino acids, such as charge and polarity (Table 2.1). Miyata and colleagues (1979) produced a simple measure of the chemical distance among amino acids by taking into account their charge, polarity, and volume. Figure 2.1 illustrates these distance relationships among the amino acids in the form of a network in which similar amino acids cluster together.

The sequences of amino acids in related polypeptides is a source of information that can be used to understand evolutionary processes. Even more informative in many cases are the sequences of the nucleic acids DNA and RNA, which provide the information used by the cell to synthesize polypeptides.

The essential building blocks of DNA (deoxyribonucleic acid) are four different types of nucleotides, each of which consists of a phosphate group, a sugar (deoxyribose), and one of four bases: adenine (A), guanine (G), cytosine (C), and thymine (T). A and G share structural similarities, and are called

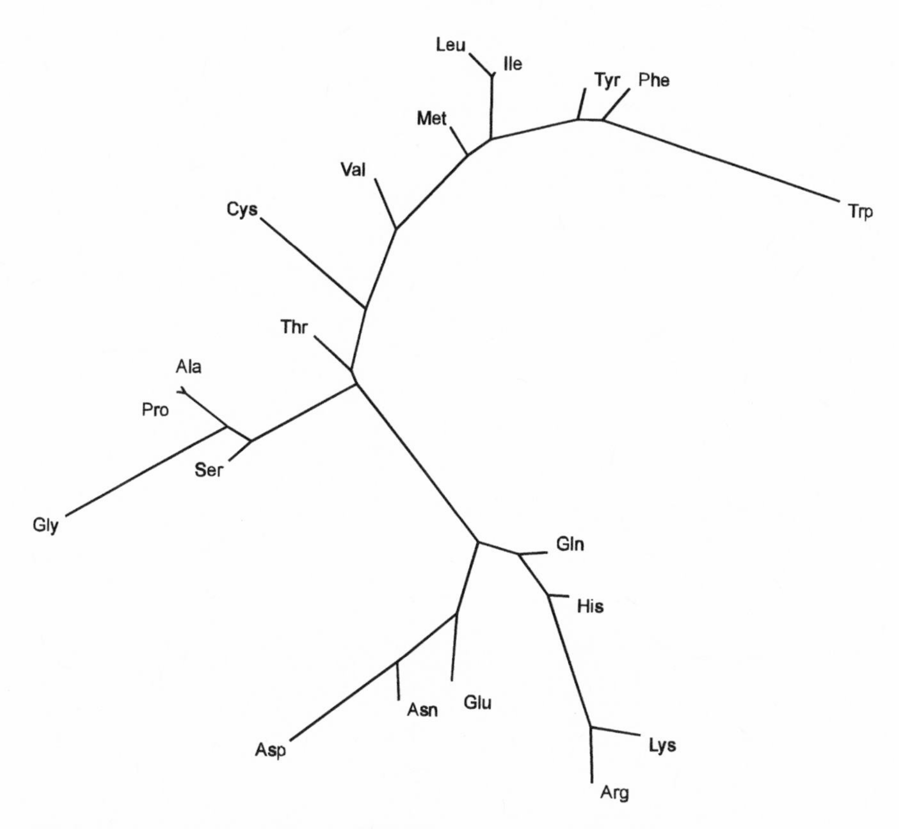

Figure 2.1. Network of amino acids based on chemical distance; the network was constructed by the neighbor-joining method of Saitou and Nei (1987).

purines; C and T resemble each other also, and are called pyrimidines. The DNA molecule is a double-stranded structure in which the bases in one strand pair with those in the other by hydrogen bonding according to the Watson-Crick rules that A pairs with T and C with G. Because the phosphate and sugar do not vary from one nucleotide to another, and because, by the rules of base pairing, the sequence of bases in one strand completely determine the bases in the other strand, a double-helical DNA molecule can be completely described as a linear sequence of the letters A, T, C, and G. The sugar and phosphate groups are linked in a way that has polarity: the phosphate links the 5′ carbon of the sugar group of one nucleotide to the 3′ carbon of the sugar group of the next nucleotide. Thus, a linear sequence of A, C, T, and G can be written in a way that indicates directionality. The convention in presenting sequences is to write them from left to right in a 5′ to 3′ direction (Figure 2.2).

RNA (ribonucleic acid) is a single-stranded molecule consisting of the same bases except that uracil (U) is substituted for T, and that the sugar is ribose. The genomes of most organisms consist of DNA, although a few viruses have RNA genomes. In cellular organisms RNA plays the role of an intermediary

```
ctgcagaaataactaggtactaagcccgtttgtgaaaagtggccaaacccataaatttggcaattacaataaaga
agctaaaattgtggtcaaactcacaaacattttattatatacattttagtagctgatgcttataaaagcaatat
ttaaatcgtaaacaacaaataaaataaaatttaaacgatgtgattaagagccaaaggtcctctagaaaaaggtat
ttaagcaacggaattcctttgtgttacattcttgaatgtcgctcgcagtgacattagcattccggtactgttggt
    M  E  D  A  K  N  I  K  K  G  P  A  P  F  Y  P  L  E  D  G  T  A  G  E
aaaATGGAAGACGCCAAAAACATAAAGAAAGGCCCGGCGCCATTCTATCCTCTAGAGGATGGAACCGCTGGAGAG
 Q  L  H  K  A  M  K  R  Y  A  L  V  P  G  T  I  A  F
CAACTGCATAAGGCTATGAAGAGATACGCCCTGGTTCCTGGAACAATTGCTTTTgtgagtatttctgtctgattt
                                      T  D  A  H  I  E  V  N  I  T  Y  A  E
ctttcgagttaacgaaatgttcttatgtttctttagACAGATGCACATATCGAGGTGAACATCACGTACGCGGAA
 Y  F  E  M  S  V  R  L  A  E  A  M  K  R  Y  G  L  N  T  N  H  R  I  V  V
TACTTCGAAATGTCCGTTCGGTTGGCAGAAGCTATGAAACGATATGGGCTGAATACAAATCACAGAATCGTCGTA
 C  S  E  N  S  L  Q  F  F  M  P  V  L  G  A  L  F  I  G  V  A  V  A  P  A
TGCAGTGAAAACTCTCTTCAATTCTTTATGCCGGTGTTGGGCGCGTTATTTATCGGAGTTGCAGTTGCGCCCGCG
 N  D  I  Y  N  E                                                     R  E
AACGACATTTATAATGAACgtaagcaccctcgccatcagaccaaagggaatgacgtatttaattttaagGTGAA
 L  L  N  S  M  N  I  S  Q  P  T  V  V  F  V  S  K  K  G  L  Q  K  I  L  N
TTGCTCAACAGTATGAACATTTCGCAGCCTACCGTAGTGTTTGTTTCCAAAAAGGGGTTGCAAAAAATTTTGAAC
 V  Q  K  K  L  P  I  I  Q  K  I  I  I  M  D  S  K  T  D  Y  Q  G  F  Q  S
GTGCAAAAAAAATTACCAATAATCCAGAAAATTATTATCATGGATTCTAAAACGGATTACCAGGGATTTCAGTCG
 M  Y  T  F  V  T  S  H  L  P  P  G  F  N  E  Y  D  F  V  P  E  S  F  D  R
ATGTACACGTTCGTCACATCTCATCTACCTCCCGGTTTTAATGAATACGATTTTGTACCAGAGTCCTTTGATCGT
 D  K  T  I  A  L  I  M  N  S  S  G  S  T  G  L  P  K  G  V  A  L  P  H  R
GACAAAACAATTGCACTGATAATGAATTCCTCTGGATCTACTGGGTTACCTAAGGGTGTGGCCCTTCCGCATAGA
 T  A  C  V  R  F  S  H  A  R
ACTGCCTGCGTCAGATTCTCGCATGCCAGgtatgtcgtataacaagagattaagtaatgttgctacacacattgt
    D  P  I  F  G  N  Q  I  I  P  D  T  A  I  L  S  V  V  P  F  H  H  G  F
agAGATCCTATTTTTGGCAATCAAATCATTCCGGATACTGCGATTTTAAGTGTTGTTCCATTCCATCACGGTTTT
 G  M  F  T  T  L  G  Y  L  I  C  G  F  R  V  V  L  M  Y  R  F  E  E  E  L
GGAATGTTTACTACACTCGGATATTTGATATGTGGATTTCGAGTCGTCTTAATGTATAGATTTGAAGAAGAGCTG
 F  L  R  S  L  Q  D  Y  K  I  Q  S  A  L  L  V  P  T  L  F  S  F  F  A  K
TTTTTACGATCCCTTCAGGATTACAAAATTCAAAGTGCGTTGCTAGTACCAACCCTATTTTCATTCTTCGCCAAA
 S  T  L  I  D  K  Y  D  L  S  N  L  H  E  I  A  S  G  G  A  P  L  S  K  E
AGCACTCTGATTGACAAATACGATTTATCTAATTTACACGAAATTGCTTCTGGGGGCGCACCTCTTTCGAAAGAA
 V  G  E  A  V  A  K  R                                                  F
GTCGGGGAAGCGGTTGCAAAACGgtgagttaagcgcattgctagtatttcaaggctctaaaacggcgcgtagCTT
  H  L  P  G  I  R  Q  G  Y  G  L  T  E  T  T  S  A  I  L  I  T  P  E  G  D
CCATCTTCCAGGGATACGACAAGGATATGGGCTCACTGAGACTACATCAGCTATTCTGATTACACCCGAGGGGGA
  D  K  P  G  A  V  G  K  V  V  P  F  F  E  A  K  V  V  D  L  D  T  G  K  T
TGATAAACCGGGCGCGGTCGGTAAAGTTGTTCCATTTTTTGAAGCGAAGGTTGTGGATCTGGATACCGGGAAAAC
  L  G  V  N  Q  R  G  E  L  C  V  R  G  P  M  I  M  S  G  Y  V  N  N  P  E
GCTGGGCGTTAATCAGAGAGGCGAATTATGTGTCAGAGGACCTATGATTATGTCCGGTTATGTAAACAATCCGGA
  A  T  N  A  L  I  D  K  D  G  W  L  H  S  G  D  I  A  Y  W  D  E  D  E  H
AGCGACCAACGCCTTGATTGACAAGGATGGATGGCTACATTCTGGAGACATAGCTTACTGGGACGAAGACGAACA
  F  F  I  V  D  R  L  K  S  L  I  K  Y  K  G  Y  Q
CTTCTTCATAGTTGACCGCTTGAAGTCTTTAATTAAATACAAAGGATATCAGgtaatgaagattttacatgcac
                      V  A  P  A  E  L  E  S  I  L  L  Q  H  P  N  I  F  D
acacgctacaatacctgtagGTGGCCCCCGCTGAATTGGAATCGATATTGTTACAACACCCCAACATCTTCGACG
A  G  V  A  G  L  P  D  D  D  A  G  E  L  P  A  A  V  V  V  L  E  H  G  K
CGGGCGTGGCAGGTCTTCCCGACGATGACGCCGGTGAACTTCCCGCCGCCGTTGTTGTTTTGGAGCACGGAAAGA
T  M  T  E  K  E  I  V  D  Y  V  A
CGATGACGGAAAAAGAGATCGTGGATTACGTCGCCAgtaaatgaattcgttttacgttactcgtactacaattct
         S  Q  V  T  T  A  K  K  L  R  G  G  V  V  F  V  D  E  V  P  K  G  L
tttcatagGTCAAGTAACAACCGCGAAAAAGTTGCGCGGAGGAGTTGTGTTTGTGGACGAAGTACCGAAAGGTCT
  T  G  K  L  D  A  R  K  I  R  E  I  L  I  K  A  K  K  G  G  K  S  K  L  *
TACCGGAAAACTCGACGCAAGAAAAATCAGAGAGATCCTCATAAAGGCCAAGAAGGGCGGAAAGTCCAAATTGTA
Aaatgtaactgtattcagcgatgacgaaattcttagctattgtaatattatatgcaaattgatgaatggtaattt
tgtaattgtgggtcactgtactattttaacgaataataaaatcaggtataggtaactaaaaa
```

Figure 2.2. An example of a eukaryotic gene, the luciferase gene of the firefly *Photinus pyralis*. In accordance with convention, the base sequence from 5′ to 3′ is written left to right. Lowercase letters indicate untranslated nucleotides, while uppercase letters indicate translated nucleotides in exons. The amino acid translation (single-letter code) is written above the nucleotide sequence.

(messenger RNA or mRNA) that carries the coding information found in DNA to the ribosomes where proteins are assembled.

In the process of transcription, the double-helical DNA molecule unravels and, by the rules of base pairing, an mRNA complementary to one strand of the DNA is formed. Thus, the sequence of the mRNA is the same as that in the other strand (the coding strand) of the DNA, except that U is substituted for T. In the case of eukaryotes, most protein-coding genes are encoded in pieces (exons), which are interrupted by noncoding segments (introns) (Figure 2.2). Introns are spliced out of the preliminary transcript to form a mature mRNA. The mRNA serves as a template for protein synthesis by means of a triplet ("three-letter") code in which each codon of three bases corresponds to an amino acid to be added to the polypeptide chain in the process of translation, or to a "stop" codon, which terminates the synthesis of the polypeptide. Each codon in the mRNA is recognized by an anticodon of a transfer RNA (tRNA) molecule, which carries the appropriate amino acid and adds it to the polypeptide chain. The anticodon recognizes the codon by base pairing, but this pairing needs only to be strictly in accord with the Watson-Crick rules for the first two bases of the codon. At the third base of the codon (the so-called wobble position) the pairing between codon and anticodon need not be strictly in accord with the Watson-Crick rules, although it is not completely unrestricted either (Osawa et al. 1992).

From the point of view of evolutionary studies, the redundancy of the genetic code is its most remarkable feature; that is, most amino acids are encoded by more than one type of codon (Table 2.1). Thus, mutations in coding regions are of two types: synonymous (which do not change the amino acid) and nonsynonymous (which do change the amino acid). Over evolutionary time, changes of these two types occur at very different rates, and comparing these rates can be used to provide information regarding the action of natural selection.

Most organisms use the same genetic code—the so-called universal genetic code, which is not actually universal (Table 2.1). In many eukaryotic organisms, the mitochondrial genome uses a somewhat modified genetic code in comparison to the nuclear genome. In addition, a few organisms have been found to have minor differences from the universal code in the case of their nuclear genes as well.

If we compare two DNA sequences that have a common ancestor, we will frequently observe that a number of differences have accumulated over evolutionary time (Figure 2.3). Some of these changes may involve insertions or deletions of one or more base pairs. The computer programs used to align DNA sequences typically infer that such events have occurred to maximize the alignment of nucleotides. Insertions and deletions together are classified as *indels*, because generally (in the absence of other information) we do not know whether an insertion in one sequence or a deletion in the other is responsible for the observed gap. Other changes involve the substitution of one nucleotide for another. The simplest statistic we can use to measure the amount of evolutionary change is the proportion of difference. We simply count the number of differences and divide that by the number of sites compared. We can compute the

```
δ1      ATTTACGGAGCTACCGGCCAGGGCTCCACTGCAGACACGGTTGTACCAGGTGCCG
δ2      .C--.......C.....----....G......G..........C.........T.

δ1      GGATCGCCGC ACACCCGAGC AAAACGTCGT CTGAG
δ2      .......... ....G..... ........C. .....
```

Figure 2.3. Example of two related sequences, showing nucleotide differences. The sequences are 5′ untranslated regions of duck δ1 and δ2 crystallin genes. "-" indicates a gap in the lower sequence (indel) postulated by the alignment; '.' indicates identity with the top sequence.

proportion of difference between two DNA sequences or between two amino acid sequences. In computing the proportion of difference the usual practice is to exclude sites at which the alignment postulates an indel event.

However, it is important to realize that the differences that can be observed between two related sequences may not represent all of the changes that have taken place since the two diverged from their common ancestor. Having only the two extant sequences to compare, we are usually not in a position to know what intermediate stages each has passed through in the time (possibly millions of years) since they diverged from their common ancestor. Changes that are no longer observable may have taken place in the past, including the following:

1. When the two sequences have changed in parallel at a particular site, both will now have the same nucleotide at that site. Thus, although two changes have taken place, we will observe no difference between the sequences, and thus assume that no change has taken place (Figure 2.4A).
2. When one sequence has changed at a given site but has subsequently changed back to the original nucleotide, again no difference between the two sequences will be observed, even though two changes have actually taken place (Figure 2.4B).
3. When two changes have occurred at one site in one sequence, we will know about only one of them, because we can observe only one difference between the extant sequences (Figure 2.4C).
4. If we do not know the ancestral sequence and have no reliable way of reconstructing it, we will count only one difference at a site where both sequences have changed from the ancestral sequence (Figure 2.4D).

A goal of molecular evolutionists has been to construct models that can be used to adjust statistically for such unknown (and usually unknowable) past events. (Such events are collectively referred to as "multiple hits," because they all involve more changes or "hits" than are actually observable.) This kind of statistical model takes the raw observed proportion of the difference between two sequences and, on the basis of certain assumptions about the pattern of nucleotide substitution, produces a corrected estimate of the number of nucleotide substitutions per site. Inevitably, this corrected estimate is higher than the raw proportion of difference, because the goal is to statistically account for

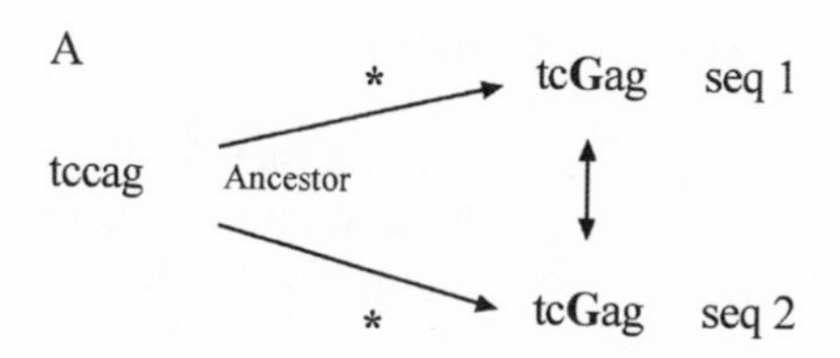

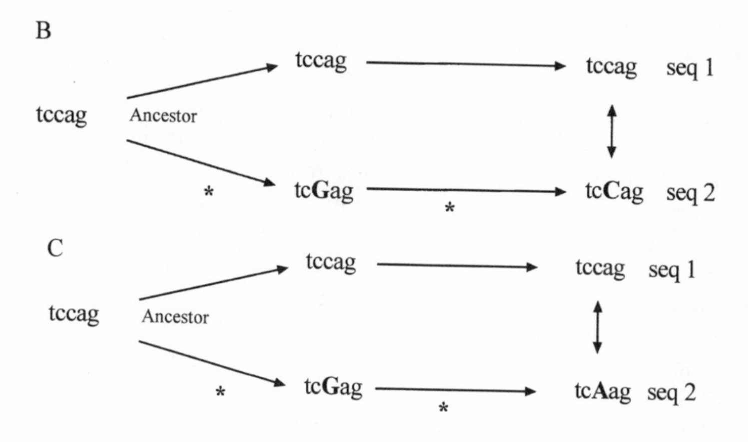

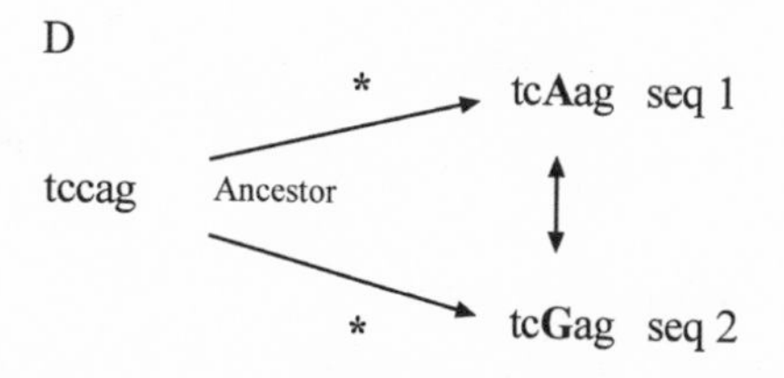

Figure 2.4. Examples of unobserved changes since the common ancestor of two nucleotide sequences (seq 1 and seq 2). (**A**) Both sequences have had a substitution of G for C at the third position, so that there is no difference between the two contemporary sequences. (**B**) Seq 2 undergoes two substitutions, one of which reverts to the ancestral nucleotide; again, no difference is observed between the two contemporary sequences. (**C**) Two substitutions have occurred in seq 2; but, because both substitutions have occurred at the same site, only one difference is observed between the contemporary sequences. (**D**) Both sequences have had substitutions at the same site, so that, again, only one difference is observed.

unobserved changes. The most complicated model would include 16 parameters—one for each of the possible types of nucleotide substitution. The simplest model would include just one parameter, on the assumption that each type of nucleotide substitution is equally likely. Models with intermediate numbers of parameters can also be constructed.

In general, the problem with a complicated model is that it will have a high error of estimation. Unless the number of nucleotides examined is extraordinarily large, the sequences compared will not be likely to contain enough information to estimate a large number of parameters with any degree of accuracy. For this reason, simple models are generally used in evolutionary studies.

The simplest of these, the one-parameter model, was first proposed by Jukes and Cantor (1969). As mentioned, this model assumes that all nucleotide substitutions are equally likely. A, for example, is assumed to be equally likely to be replaced by C, T, or G, and so forth. This assumption is probably never strictly true; but it is close enough to the truth in many cases that Jukes and Cantor's model can be used. Furthermore, if the number of nucleotide substitutions per site is low (fewer than about 0.25), the effect of biases in the nucleotide substitution pattern is often so slight that a more complicated model is unwarranted (Kumar et al. 1993).

Jukes and Cantor's model yields a straightforward expression for the number of nucleotide substitutions per site (d) as a function of the proportion of nucleotide difference (p):

$$d = -3/4 \ln (1 - 4/3p) \tag{2.1}$$

This estimate is frequently referred to in the literature as the "Jukes-Cantor distance." Its variance is given by

$$\mathrm{var}(d) = [9p(1 - p)]/(3 - 4p)^2 n \tag{2.2}$$

where n is the number of nucleotide sites compared (Nei 1987, p. 66).

Note that when $p = 3/4$, expression 2.1 becomes undefined because it would involve taking the logarithm of zero. It makes good sense for this expression to be undefined when $p = 3/4$. Assuming equal usage of all four nucleotides, two random sequences are expected to be identical at one-fourth of all positions (i.e., $p = 3/4$). Therefore, there is no evidence of an evolutionary relationship between two nucleotide sequences that are 75% different. As p approaches 3/4, the value of d gets very large, and its standard error gets very large. In these circumstances, the two sequences compared are said to be saturated with changes; and no meaningful estimation of the number of nucleotide substitutions per site can be made. However, in the case of protein-coding sequences, it is sometimes true that meaningful evolutionary information can be obtained from the comparison of amino acid sequences even when the DNA sequences are at or near saturation.

The most commonly observed deviation from Jukes and Cantor's model is a bias in favor of transitions. A transition is a mutation involving replacement of a purine by a purine or of a pyrimidine by a pyrimidine. A transversion, on the other hand, involves replacement of a purine by a pyrimidine or vice versa.

Mitochondrial genomes of animals often have been found to show pronounced transitional biases. For example, the control region of the mitochondrial DNA of humans shows a ratio of transitions to transversions that is greater than 9 : 1 (Kocher and Wilson 1991). For such cases, Kimura (1980) proposed a simple two-parameter model, with separate transitional and transversional rates. For details of the estimation procedure by this and other models, see Nei (1987) and Kumar et al. (1993).

Synonymous and Nonsynonymous Substitution

In protein-coding sequences, the most striking difference among nucleotide sites with respect to the rate of substitution is the difference between sites at which a mutation will change the amino acid and sites at which a mutation will not change the amino acid. This difference among sites is, of course, a consequence of the redundancy of the genetic code (Table 2.1). Sites in protein-coding sequences can be characterized as follows:

1. A *nondegenerate site* is one at which any possible mutation will change the amino acid. All second positions of codons are nondegenerate sites, as are most first positions (Table 2.1). In the universal genetic code, the third positions of methionine (ATG) and tryptophan (TGG) codons are also nondegenerate sites (Table 2.1).
2. A *fourfold degenerate site* is one at which any possible mutation will not change the amino acid. For example, GGT, GGC, GGA, and GGG all encode glycine (Table 2.1). Thus, whatever nucleotide is in the third position of a glycine codon (GG-), it still encodes glycine.
3. At a *threefold degenerate site*, any change among three of the four possible nucleotides will not change the amino acid encoded, while mutation to the fourth possible nucleotide does result in a different amino acid. In the universal genetic code, there is only one threefold degenerate site—the third position of an isoleucine codon. ATT, ATC, or ATA all encode isoleucine, but ATG encodes methionine (Table 2.1).
4. At a *twofold degenerate site*, mutation between two of the four possible nucleotides will not change the amino acid, but mutation to either of the other two nucleotides will change the amino acid. Several third positions and a few first positions are twofold degenerate sites. For example, both AAA and AAG encode lysine but AAT or AAC encode asparagine (Table 2.1).

For many types of evolutionary study, particularly for studies of adaptive evolution at the molecular level, it is useful to obtain separate estimates of the number of synonymous nucleotide substitutions per synonymous site (designated d_S) and of the number of nonsynonymous nucleotide substitutions per nonsynonymous site (designated d_N). A number of methods have been proposed to estimate these quantities. I will discuss a simple method proposed by Nei and Gojobori

(1986). The key to this method is the counting of synonymous and nonsynonymous sites.

In general, the genetic code affords fewer opportunities for synonymous changes than for nonsynonymous changes. Furthermore, the likelihood of either type of mutation is highly dependent on amino acid composition. For example, a protein containing a large number of leucines will contain many more opportunities for synonymous change than will a protein with a high number of lysines. This is true because the third position of four of the leucine codons (CT-) are fourfold degenerate sites, and the first positions of four of the codons (CTA, CTG, TTA, and TTG) are twofold degenerate sites (Table 2.1). Thus, at a typical leucine codon, there are several possible mutations that will not change the amino acid. By contrast, the only opportunity for a synonymous change at a lysine codon is one of three possible mutations at the third position (Table 2.1). Estimating the number of synonymous substitutions per synonymous sites and the number of nonsynonymous substitutions per nonsynonymous sites is a way of taking into account the relative likelihood of the two types of change in a given sequence.

In counting synonymous and nonsynonymous sites, dealing with both nondegenerate sites and fourfold degenerate sites is straightforward. A nondegenerate site is always a nonsynonymous site, because any mutation there will be nonsynonymous. A fourfold degenerate site, on the other hand, is a synonymous site because any mutation there will be synonymous. A problem arises, however, in the case of twofold and threefold degenerate sites because at these sites some mutations are synonymous and some are nonsynonymous. Nei and Gojobori's (1986) method deals with such sites by counting fractional sites. Consider AAA, a lysine codon. The first two positions in the codon are nondegenerate and thus are counted as nonsynonymous sites. At the third position, one possible mutation (to AAG) is synonymous, while the other two (to AAT and AAC) are nonsynonymous (Table 2.1). Nei and Gojobori's method counts the third position of the AAA codon as one-third of a synonymous site and two-thirds of a nonsynonymous site, reflecting the fact that one-third of possible mutations at that site are synonymous and two-thirds nonsynonymous.

In comparing two sequences (which we can designate sequence 1 and sequence 2, respectively), we count the number of synonymous sites in each sequence (N_{S1} and N_{S2}, respectively) and the number of nonsynonymous sites in each sequence (N_{N1} and N_{N2}, respectively). Then we compute the unweighted averages:

$$N_S = (N_{S1} + N_{S2})/2 \tag{2.3}$$

and

$$N_N = (N_{N1} + N_{N2})/2 \tag{2.4}$$

Next, it is necessary to count the numbers of synonymous and nonsynonymous differences between the two sequences. Some examples are illustrated in Figure 2.5. In the first codon shown, the difference between the two sequences is obviously a nonsynonymous difference, because it changes the amino acid (from

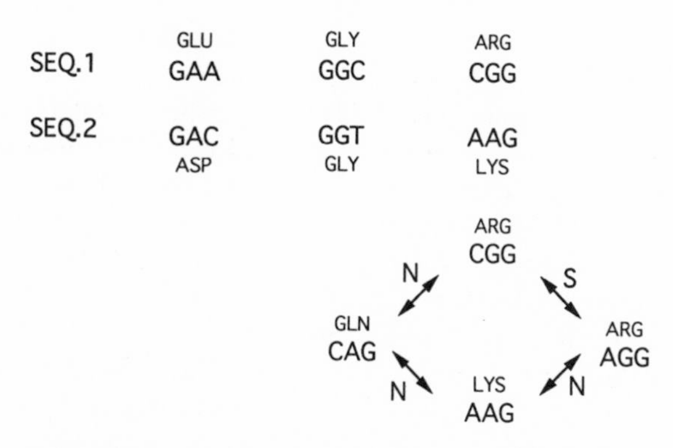

Figure 2.5. Synonymous and nonsynonymous mutations in coding regions. (*Top*) The first codon provides an example of a nonsynonymous mutation; the second codon a synonymous mutation; and the third codon a case of two differences in a single codon. (*Bottom*) Possible pathways relating the two codons at the third position, showing synonymous (S) and nonsynonymous (N) changes required by each pathway.

glutamic acid in sequence 1 to aspartic acid in sequence 2) (Figure 2.5). In the second codon, the one difference is clearly synonymous, because it does not change the amino acid (Figure 2.5). In the third codon shown, however, counting changes is less straightforward, because there are two differences between sequence 1 (CGG) and sequence 2 (AAG) (Figure 2.5).

The way that the method deals with codons having two (or three) differences between the two sequences being compared is to average the differences that would occur along all possible mutational pathways connecting the two sequences. In the case of the third codon of the sequences compared in Figure 2.5, there are two such pathways. One (CGG-CAG-AAG) involves two nonsynonymous differences, while the other (CGG-AGG-AAG) involves one synonymous difference and one nonsynonymous difference (Figure 2.5). Averaging the two pathways, we get $(2 + 1)/2 = 1.5$ nonsynonymous differences and $(0 + 1)/2 = 0.5$ synonymous differences. A similar approach is taken in the more complicated case when there are three differences between two codons being compared. Note that in averaging pathways, we need make no assumption regarding the direction of evolution or which sequence was ancestral. Thus, in the pathway CGG-CAG-AGG of Figure 2.5, any of these three codons might have been found in the common ancestor of the two sequences being compared.

Because Nei and Gojobori's method uses the unweighted average of pathways, it assumes that each possible pathway is equally likely. Because protein evolution is generally conservative (see Chapter 3, The Dynamics of Neutral Evolution), this assumption may sometimes be incorrect. In the third codon of Figure 2.5, for example, the pathway CGG-AGG-AAG might actually be more likely than the pathway CGG-CAG-AAG, because the latter pathway involves a residue charge change while the former pathway does not. There are some methods of estimating synonymous and nonsynonymous substitutions that attempt to correct for the difference in likelihood between different pathways by using a weighted average, where the weights are derived from an empirical matrix of amino acid replacements in a large data set of evolutionarily conserved proteins (e.g., Li et al. 1985). However, when we are studying positive Darwinian selection, which acts to favor diversity at the amino acid level, the pattern of amino acid change may be quite different from that seen in conserved proteins. Natural selection may actually favor nonconservative amino acid changes, such as those involving charge changes (e.g., Hughes et al. 1990). Thus, for the study of adaptive evolution, an unweighted pathway method seems preferable.

When the numbers of synonymous (M_S) and nonsynonymous (M_N) differences between a pair of sequences have been calculated, then we can compute the proportion of synonymous differences per synonymous site (p_S) by the following:

$$p_S = M_S/N_S \tag{2.5}$$

Similarly, we can compute the proportion of nonsynonymous differences per nonsynonymous site (p_N) by:

$$p_N = M_N/N_N \tag{2.6}$$

To correct p_S and p_N for multiple hits, Nei and Gojobori (1986) used Jukes and Cantor's (1969) formula (equation 2.1). The number of synonymous nucleotide substitutions per site (d_S) is thus given by

$$d_S = -3/4 \ln [1 - 4/3(p_S)] \tag{2.7}$$

And the number of nonsynonymous substitutions per nonsynonymous site (d_N) is given by

$$d_N = -3/4 \ln[1 - 4/3(p_N)] \tag{2.8}$$

The variances of d_S and d_N are obtained from equation 2.2. In computing var (d_S), p_S and N_S are substituted for p and n, respectively, in equation 2.2. Likewise, in var (d_N), p_N and N_N are substituted for p and n, respectively.

The estimates of d_S and d_N will become unreliable as p_S and p_N approach 0.75. As mentioned previously, under Jukes and Cantor's model, two sequences that are 25% identical (75% different) show no evidence of evolutionary relationship, because this degree of similarity is expected for two randomly chosen sequences. In most actual coding sequences, synonymous sites evolve much more rapidly than do nonsynonymous sites (see Chapter 3, The Dynamics of Neutral Evolution). Thus, it is often the case that synonymous sites become saturated long before nonsynonymous sites do, and that d_S between two sequences cannot be estimated but d_N can be. For example, in the case of nuclear genes of vertebrates, d_S estimates are usually unobtainable between sequences that last shared a common ancestor 150–200 million years ago.

Other widely used methods of estimating synonymous and nonsynonymous substitutions are those of Li et al. (1985) and of Li (1993). The authors of these papers used the symbols K_S and K_A in place of d_S and d_N, respectively. Aside from slight differences in the method of computation, these different symbols have an essentially equivalent meaning.

Li (1993) observed that the way of counting fractional synonymous and nonsynonymous sites used by Nei and Gojobori's (1986) and similar methods may underestimate the likelihood of synonymous substitution at twofold degenerate sites if there is a bias in favor of transitions. This would occur because, at all twofold degenerate sites, transitional mutations are synonymous, while transversional mutations are nonsynonymous (Table 2.1). Instead, he proposed a method that estimates the fraction of synonymous and nonsynonymous sites at twofold degenerate sites from the observed frequencies of transitional and transversional differences at these sites in the data. However, there are two problems with such an approach. First, estimating such a quantity from the data itself will be subject to large stochastic errors because the number of twofold degenerate sites in a sequence will typically be small unless the sequences compared are very long. Second, the observed frequency of transitional differences at twofold degenerate sites will not give a true indication of the extent of transitional mutational bias. Because any transversional difference at a twofold degenerate site will change the amino acid, such a mutation will often be deleterious to protein structure. Thus, it is likely to be eliminated from the population by natural selection (by so-called purifying selection; see Chapter 3, The Dynamics of Neutral Evolution). Thus, even if there is no mutational bias in favor of transitions, twofold degenerate sites will often show a strong preponderance of transitional (and, therefore, synonymous) differences. A better way of estimating transitional bias would be to examine fourfold degenerate sites, because at these sites all mutations are synonymous. However, in this case, the problem of stochastic error would also still be present because of the limited number of sites.

The comparison of rates of synonymous and nonsynonymous nucleotide substitution can be a powerful tool in the study of adaptive evolution. As mentioned, in the case of most genes d_S exceeds d_N. In contrast, when natural selection favors changes at the amino acid level, d_N can exceed d_S. In Chapters 4–7 I will discuss cases where the comparison of d_S and d_N has been used to test for natural selection in a variety of systems. The hypothesis that $d_S = d_N$ can be tested by a z-test, which assumes approximate normality of the distribution of these two parameters. The test statistic is

$$z = (d_N - d_S)\ [\text{var}\ (d_N) + \text{var}\ (d_S)]^{-0.5} \tag{2.9}$$

Note that, if there is a mutational bias toward transitions at twofold degenerate sites, Nei and Gojobori's (1986) method will slightly over-estimate d_S. Thus, this method provides a conservative test for natural selection. This test is valid unless the number of substitutions is very small (Zhang et al. 1997).

Nei and Jin (1989) and Ota and Nei (1994a) presented methods for estimating the variances of mean d, d_S, or d_N for all pairwise comparisons within or between sets of sequences. The estimation of these variances is not straightforward because of evolutionary relationships among the sequences compared. Thus, for example, d_S between two related sequences is not independent of d_S between each of these sequences and a third related sequence. The methods of Nei and Jin (1989) and of Ota and Nei (1994a) take into account the covariance among estimates of nucleotide substitutions per site.

Conservative and Radical Nonsynonymous Differences

In some cases, natural selection may act to favor amino acid replacements that change certain properties of amino acids. For example, under certain circumstances, changes in amino acid residue charge may be selectively favored. Hughes et al. (1990) proposed a method of testing for such selection. By analogy with Nei and Gojobori's (1986) method, this method divides nonsynonymous sites into conservative sites and radical (nonconservative) sites with respect to some qualitative amino acid property of interest, such as charge, polarity, and so forth. For example, if the property of interest is residue charge, then a site at which all possible mutations will lead to a change of residue charge is a radical site. An example would be the first position of the aspartic acid codon GAC (Table 2.1). Aspartic acid is a negatively charged residue, but any mutation at the first position of GAC will cause a charge change, either to the positively charged amino acid histidine (CAC) or the neutral amino acids tyrosine (TAC) or asparagine (AAC). As with Nei and Gojobori's (1986) method, fractional sites are counted when some mutations at a site will change the property of interest, while others will not.

This method is used to compute the proportion of conservative nonsynonymous substitutions (p_{NC}) and the proportion of radical nonsynonymous substitutions (p_{NR}). These proportions are not corrected for multiple hits because in this case Jukes and Cantor's (1969) or more complex models are not really applicable. When $p_{NC} = p_{NR}$, nonsynonymous differences occur at random with respect to the property of interest (Figure 2.6). When $p_{NC} > p_{NR}$, there are numerous nonsynonymous differences, but these have occurred in such a way as to conserve the amino acid residue property of interest (Figure 2.6). For example, with respect to residue polarity, $p_{NC} > p_{NR}$ is often seen in comparisons of closely related transmembrane regions. In transmembrane regions, nonpolar, hydrophobic residues are favored because these provide an anchor in the phospholipid bilayer of the cell membrane. Often there are frequent amino acid changes in transmembrane regions, but these occur in such a way as to preserve the overall hydrophobic character of the domain.

On the other hand, when $p_{NR} > p_{NC}$, nonsynonymous differences occur in such a way as to change the amino acid property of interest (Figure 2.6). When p_{NR} is significantly greater than p_{NC}, this is evidence of a pattern of amino acid replacement that has changed the property of interest to a greater extent than expected under random substitution. Such a nonrandom pattern of amino acid replacement, in turn, is suggestive of natural selection favoring changes in the residue property. Note that, because this method examines nonsynonymous differences only, it can sometimes be used in cases where estimation of d_S would be unreliable because the sequences studied diverged in the distant past.

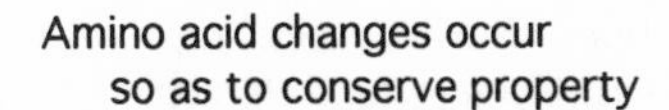

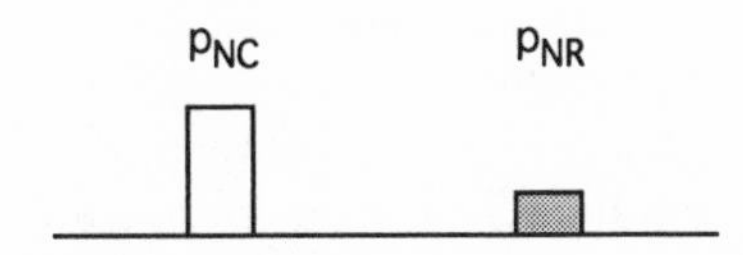

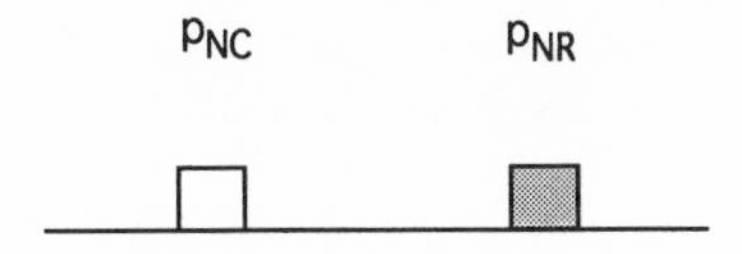

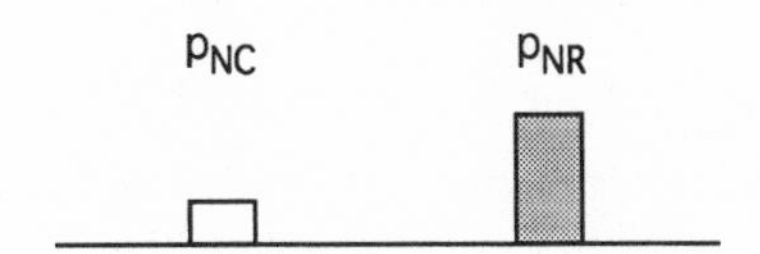

Figure 2.6. Possible relationships between the proportion conservative (p_{NC}) and radical (p_{NR}) nonsynonymous differences with respect to some amino acid property of interest.

Phylogenetic Tree Reconstruction

One of the most useful approaches to the study of evolution is the reconstruction of evolutionary relationships among homologous nucleotide sequences—that is, among sequences that are descended from a common ancestral sequence. These relationships are commonly represented by means of a diagram known as a phylogenetic tree. The development and refinement of methods for the reconstruction of phylogenetic trees has been an active area of research in biology and biological statistics for the past three decades. Unfortunately, it has also been an area of considerable controversy—often needlessly so. The reader is referred elsewhere (Nei 1987; Li 1997) for a detailed discussion of the various methods of phylogenetic reconstruction that have been proposed. Here, I offer only a brief overview of the problem.

It is important to emphasize that any method of phylogenetic analysis represents an attempt to reconstruct evolutionary relationships that are, in most cases, unknown to us. For any set of related sequences, the "true tree" represents the evolutionary relationships among them with 100% accuracy. However, in general, we do not know the true tree. Any tree we construct represents an estimation of the true tree. However, it is desirable to use a method of reconstruction that has a high probability of recovering the true tree.

An important distinction in phylogenetic analysis is that between gene trees and species trees. A species tree reconstructs the relationships among the members of a set of species (or sometimes of subspecies or populations within species). A gene tree reconstructs the relationships among a set of related genes. One or more gene trees may be used to infer a species tree, but a particular gene tree need not correspond to the species tree. The following factors may cause the gene tree not to correspond to the species tree (Nei 1987):

1. When two or more species are descended from an ancestral species that is polymorphic at a locus, different alleles may become fixed in the descendant species; but the relationship among these alleles may not correspond to the relationship among the species, because the alleles predate speciation. Ordinarily, this is only a problem when the species studied are closely related, because selectively neutral polymorphism generally do not last long (see Chapter 3, The Dynamics of Neutral Evolution). It is more of a problem when natural selection acts to maintain a polymorphism, in which case polymorphism can last much longer (see Chapter 4, Trans-Species Polymorphism at MHC Loci).
2. The genes analyzed may be paralogous rather than orthologous. Two genes in different species are said to be orthologous if they are descended without gene duplication from a gene in the two species' common ancestor (Figure 2.7). In contrast, two genes are said to be paralogous if the two genes are descended from different daughter genes that arose as a result of a gene duplication event (Figure 2.7). When paralogous genes rather than orthologous genes are compared, the gene tree is not expected to be the same as the species tree (Figure 2.7). Sometimes, researchers may be uncertain whether the genes they analyze are orthologous or paralogous. Consider the case where a given gene family is well known in one species and all of its members have been mapped and sequenced. If there is a second species whose genome is not well known, it may be uncertain how many members of the gene family are present. If a sequence is obtained from the latter species, it may not be clear to which gene in the former species the sequence corresponds. In this case, a phylogenetic analysis of genes will be necessary before it will be possible to use the sequences in a phylogenetic analysis of species.

Figure 2.8 shows an illustration of this problem in the case of a small gene family known as CD1. CD1 are related to class I major histocompatibility complex molecules (Chapter 4), but their function is unknown. In the human

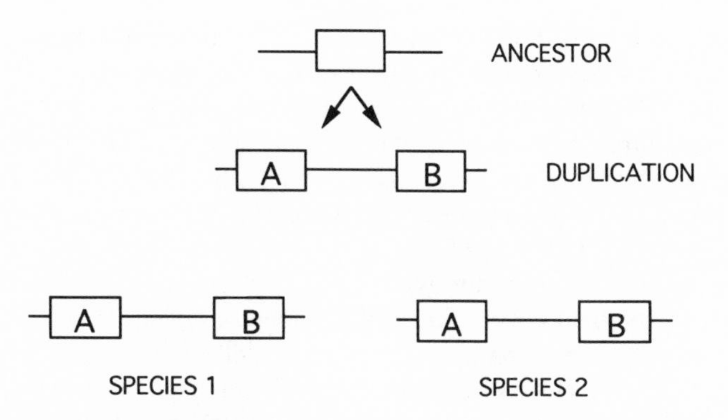

Figure 2.7. Orthologous and paralogous relationships between genes. After gene duplication in the common ancestor of species 1 and species 2 gives rise to two genes (A and B), A in species 1 is orthologous to A in species 2, and B in species 1 is orthologous to B in species 2. But A in species 1 is paralogous to B in species 2, and B in species 1 is paralogous to A in species 2.

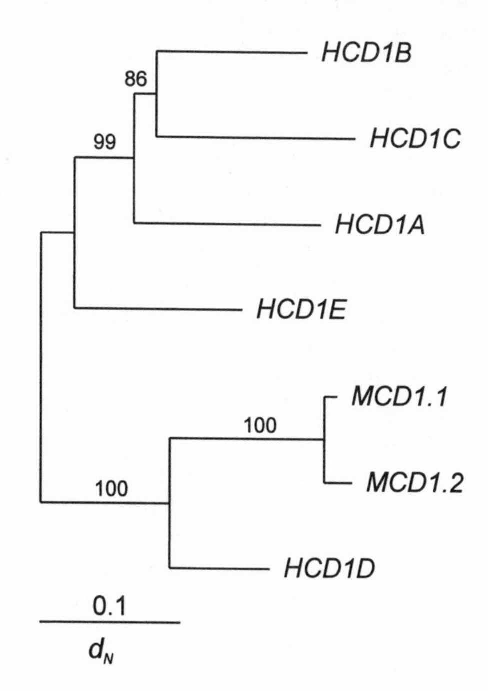

Figure 2.8. Phylogenetic tree of human (*H*-) and mouse (*M*-) *CD1* genes. The tree was constructed on the basis of nonsynonymous nucleotide substitutions per site (d_N) by the neighbor-joining method.

genome, there are five CD1 genes (*HCD1A-E*), whereas the mouse has only two (*MCD1.1* and *MCD1.2*). The phylogenetic tree suggests that the two mouse genes are orthologous to the human HCD1D, and that the mouse lacks orthologues of the other four human genes (Figure 2.8). *MCD1.1* and *MCD1.2* evidently arose as a result of gene duplication after the divergence of the rodent lineage (leading to the mouse) from the primate lineage (leading to the human) (Hughes 1991a).

Controversies regarding the methods of phylogenetic reconstruction have their origin in the field of systematics, the branch of biology that deals with the classification of species of organisms. Until the middle of this century, systematists largely approached their task in an intuitive and often haphazard fashion, relying on a variety of types of evidence to infer evolutionary relationships among species but lacking an objective, repeatable method that could be applied consistently to different groups of organisms.

Feeling that traditional systematics was more an art than a science, a number of biologists attempted to introduce objectivity into the classification of organisms. One approach was called numerical taxonomy or phenetics (Michener and Sokal 1957; Sneath and Sokal 1973). Using this approach, an investigator would obtain measurements on as many quantifiable traits as possible for the organisms in which he or she was interested. From these measurements, pairwise distances among the species were then computed. These distances, in turn, were used to construct a phylogenetic tree; a number of algorithms for constructing phylogenetic trees from such distance matrices were developed by numerical taxonomists.

However, many systematists were troubled by certain aspects of this approach. The pheneticists emphasized that their goal was to obtain an objective measure of overall similarity, and that this goal would best be achieved if all characters examined were weighted equally (Sneath 1961). Because systematists were accustomed to believe that certain characters conveyed greater information about evolutionary relationships than did others, they were reluctant to accept the idea of equal weighting. Confidence in phenetic methods was further shaken by evidence that the classification is highly dependent on both the characters examined and the exact methods used for both distance estimation and clustering (Minkoff 1965). Further, pheneticists in their writings often expressed skepticism regarding our ability to reconstruct the evolutionary relationships among species. Rather, they argued that the task of the systematist is to present a workable classification based on overall phenotypic resemblance (hence the term *pheneticist*). Such a classification, they felt, was very likely to reflect evolutionary relationships—but could not be guaranteed to do so.

A quite different approach to systematics was taken by the German entomologist Willi Hennig (1950, 1965), founder of the school known as "phylogenetic systematists" or "cladists." Hennig emphasized the use of discrete characters in systematics, to each of which a primitive (or ancestral) and a derived state could be assigned (Figure 2.9). He held that species should be grouped together on the basis of shared derived characters (Figure 2.9). Cladists believe that the polarity of characters (i.e., which state is primitive and which is derived) can

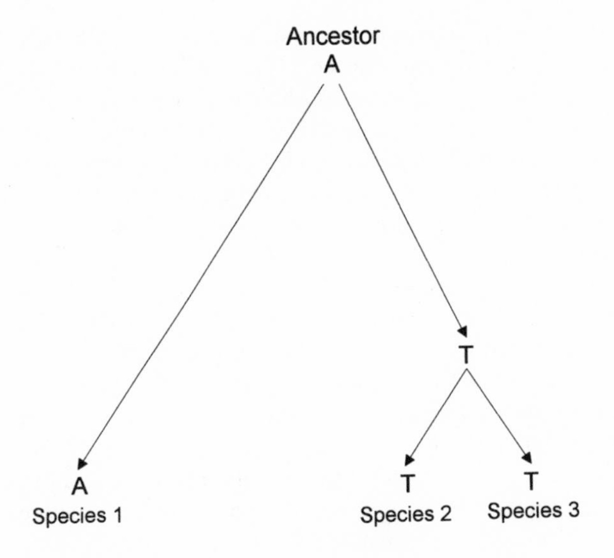

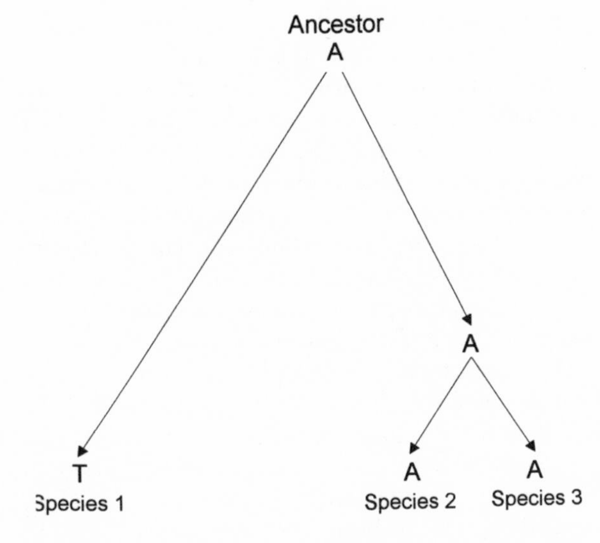

Figure 2.9.
(*Top*) Example of a shared derived character: T, shared by species 2 and species 3. (*Bottom*) Example of a shared ancestral character: A, shared by species 2 and species 3.

often be determined by comparison with an outgroup, where an outgroup is a species or group of species known to be distantly related to the group of species whose interrelationships are of primary interest.

One problem with this approach is that different characters may tell different stories. A common cause of such differences is convergent or parallel evolution. The term *convergent evolution* is used by morphologists to describe a case where two taxa have evolved similar morphological structures starting from very different ancestral states. For example, both fishes and whales have forelimbs that are fin-like and function to propel the animal through water, but these structures evolved independently in the two groups and in very different ways. If we were to consider the presence of fin-like forelimbs to be a shared derived character, we might classify fishes and whales together (as Aristotle did). However, other characters (such as the presence of hair and mammary glands) reveal that whales are, in fact, mammals, and are not closely related to fishes. Thus, we conclude that whales have evolved a fish-like body plan convergently in response to the demands of an aquatic environment. Similarly, parallel evolution is said to occur when two species independently evolve a similar morphology starting from essentially the same ancestral state. Although the concepts of convergent and parallel evolution were originally developed in the context of morphological evolution, they are readily transferred to molecular evolution (Figure 2.10).

When different characters support different phylogenies, cladists prefer the phylogeny that is most parsimonious—that is, the phylogeny that assumes the smallest possible number of changes of state in the evolutionary process. No theoretical justification has been given for this preference, other than the intuitive sense common to all scientific endeavors that simpler explanations are more likely to be true than are complicated explanations. Computer algorithms have been developed for finding the most parsimonious phylogenetic tree from both morphological and molecular data. These methods—collectively referred to as maximum parsimony (MP) methods—choose the tree that is supported by the greatest number of the characters chosen for analysis by the investigator.

Thus, the choice of characters for analysis will have a large influence on the outcome of the analysis; and reexamination of individual cases has often led to strikingly different conclusions (Mayr 1981; Feduccia 1996). One major source of difficulty has been lack of independence among the characters examined, especially when convergent evolution is also present. One instructive example involves loons, grebes, and a group of extinct birds called hesperornithiforms. All three of these groups of birds share an unusual locomotor adaptation known as foot-propelled diving. On the basis of a cladistic analysis, Cracraft (1982) argued that all three groups of birds are related because they share a suite of derived characters. However, molecular data do not support a close relationship of loons and grebes (Sibley and Ahlquist 1990). (Of course, we have no molecular data for the extinct hesperornithiforms.) In fact, it appears that the shared derived characters that Cracraft used to infer a close relationship of these taxa are all adaptations for foot-propelled diving (Feduccia 1996). Thus, the characters are not independent, but rather are likely to occur together in any bird adapted to this mode of locomotion, which appears to have evolved convergently in these three groups of birds (Feduccia 1996). Counting each as a separate character led to an inflated estimate of the amount of support for grouping the three taxa together.

The conflict between pheneticists and cladists properly belongs to the era of morphological systematics—an era that is now effectively at an end. The availability of molecular data has revolutionized the field and made many old controversies obsolete. Unfortunately, some systematists have continued to view the field from the perspective of old controversies that are no longer relevant. For example, some cladists are adamantly opposed to the use of any sort of

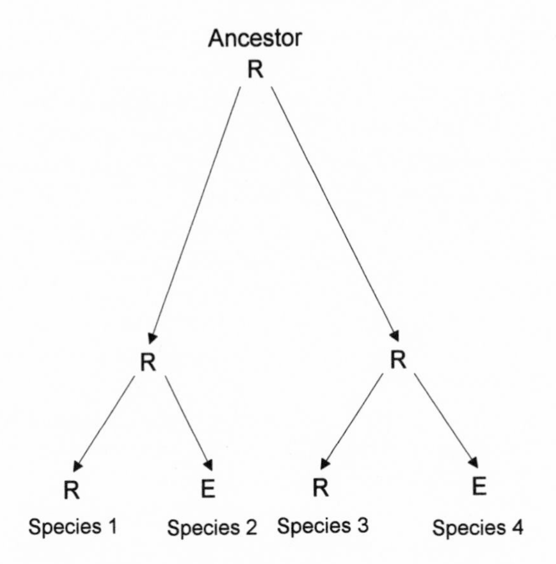

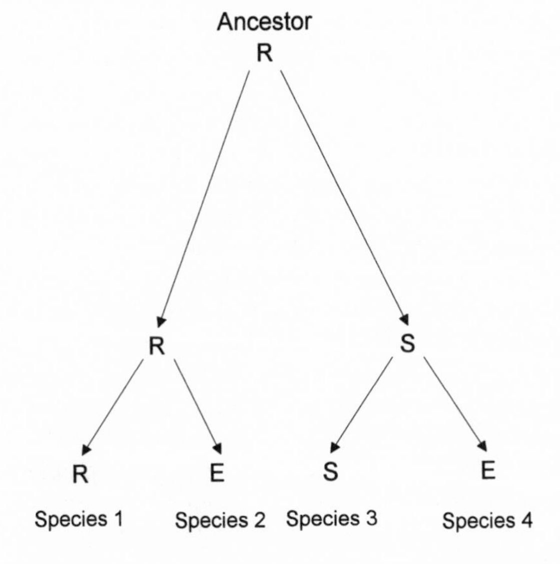

Figure 2.10.
(*Top*) An example of parallel evolution at the amino acid level: E evolves in parallel (independently) in species 2 and in species 4. (*Bottom*) An example of convergent evolution at the amino acid level: species 2 and species 4 each evolve E by replacement of a different ancestral residue.

phylogenetic analysis that utilizes a matrix of distances among sequences. There are several such methods available—some of them developed by pheneticists for analysis of phenetic distances and others developed more recently for use with molecular data. In applying such "distance matrix" methods to DNA sequence data, investigators use as "distances" estimates of the number of nucleotide substitutions per site based on Jukes and Cantor's (1969) model or another appropriate model. Thus, it is really not fair to compare such molecular distance matrix methods with the use of phenetic distances and to tar both types of analysis with the same brush. Phenetic analysis computes a distance based on a large number of arbitrarily chosen phenotypic characters whose genetic basis is unknown, and the exact meaning of such a distance in evolutionary terms is unclear. Critiques of this approach thus do not apply in the case of distance estimates based on explicit models of the process of nucleotide substitution.

The most widely used distance matrix method of phylogenetic reconstruction for molecular data is the neighbor-joining (NJ) method (Saitou and Nei 1987). The NJ method provides a rapid, heuristic method for finding a tree that is close to or (sometimes) identical to the minimum-evolution tree, where the minimum-evolution tree is one in which the total sum of the branch lengths is at a minimum (Rzhetsky and Nei 1992a). There is theoretical justification for expecting that the minimum-evolution tree will be the true tree. As long as the distance estimates are unbiased, the expected value of the sum of the branch lengths is smallest for the true tree (Rzhetsky and Nei 1992b). Computer simulations have shown that the NJ method generally outperforms other distance methods, especially when the rate of evolution is different in different branches of the tree (Saitou and Nei 1987; Saitou and Imanishi 1989).

The MP method seems not to perform quite as well as the NJ method, particularly when the degree of sequence divergence is substantial (Saitou and Imanishi 1989). Presumably, the reason for this is that the MP method incorporates no correction for multiple hits. Nonetheless, when applied judiciously, the MP method does quite well in the case of molecular data; indeed, MP does much better in the case of molecular data than in the case of morphological data.

As illustrated above with the example of diving birds, the factor most likely to lead to an erroneous conclusion with the MP method is convergent evolution, especially when it involves a number of characters simultaneously. This factor is much less of a problem with molecular data than with morphological or other phenotypic data. Convincing examples of convergent evolution at the level of the molecular sequence are quite rare, and in most known cases, the number of sites involved is quite small (Doolittle 1994; Zhang and Kumar 1997; Chapter 6, Convergent Sequence Evolution). Thus, even an undetected case of convergent evolution may not lead to seriously misleading conclusions in MP analysis, as long as the sequence analyzed is reasonably long.

In the case of both MP and distance matrix methods, the most common error seen in published analyses is the failure to take into account saturation of synonymous sites in the comparison of distantly related coding sequences. In such a case, NJ and other distance matrix methods are frequently applied to the Jukes-Cantor distance (or some similar distance) based on all nucleotide sites. This approach is unreliable, because in such a case synonymous sites actually convey no evolutionary information. It would be preferable to base the phylogeny on d_N or on an estimate of the number of amino acid replacements per site (see Nei 1987; Kumar et al. 1993). Likewise, MP trees are often constructed on the basis of all nucleotide positions even when synonymous sites are saturated. In this case, chance resemblances at synonymous sites may be interpreted as shared derived characters. When synonymous sites are at or near saturation, MP trees should be based either on just the first and second codon positions (where most mutations are nonsynonymous) or on amino acid sequences.

When phylogenetic trees are based on amino acid sequences, one may use the uncorrected proportion of differences or some amino acid distance that corrects for multiple amino acid replacements per site. The most commonly used of these assumes that the process of amino acid replacement follows a Poisson distribution (Nei 1987). As a consequence, it is assumed that the probability of replacement is the same for all sites. Although this assumption is unlikely to be true for real sequences, the error resulting from violations of this assumption is generally small, unless the differences among sites with respect to the rate of replacement are very pronounced (Nei 1987). In the latter case, a correction based on the gamma distribution is preferable (Ota and Nei 1994b).

In practice, both NJ and MP methods give very similar results. Another method, the maximum likelihood (ML) method, is less frequently used because it takes a great deal of computer time. When used with an appropriate substitution model, it also generally yields results similar to NJ and MP. Phylogenetic trees presented in this book have been constructed by the NJ method, but in most cases the MP and/or ML methods were also applied to the same data and produced similar reconstructions.

One attractive aspect of the MP method is that it permits reconstruction of changes that are hypothesized to have taken place in the past, and thus of ancestral molecular sequences. However, because it does not correct for multiple changes and assumes equal likelihood of all possible substitutions, the MP method often does not provide good resolution of ancestral sequences. Methods based on maximum likelihood, such as that of Yang et al. (1995), provide a better resolution. Yang et al.'s (1995) method makes it possible to reconstruct ancestral sequences on the basis of a given substitution model and a phylogenetic tree, which may be reconstructed by any method. By this method, one can estimate the probability of a given reconstruction, assuming the substitution model and the phylogenetic tree.

Because a finite number of sites is examined, any tree-making method may yield an incorrect tree due to chance. Thus, it is desirable to have some sort of statistical test of the reliability of branching patterns within a tree. For NJ trees, two such tests are in common use. The first involves "bootstrapping" (Felsenstein 1985). Bootstrapping involves creating pseudosamples by random selection of sites from the data set, with replacement. This process is repeated a large number of times (say, 1000), and a phylogenetic tree is constructed for each pseudosample. The percentage of pseudosamples supporting a given clustering pattern provides an index of the support for that pattern. Bootstrapping can also be applied to MP trees. The second way of testing NJ trees involves estimating the standard error of internal branch lengths in the tree (Rzhetsky and Nei 1992a). The ratio of the branch length to its standard error is a z-statistic for testing the null hypothesis that the branch length is zero.

3

The Neutral Theory

In an influential book called *The Structure of Scientific Revolutions* (1962), the historian and philosopher of science Thomas Kuhn presented a model of how changes in scientific theory take place. Kuhn calls a major theoretical change a "scientific revolution" or a "paradigm shift." According to Kuhn, in between revolutions (in what Kuhn calls "normal science"), a field of science is dominated by a general theory, or "paradigm," that is accepted by virtually all scientists working in that field. It may happen that observations begin to accumulate that are not easily explained by the dominant paradigm. The tendency of most scientists is to explain away such inconsistencies or anomalies—or even to ignore them, to push them under the rug. However, when a sufficiently large mass of data contradicting the dominant theory has accumulated, a crisis ensues. Eventually, a new paradigm is put forward, which explains both the recently accumulated observations as well as previous knowledge in the field. There follows a period of controversy in which the new theory is subjected to rigorous testing. If the predictions of the new theory are verified, then it becomes the new paradigm for the field, and a new period of "normal science" begins.

Kuhn's model was based on his interpretation of certain events in the history of physics, and its general applicability to the biological sciences can be questioned. The growth of biology has often been rather haphazard, and biologists in general tend to shy away from overarching paradigms. But, at least in the rather specialized fields of population genetics and evolutionary biology, the past three decades have witnessed a minor sort of scientific revolution that fits the Kuhnian model fairly well. This scientific revolution involved the development of the neutral theory of molecular evolution by Motoo Kimura and others, the controversies between opponents of the theory ("selectionists"), and its proponents ("neutralists"), culminating in the widespread tacit acceptance of the basic postulates of the neutral theory by most evolutionary biologists.

Here I review some of the data that led to proposal of the neutral theory.

Then I discuss the basic tenets of the theory in an essentially qualitative way. The mathematical theory of the dynamics of neutral alleles in populations has been developed elsewhere (Kimura 1983; Nei 1987) and is beyond the scope of this book. However, this theory yields some very simple mathematical results, which have important implications for evolutionary biology. I discuss a few of these, with particular emphasis on the insights these have to offer for the study of adaptive evolution.

Background of the Neutral Theory

In his 1968 paper that, along with that of King and Jukes (1969), launched the neutral theory, Kimura drew attention to two anomalies that were not easily explained by evolutionary theory widespread at that time. By the 1950s, most writings on evolutionary biology had come to emphasize the role of natural selection as the dominant force in evolution and to relegate other factors to a relatively minor role. The Neo-Darwinian synthesis had revived interest in natural selection in the 1920s and 1930s, but the original Neo-Darwinians had not placed so overwhelming an emphasis on natural selection. Sewell Wright, in particular, accorded an important role to genetic drift; that is, random change in gene frequency due to sampling of gametes in a finite population (Wright 1931). Later, however, Wright (e.g., 1949), in concert with other Neo-Darwinians, emphasized the pervasiveness of natural selection. Indeed, by the late 1940s and early 1950s there occurred what Gould (1983) has called a "hardening" of the Neo-Darwinian synthesis, with an increasingly selectionist bias.

Natural selection was seen to be at work in two major ways: (1) the maintenance of polymorphisms in populations (balancing selection); and (2) the substitution of one allele for others; that is the fixation of new mutations (directional selection). Virtually all genetic polymorphisms were held to be balanced polymorphisms. Populations of organisms were believed to be very large, so that genetic drift was unlikely to be an important factor. Thus, when a new mutant allele appeared in a population, if that allele conferred a selective advantage on its bearer (even a slight advantage), that allele would eventually become fixed at that locus; in other words, it would reach a population frequency of 100%. Because of the large population size, it was not thought likely that such advantageous allele would be lost due to genetic drift, even when rare; and the possibility that drift would ever lead to fixation of a selectively neutral or deleterious allele was discounted.

Kimura (1968), who had already made important contributions to the mathematical modeling of genetic drift (e.g., Kimura 1955), argued that data that had recently become available from protein sequencing were inconsistent with a major role for selection in most change at the molecular level over evolutionary time. His calculations predicted that neutral or nearly neutral mutations were occurring at a much higher rate than previously supposed. A high neutral mutation rate might, in turn, explain the unexpectedly high levels of polymorphism that were revealed by studies of natural populations using the technique of protein electrophoresis (e.g., Lewontin and Hubby 1966), without the need for balancing selection at most polymorphic loci. Here, I will briefly outline Kimura's arguments.

In the late sixties, a small number of amino acid sequences of orthologous proteins were available for different species of mammals. Averaging the available data, Kimura (1968) estimated that, on average, one amino acid substitution occurs every 28×10^6 years in a polypeptide 100 residues in length. This may not seem like a very rapid rate of evolution, but Kimura argued that, in fact, such a rate is much too rapid to be accounted for by natural selection.

To show that the observed rate of molecular evolution was too high to be explained by natural selection, Kimura relied on the concept of the "cost of selection" or "substitutional load" developed by Haldane (1957). Haldane's theory was based on the observation of plant and animal breeders that it is often impossible to select for a number of different characters simultaneously. One reason for this difficulty may be negative genetic correlation among characters; for example, the same gene or genes that cause a high value of one character (such as egg production in poultry) might cause a low value of another character (such as body weight). However, even when there is no genetic correlation among the traits selected for, simultaneous selection on a number of traits can pose difficulties. If the breeder chooses for breeding only those individuals that have high scores on a number of traits, he will soon find that he is breeding from a very small fraction of his stock.

The same phenomenon is expected to occur in populations subject to natural selection as well. Haldane (1957) took an example from the phenomenon of industrial melanism in pepper moths. In this moth a mutation causing dark pigment (melanism) was favored in areas where tree trunks were darkened as a result of human industrial activity; in these areas it was estimated that moths having the original grey color of the species, because they stood out on soot-blackened tree trunks, suffered a 50% mortality due to predation by birds. Haldane asked what would happen if an equally high level of natural selection were being applied simultaneously at 10 independent genetic loci; that is, in addition to the melanic allele at the locus controlling body color, suppose that there were equally favored new mutant alleles at nine other independently assorting loci. The result would be that only one out of every 2^{10} (or 1,024) individuals having the original genotype at all 10 loci would survive. The population would be very likely to go extinct before all 10 favored alleles could become fixed. Using a simple model, Haldane showed that, without an undue strain on the population, a new gene can be substituted only about once every 300 generations.

At the molecular level, however, Kimura pointed out that substitutions must be occurring much more rapidly than that, given the estimate of one amino acid substitution per 28 million years in a 100-residue polypeptide. Kimura assumed a genome size of 4×10^9 base pairs for mammals. A 100 codon coding region includes 300 base pairs, and because about 20% of nucleotide substitution is synonymous, he estimated that one amino acid replacement corresponds to about 1.2 base pair substitutions. From these values he computed the time it would take for one nucleotide substitution to occur in the genome as

$$(28 \times 10^6) \div (4 \times 10^9/300) \div 1.2 = 1.8 \text{ years}$$

This is a far more rapid rate than once every 300 generations. In fact, three years is close to the generation time for an average mammal; so these calculations suggest that nucleotide substitutions accumulate in the mammalian genome more rapidly than once per generation—a far too rapid rate to be explained by natural selection.

In fact, we can now see that Kimura considerably underestimated the rate of nucleotide substitution. Because at the time only amino acid sequences were available, Kimura was evidently unaware of the dramatic difference between evolutionary rates at synonymous and nonsynonymous sites that has since become evident from comparisons of nucleotide sequences (see next Section). Furthermore, he was unaware of the large quantities of noncoding DNA that exist in the genomes of higher organisms.

Using a more recent estimate of 3.3×10^9 base pairs in a mammalian genome (Miklos and Rubin 1996) and assuming that there are about 70,000 protein-coding genes with an average length of 400 codons, we can estimate that about $70{,}000 \times 1200 \times 0.75$ or 0.063×10^9 nucleotide sites are nonsynonymous sites, whereas the remaining 3.237×10^9 are either synonymous sites within codons or are sites in noncoding DNA. Some sites in noncoding regions are functionally important, and thus evolve at a low rate, but these sites are very few in number. Ignoring these latter and assuming that most noncoding sites evolve at about the same rate as synonymous sites in coding regions, we can use comparisons between the rat and mouse (Hughes and Yeager 1997a) to provide estimates of average rates of evolution at nonsynonymous (0.8×10^{-9} substitutions/site/year) and at synonymous/noncoding sites (2.5×10^{-9} substitutions/site/year). For the whole genome, we expect $(3.237 \times 10^9 \times 2.5 \times 10^{-9}) + (0.063 \times 10^9 \times 0.8 \times 10^{-9}) = 8.14$ substitutions per year. This means a time of about 0.12 years per substitution, or one substitution every month and a half. Clearly, this is much faster than one substitution per generation, even for small rodents. If it is true, as some data seem to indicate (e.g., Li 1997), that the rate of molecular evolution is somewhat higher in rodents than in primates, these values would be somewhat lower in the case of humans than in the case of the mouse and rat. Nonetheless, they would still be far too fast to be accounted for by selection.

A similar argument can be made regarding the extent of polymorphism observed in natural populations; namely, there is too much polymorphism for all of it to be balanced polymorphism maintained by natural selection. The best-studied form of balancing selection is overdominant selection or heterozygote advantage. Overdominant selection entails a cost that can be easily quantified for simple models. Consider, for example, a single locus with two alleles. Let the heterozygote have the highest fitness, and let each homozygote be 1% less fit than the heterozygote—a very slight fitness advantage for the heterozygote. We can interpret this to mean that a homozygote will have 1% less chance of surviving to reproduce than will a homozygote. Assuming that, at least with regard to the effects of this particular locus, a heterozygote has 100% chance of surviving to reproduce, if two heterozygotes mate, their offspring will have on average [1/2 (100%) + (1/2)(99%)] or 99.5% chance of surviving to reproduce, because one-half of their offspring are expected to be heterozygous and one half to be homozygous. This is where the cost of selection comes in: even a mating between two individuals of the fittest genotype (the heterozygote) will have offspring that are, on average, less fit than themselves.

Thus, in the premolecular era, evolutionary biologists were aware that overdominant selection imposes a cost, yet they generally believed that virtually all polymorphisms in natural populations are selectively maintained. This would not pose a problem if polymorphisms were relatively few, but early electrophoretic surveys showed a surprisingly high level of polymorphism. For example, Lewontin and colleagues, in examining several geographically separated populations of the fruit fly *Drosophila pseudoobscura*, found that 10 of 28 loci (36%) were highly polymorphic in more than one of the populations studied (Lewontin and Hubby 1966; Lewontin 1974). Because these loci were chosen for analysis simply because methods were available to detect their gene products, it was assumed that they represented a random sample of the protein-coding loci. If so, evolutionary biologists were faced with the disturbing possibility that as many as 36% of loci in the genomes of multicellular eukaryotes might be polymorphic.

Why this possibility was disturbing, given the widespread belief in the role of selection in maintaining polymorphism, is easily seen by applying the simple model seen above. Assuming that *D. pseudoobscura* has 12,000 protein-coding loci (Miklos and Rubin 1996), if 36% of them are polymorphic, there are 4,320 polymorphic loci in this species. Assume further that at each of these loci there are two alleles, that in each case the fitness of the homozygote is 1% less than that of the heterozygote, and that each locus acts independently in its effects on fitness. Individuals heterozygous at all 4,320 loci will presumably be rare, but this genotype will have the highest fitness. In a mating between two such multiple heterozygotes, the average chance of their offspring surviving to reproduce will be 0.995^{4320} or about 4×10^{-10}. These individuals would have to produce more than 1 billion eggs to assure that even one would survive! This sort of reasoning led Kimura and others to conclude that natural polymorphism was far too extensive for all but a small fraction of it to be selectively maintained.

The Dynamics of Neutral Evolution

The lines of reasoning outlined in the previous section led Kimura to propose the so-called neutral theory of molecular evolution, which has two main tenets: (1) Most polymorphisms are selectively neutral, and allelic frequencies change from one generation to the next as a result of random genetic drift. (2) Most nucleotide substitutions occurring over evolutionary time result from the chance fixation of allelic variants by genetic drift. Thus, the neutral theory sees both current-day polymorphism and change over evolutionary time as two aspects of the same phenomenon.

It is important to realize that Kimura (and subsequent "neutralists") never denied the importance of natural selection in adaptive evolution. They merely claimed that adaptively favorable mutations fixed as a result of positive selection constitute a relatively small proportion of all nucleotide substitutions. Nor did they deny that some polymorphisms are maintained by balancing selection. A few examples of such selection were already well documented—such as the sickle cell polymorphism in humans, which is maintained in regions where malaria is present because heterozygotes are resistent to the malaria parasite. But such cases were expected to constitute a minority of polymorphisms.

Both before and after his proposal of the neutral theory, Kimura developed an extensive body of population genetics theory regarding the dynamics of neutral alleles in finite populations (for a summary, see Kimura 1983). Most of this theory is based on a type of equation called a diffusion equation, commonly used to model stochastic processes (Kimura 1964). Although the mathematics involved in formulating and solving these equations are fairly advanced, fortunately this approach yields a number of very simple results that have great importance for understanding molecular evolution. Here, I will discuss some of these results and examine their relevance to the analysis of molecular sequence data.

A key concept in the mathematical study of genetic drift is that of effective population size (N_e). Because genetic drift relates to chance fluctuations in gene frequency due to the effects of population size, it is not surprising that population size will have a big effect on the process of drift. Drift is most easily modeled in an idealized population having the following characteristics: a constant population size from one generation to the next; random mating; an equal number of each sex contributing to the gene pool; and nonoverlapping generations. In this idealized case, the relationship between population size and genetic drift is straightforward, with the effects of drift being greater in smaller populations. In real populations, of course, one or more of the conditions for such an idealized population will most likely be violated. The concept of effective population size represents a way of correcting for such violations. For a given real population, N_e is defined as the size of an idealized population having the same characteristics (with regard to genetic drift) as the real population. N_e is typically lower than the census count of the population—often a great deal lower. Simple formulas are available to compute N_e in the case of fluctuating population size, unequal sex ratio, and overlapping generations (e.g., Nei 1987; Li 1997). In practice, more than one such factor may be operating, but in many cases there is one primary factor that has the most substantial impact on N_e.

The effect of N_e on the rate of genetic drift can be seen in a simple equation for the mean time to fixation (t) (in generations) of those neutral mutations that eventually become fixed:

$$t = 4N_e \tag{3.1}$$

If two populations have the same generation time but different N_e, the average time to fixation (in years) of neutral alleles will be faster on average for the population with the smaller N_e.

The simplest prediction of the neutral theory concerns the rate of substitution of neutral alleles. Let K_0 be the rate of substitution of neutral alleles (per generation) and u the neutral mutation rate per generation (i.e., the rate of production of new, selectively neutral mutations). Then

$$K_0 = u \quad (3.2)$$

As long as the neutral mutation rate remains constant, so does the rate of substitution (that is, of fixation) of neutral alleles. The same relationship can be expressed still more revealingly as follows:

$$K_0 = u_T f_0 \quad (3.3)$$

where u_T is the total mutation rate, and f_0 is the fraction of all mutations that are selectively neutral. Note that, although the effective population size is an important factor in predicting many aspects of genetic drift, equation 3.3 holds true regardless of N_e.

To express K_0 in years rather than in generations, one can use

$$K_0 = (u_T/g)\ f_0 \quad (3.4)$$

where g is the generation time in years (Kimura 1987). For a given set of homologous genes, the rate of evolution is often remarkably constant over time, even for organisms of very different generation times; this phenomenon has been referred to as the "molecular clock." In fact, when examined closely, molecular clocks turn out to tick far less regularly than has sometimes been supposed (see Gillespie 1991; Li 1997). Nonetheless, the roughly clock-like nature of molecular evolution can be explained by equation 3.4 if there is a strong positive correlation between u_T and g.

The basic idea behind equations 3.3 and 3.4 is used very extensively by biochemists and molecular biologists, most of whom are blissfully unaware that they are using a concept derived from population genetics, still less that they are using one of the key ideas of the neutral theory. For example, it is a standard practice to align a number of related protein sequences and to look for regions that show few differences among sequences. Regions with few differences, called "conserved" regions, are routinely proposed as likely candidates for functionally important regions of the protein. The reasoning is that a functionally important part of the protein is less likely to change over the course of evolution than are functionally less important parts. By now, a great many cases of proteins of known structure and function are known, and this generalization holds true for most cases.

For example, Figure 3.1 shows aligned sequences of some copper-binding proteins of arthropods: (1) hemocyanin, which is used for oxygen transport in crustaceans and in arachnids; and (2) phenoloxidase, which plays a key role in the sclerotization (hardening) of cuticle in insects and other arthropods (Voit and Feldmaier-Fuchs 1990; Fujimoto et al. 1995; Hall et al. 1995; Kawabata et al. 1995). The portion of the sequence shown in Figure 3.1 is the second of three protein domains, and this domain is known to contain the two copper-binding sites (A and B) (Gaykema et al. 1994; Fujimoto et al. 1995). Several amino acids in the two copper-binding sites are conserved in all sequences

```
                                                           CuA                        50
Signal crayfish proPO          ATTPLVIEYGPEFANTNQKAEHRV | SYWREDFGINSHHWHWHLVYPIEMN-
Drosophila proPO               SGARVHVDIPQNYTASDREDEQRL | AYFREDIGVNSHHWHWHLVYP-TTGP
Hornworm proPO                 TIPRTPIIIPRDYTATDLEEEHRL | AYWREDLGINLHHWHWHLVYPFSASD
Tarantula hemocyanin A         DESDIIVDVKD--TGNILDPEYKL | AYFREDIGVNAHHWHWHVVYPSTYDP
Tarantula hemocyanin E         DQ-DISVHVVE--TGNILDEEYKL | AYFKEDVGTNAHHWHWHIVYPATWDP
Penaeid shrimp hemocyanin      KQKQTPGKFKSSFTGTKKNPEQRV | AYFGEDIGLNTHHVTWHMEFPFWWND
Pacific crab hemocyanin 6      KMTQTAAKIESHFTGSKSNPEQRV | AYFGEDIGMNTHHVTWHLEFPFWWDD

Signal crayfish proPO          ---VN-RDRKGEL-FYYMHQQMVARY | DWERLSVNLNRVEKLENWRVPIPD
Drosophila proPO               TEVVN-KDRRGEL-FYYMHHQILARY | NVERFCNNLKKVQPLNNLRVEVPE
Hornworm proPO                 EKIVA-KDRRGEL-FFYMHQQIIARY | NCERLCNSLKRVKKFSDWREPIPE
Tarantula hemocyanin A         AFFGKVKDRKGEL-FYYMHQQMCARY | DCERLSNGLNRMIPFHNFNEPL-G
Tarantula hemocyanin E         AFMGRMKDRKGEL-FYYMHQQMCARY | DCERLSNGMRRMIPFSNFDEKL-E
Penaeid shrimp hemocyanin      AYG-HHLDRKGEN-FFWIHHQLTVRF | DAERLSNYLDPVGELQ-WNKPIVD
Pacific crab hemocyanin 6      AHENHHIERKGESCSSWVHHQLTVRF | DAERLSNYLDPVDELH-WDDVIHE
                                                                                 150
Signal crayfish proPO           GYFSKLTANNSGRPWGTRQDNTFIKDFRRNDAGLDFIDISDMEIWRSRLM
Drosophila proPO                GYFPKILSSTNNRTYPARVTNQKLRDVDR--HDG-RVEISDVERWRDRVL
Hornworm proPO                  AYYPKLDSLTSARGWPPRQAGMRWQDLKRPVDGL-NVTIDDMERYRRNIE
Tarantula hemocyanin A          GYAAHLTHVASGRHYAQRPDGLAMHDVRE-------VDVQDMERWTERIM
Tarantula hemocyanin E          GYSAHLTSLVSGLPYAFRPDGLCLHDLKD-------IDLKEMFRWRERIL
Penaeid shrimp hemocyanin       GFAPHTTYKYGGQ-FPARPDNVKFEDVD------DVARIRDMVIVESRIR
Pacific crab hemocyanin 6       GFDPQAVYKYGGY-FPSRPDNIHFEDVD------GVADVRDMLLYEERIL
                                                                          CuB  200
Signal crayfish proPO          DAIHQGYMLNRNGERVPLSDNVTTGKRGIDILGDAFEADAQLSP | NYLFYG
Drosophila proPO               AAIDQGYVEDSSGNRIPLDEV-----RGIDILGNMIEASPVLSI | NYNFYG
Hornworm proPO                 EAIATGNVILPDKSTKKLD---------IDMLGNMMEAS-VLSP | NRDLYG
Tarantula hemocyanin A         EAIDLRRVISPTGEYIPLDEE-----HGADILGALIESS-YESK | NRGYYG
Tarantula hemocyanin E         DAIDSGYYIDNEGHQVKLDIV-----DGINVLGALIESS-FETK | NKLYYG
Penaeid shrimp hemocyanin      DAIAHGYIVDSEGKHIDISN-----EKGIDILGDIIESS-LYSP | NVQYYG
Pacific crab hemocyanin 6      DATAHGYV-RINGQIVDLRN-----NDGIDLLGDVIESS-LYSP | NPQYYG
                                                                                 250
Signal crayfish proPO          DLHNTGHVLLAFCHDNDNSHREEIGVMGDSATALRDPVFYRWHKFVDDIF |
Drosophila proPO               NLHNEGHNIISFAHDPDYRHLEDFGVMGDVTTAMRDPIFYRWHGFIDTVF |
Hornworm proPO                 SIHNNMHSFSAYMHDPEHRYLESFGVIADEATTMRDPFFYRVHAWVDDIF |
Tarantula hemocyanin A         SLHNWGHVMMAYIHDPDGRFRETPGVMTDTATSLRDPIFYRYHRFIDNVF |
Tarantula hemocyanin E         SLHNWGHVMMARLQDPDHRFNENPGVMSDTSTSLRDPIFYRYHRFIDNIF |
Penaeid shrimp hemocyanin      ALHNTAHIVLGRQGDPHGKFDLPPGVLEHFETATRDPSFFRLHKYMDNIF |
Pacific crab hemocyanin 6      ALHNTAHMMLGRQGDPHGKFDLPPGVLEHFETATRDPAFFRLHKYMDNIF |

Signal crayfish proPO          QEYKLTQ---PPYTMED
Drosophila proPO               NKFKT---RLNPYNAGE
Hornworm proPO                 QSFKEAPHNVRPYSRSQ
Tarantula hemocyanin A         QEYKKT---LPVYSKDN
Tarantula hemocyanin E         QKYIAT---LPHYTPED
Penaeid shrimp hemocyanin      KEHKDN---LPPYTKAD
Pacific crab hemocyanin 6      RKHKDS---LPPYTKEE
```

Figure 3.1. Alignment of the central region of arthropod prophenoloxidases (proPO) and hemocyanins, showing the two copper-binding domains CuA and CuB.

illustrated (Figure 3.1). Particularly notable among these are histidine residues, which are known to form the ligands for copper (Gaykema et al. 1984). In addition to these histidines, numerous other residues in the copper-binding sites are conserved in all sequences analyzed. Outside the copper-binding sites, there are some residues conserved in all sequences, but conserved residues constitute a much lower proportion of all residues outside the copper-binding sites than in these sites.

The by now commonplace observation that functionally important amino acid residues tend to be conserved over evolutionary time can be explained in a straightforward way on the basis of equation 3.3. Consider a nonsynonymous nucleotide site that forms part of a codon encoding an important amino acid residue; for example, the second position of a histidine codon (CAT or CAC) encoding one of the copper-binding residues of an arthropod hemocyanin. Because any change at this position will change the amino acid and thus destroy the function of the protein, it will presumably be deleterious to the animal's fitness. Thus, f_0, the proportion of mutations that are neutral at the site in question, will be zero (or very nearly zero). As a consequence, equation 3.3 predicts that the rate of nucleotide substitution at that site will be zero (or very nearly zero). By contrast, there may be many nucleotides—particularly those encoding residues outside the copper-binding sites—at which all or most possible mutations will have no real effect on the protein's function, and thus will not affect the organism's fitness. At such sites, f_0 will presumably be quite high, though perhaps never approaching 1.

Finally, there are likely to be quite a few amino acid positions where it does not matter which exact residue is present as long as the residue is of a certain type. There are numerous positions at which the same type of residue (e.g., hydrophobic) is conserved in all of the sequences aligned in Figure 3.1. For instance, at position 41 in Figure 3.1 (a position within the copper A-binding site), the residues isoleucine, leucine, and methionine occur, all of which are hydrophobic. Consider shrimp hemocyanin, which has methionine (encoded by ATG) at that position. Mutations to CTG or TTG (leucine), to ATA (isoleucine), or perhaps to GTG (valine) may all be selectively neutral, because all of these are hydrophobic residues. But a mutation to AGG, encoding the positively charged residue arginine, might be deleterious. At such codons, the rate of evolution is expected to be intermediate between those that are strongly constrained due to an important functional role and those that are completely unconstrained.

We can quantitatively test predictions of this sort by estimating the numbers of nonsynonymous nucleotide substitutions per site (d_N) separately for different gene regions. Consider the arthropod hemocyanins and phenoloxidases aligned in Figure 3.1. Table 3.1 shows mean d_N values for all pairwise comparisons

Table 3.1. Mean number of nonsynonymous substitutions per 100 sites ($d_N \pm$ SE) in pairwise comparisons of hexamerin family members

	proPO and hemocyanin (N = 9)	Arylphorin subfamily (N = 25)
Domain I	83.7 ± 4.7***	53.2 ± 2.6
Domain II		
CuA	30.8 ± 3.6	55.4 ± 3.7
CuB	46.2 ± 4.1**	88.3 ± 6.0***
Remainder	61.9 ± 2.7***	73.5 ± 3.8***
Domain III	65.7 ± 2.6***	61.1 ± 1.8

Tests of the hypothesis that d_N equals that in CuA: $^{**}p < .01$; $^{***}p < .001$.

among these sequences, computed separately for the following regions: domain I; the copper-binding site A; the copper-binding site B; the remainder of domain II (excluding the two copper-binding sites); and domain III. Mean d_N is lowest in the copper-binding site A, being less than half the values for regions outside the copper-binding sites (Table 3.1). The next lowest values is in the copper-binding site B, which is also lower than those for the remainder of the molecule (Table 3.1). The results support the hypothesis that the proportion of neutral nonsynonymous mutations in the copper-binding sites is lower than elsewhere in the gene.

Because the copper-binding sites correspond to small numbers of codons (Figure 3.1), it seems unlikely that the actual mutation rate is lower in these codons than in surrounding portions of the gene. If so, as many mutations must occur at a typical nonsynonymous site in the copper-binding region as elsewhere in the gene. What happens to these mutants? The neutral theory predicts that the vast majority of them are selectively deleterious and are eliminated quickly by natural selection. This kind of natural selection is called conservative or purifying natural selection, to distinguish it from positive, Darwinian selection, as discussed later in this book. What purifying selection does is to eliminate deleterious mutations, whereas positive selection favors adaptively advantageous mutations. When molecular biologists speak of a protein region as being "under strong selection," they are generally thinking of purifying selection rather than positive selection. Indeed, every known protein-coding gene shows some evidence of purifying selection, while cases of positive selection at the molecular level are relatively rare.

Sometimes, a change in a protein's function over evolutionary time leads to a relaxation of purifying selection at nonsynonymous sites. When this happens, equation 3.3 predicts that f_0 at these sites will increase and, thus, the rate of nonsynonymous evolution will increase. The gene family that includes the arthropod hemocyanins and phenoloxidases provides an interesting test of this prediction. This family also includes a group of insect hemolymph proteins called arylphorins and hexamerins; these show evidence of an evolutionary relationship to hemocyanins and phenoloxidase (Figure 3.2), but they do not bind copper (Willott et al. 1989; Jones et al. 1993; Jamroz et al. 1996). When these proteins are aligned with their copper-binding relatives, regions homologous to the copper-binding sites are clearly observed (data not shown). Because some of these show histidines at the same positions as in the copper-binding proteins, it seems likely that the common ancestor of the family was a copper-binding protein.

In any event, in the arylphorins and hexamerins, there is much less evidence of purifying selection on these regions than in the hemocyanins and phenoloxidases (Table 3.1). Here, mean d_N in the region corresponding to copper-binding site A is very similar to that in domains I and II, while somewhat lower than in domain II outside of the copper-binding sites (Table 3.1). Furthermore, mean d_N in copper-binding site B is actually higher than that in other regions examined (Table 3.1). Thus, these calculations provide an example of the relaxation of purifying selection in the absence of a functional constraint, and thus support

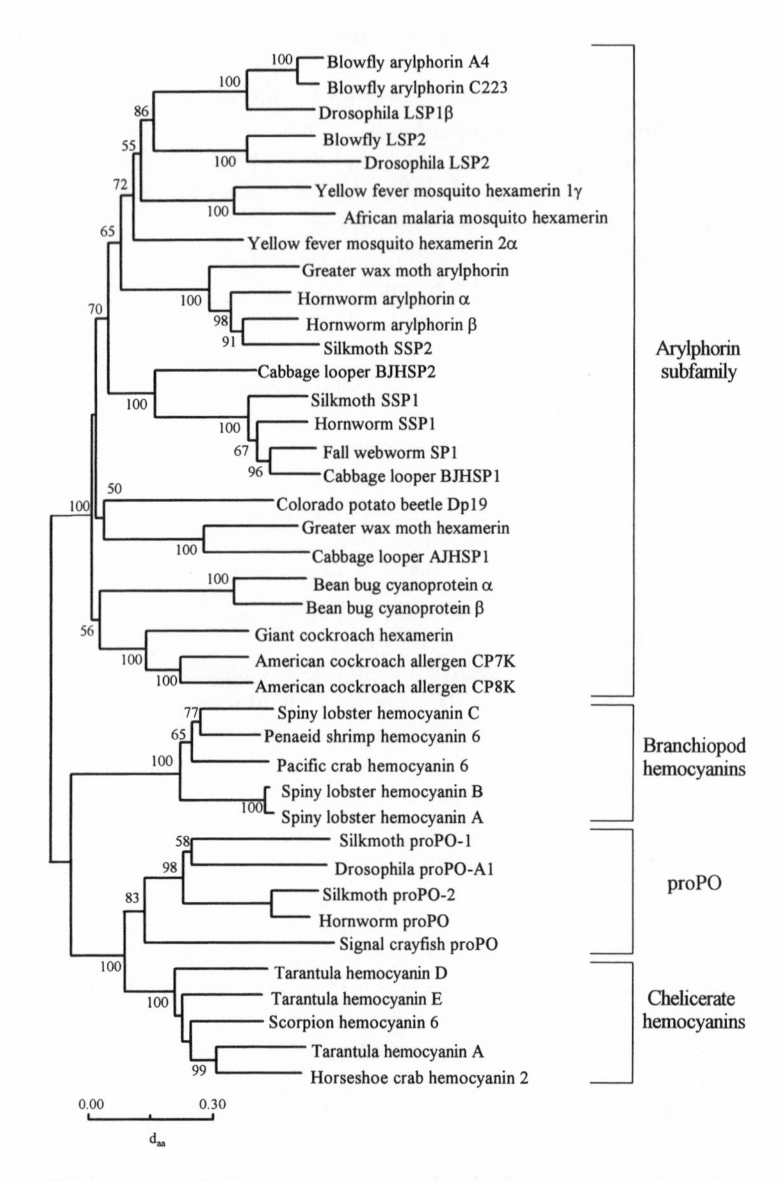

Figure 3.2. Phylogeny of arthropods prophenoloxidases (proPO) and related molecules, constructed by the neighbor-joining method on the basis of the Poisson-corrected amino acid distance (d_{aa}). Numbers on the branches are percentages of 1,000 bootstrap samples that support the branch; only values >50% are shown.

the predictions of the neutral theory. (For a summary of some classic early studies reaching similar conclusions, see Kimura 1983.)

Another important application of equation 3.3 is to the comparison between rates of evolution at synonymous and at nonsynonymous nucleotide sites. In multicellular eukaryotes, it is generally believed that different codons encoding the same amino acid are generally selectively equivalent or nearly so (see Chapter 8, Codon Usage in Multicellular Organisms, and Natural Selection on Codon Usage). If so, synonymous mutations must be selectively neutral or nearly so; and f_0 at synonymous sites must be very close to 1. By contrast, as mentioned earlier, nonsynonymous sites are a mixed bag with regard to the proportion of mutations that are neutral. At some nonsynonymous sites—those crucial to encoding functionally important amino acids—f_0 must be close to 0. At others, f_0 may approach 1, and at the majority of nonsynonymous sites f_0 is probably somewhere in between 0 and 1, because some mutations will be deleterious but not others. As a result, the average f_0 for nonsynonymous sites over a whole protein or a whole protein domain will typically fall somewhere between 0 and 1, its exact value depending on the overall level of functional constraint on the region analyzed.

One consequence of these considerations is the prediction that for most protein-coding genes d_S should exceed d_N. When the first DNA sequence data were becoming available, Kimura (1977) showed that they supported this prediction. By now, a great many comparisons have been made between homologous genes, and in the vast majority d_S exceeds d_N, as predicted by the neutral theory (e.g., Li et al. 1985; Wolfe and Sharpe 1993).

For example, Figure 3.3 shows the results of comparisons between 42 pairs of orthologous genes of the mouse and rat. The following quantities were estimated: the number of nucleotide substitutions per site (d) in introns; d_S in

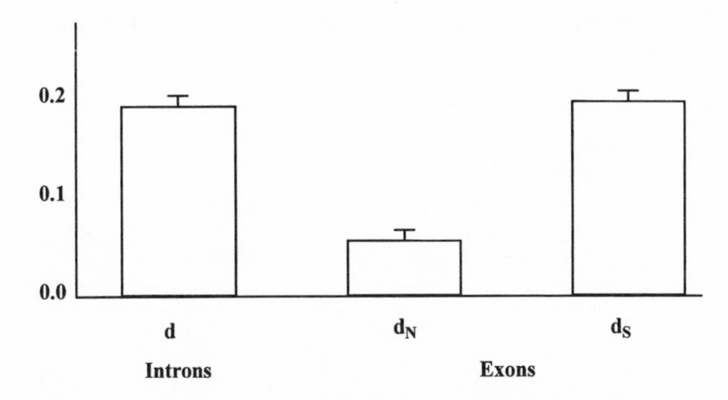

Figure 3.3. Mean numbers of nucleotide substitutions per site (d) in introns and mean numbers of nonsynonymous (d_N) and synonymous (d_S) nucleotide substitutions in exons for comparison of 42 genes between mouse and rat (data from Hughes and Yeager 1997a, 1998a).

exons; and d_N in exons. Mean d in introns and mean d_S in exons were remarkably similar, and both were about three times as high as mean d_N in exons (Figure 3.3). These results show that f_0 is about the same for sites in introns and for synonymous sites in exons. Given the overall lack of functional constraint in introns, it seems likely that both d in introns and d_S in exons reflect the mutation rate; in other words, that f_0 is close to 1 in both cases.

If d_S reflects the mutation rate, comparing d_S and d_N for different regions can provide an additional test of the hypothesis that a reduced d_N value is evidence of heightened functional constraint at the amino acid level. In the example of phenoloxidases and hemocyanins discussed above, I argued that the mutation rate in the copper-binding sites is probably no lower than in surrounding gene regions. In this case, however, all the genes studied were quite distantly related; so that synonymous sites were nearly or completely saturated with changes. Thus, in this case, no rigorous test of the hypothesis that the mutation rate is constant over different gene regions was possible. When more closely related genes are examined, synonymous sites can provide a test of the hypothesis that the mutation rate is uniform over the length of a gene.

Table 3.2 shows comparisons of rates of synonymous and nonsynonymous nucleotide substitution in the comparison between human and mouse platelet-derived growth factor receptor-α (PDGFRα). The mature molecule is a cell-surface receptor having five immunoglobulin-related extracellular domains: two C2-like and one V-like; a transmembrane domain; and a cytoplasmic domain. There is also a signal peptide, which is cleaved from the mature molecule. Values of d_S values are relatively constant across the portions of the gene corresponding to the different domains (Table 3.2). The largest d_S value (that for domain C4) is about 66% greater than the smallest (that for the signal peptide); but none of the differences are statistically significant. Thus, we cannot conclude that mutation rates differ significantly among gene regions and attribute-observed differences to chance fluctuation. By contrast, although d_N is always much lower than d_S, there are striking, statistically significant differences among d_N values for

Table 3.2 Numbers of synonymous (d_S) and nonsynonymous (d_N) nucleotide substitutions per 100 sites (±SE) in different domains of human and mouse PDGFRα genes

Domain	d_S	d_N
Leader	51.3 ± 22.1	17.9 ± 6.9
C1	61.4 ± 14.5	7.4 ± 1.9
C2	75.2 ± 17.9	8.4 ± 2.2
C3	63.1 ± 13.9	7.3 ± 1.9
C4	85.1 ± 19.3	8.3 ± 2.0
V	61.7 ± 12.5	11.9 ± 2.3
Transmembrane	73.0 ± 27.8	4.9 ± 3.1
Cytoplasmic	71.5 ± 6.8	0.8 ± 0.3

different regions (Table 3.2). The highest d_N value (that for the signal peptide) is over 22 times as great as the lowest (that for the cytoplasmic domain) (Table 3.2).

As mentioned, the neutral theory of molecular evolution was very controversial when first proposed in the late 1960s. This controversy continued throughout the 1970s with little hint of a resolution. At this time the data available for testing the theory came from surveys of polymorphism in populations by means of protein electrophoresis and from a relatively limited number of amino acid sequences. The availability of techniques for rapid sequencing of DNA changed the picture quickly. DNA sequence data were found to fit the predictions of the neutral theory quite well—in particular, the prediction that d_S should exceed d_N in most coding regions (Kimura 1977). Furthermore, the discovery of introns and of other extensive noncoding regions in the genomes of higher organisms provided examples of numerous nucleotide sites at which even the most ardent selectionist would be hard pressed to argue the action of strong natural selection. As a result, since the 1980s, in a very real sense there has been a paradigm shift, as the neutral theory has become the predominant paradigm for explaining evolutionary change at the molecular level.

Neither Kimura nor other "neutralists" denied that certain nucleotide substitutions occur as a result of positive selection. They merely asserted that such substitutions are rare relative to those resulting from fixation of selectively neutral variants as a result of genetic drift. For those interested in studying adaptive evolution at the molecular level, then, it is important to be able to distinguish cases where positive selection has been operating from the background level of neutral evolution.

Adaptive Evolution

Classical models of natural selection generally assumed an infinitely large population. This mathematically convenient assumption produced simple results applicable to very large populations but not to the small populations occurring in many species. In an infinite population, a selectively advantageous mutant allele will eventually reach fixation even if the selective advantage is very slight, although the process of reaching fixation may take a very long time. In finite populations, not all advantageous mutants will reach fixation because they may be lost due to genetic drift.

The mathematical theory of neutral evolution permits a straightforward expression for the probability of fixation of advantageous mutants in the case of the genic selection. The term "genic selection" refers to the case of a gene with intermediate dominance in terms of its fitness affects. Thus, for an advantageous mutant allele, a heterozygote having one copy of the gene will have a selective advantage s over the wild type, while a homozygote for the mutant will have a selective advantage $2s$. This, of course, is only one type of positive selection, but it is mathematically the most tractable.

In a genic selection model, a mutant with selective advantage s and initial frequency q will have the following probability of fixation (Pf):

$$Pf = \frac{1 - e^{-4N_e sq}}{1 - e^{-4N_e s}} \tag{3.5}$$

This same equation can be used to derive Pf for a neutral mutation by letting s approach zero. In this case $Pf = q$. In the case of a new mutation occurring on one chromosome in a diploid population, $q = 1/2N$. Thus, for a neutral mutation in such a population, Pf is $1/2N$.

If $4N_e s \gg 1$, then expression (3.5) becomes approximately

$$Pf = 2s(N_e/N) \tag{3.6}$$

This probability will be quite low if s is low. In fact, an advantageous mutant will behave essentially like a neutral mutant if $s < 1/2N_e$ (Kimura 1983).

The rate of substitution of selectively advantageous mutants (K_A) is given by

$$K_A = 4N_e s_A f_A u_T \tag{3.7}$$

where s_A is the average selective advantage of these mutants and f_A is the frequency of mutants that are advantageous. Comparing equations 3.3 and 3.7, we see that the main differences in the case of advantageous mutants is that both effective population and average selective advantage play a role. For a given s_A and f_A, the rate of fixation of advantageous mutants will be lower in a population with smaller effective population size; the reason for this is that, in a small population, even an advantageous mutant has some chance of drifting to extinction before it becomes fixed.

It might be thought that K_A would often be greater than K_0 if N_e is large (e.g., 1,000 or more) and s_A not very small (say, greater than 10%). Because most, if not all, advantageous mutants will be nonsynonymous, it might thus be expected that d_N should exceed d_S. However, as mentioned, this is not usually the case. The most reasonable explanation for this fact is that over all coding regions f_A is much, much lower than f_0.

As mentioned, the fact that d_S generally exceeds d_N can be interpreted as evidence that f_0 is much lower at nonsynonymous sites than at synonymous sites. In the case of two related pseudogenes that are both descended from an ancestral pseudogene, we expect that d_S and d_N will be equal, because in this case, f_0 should equal 1 at all sites. (This will not necessarily be true when we compare a pseudogene to a functional relative, because purifying selection will still be occurring on the functional gene.) However, when natural selection is actually acting to favor amino acid substitutions in some protein region, thus dramatically increasing f_A in that region, then we may actually see $d_N > d_S$. As far as I know, Hill and Hastie (1987) were the first to apply this logic to testing for positive Darwinian selection at the DNA level. As covered in subsequent chapters, this has proven to be a very powerful method of testing for positive selection. It is a conservative test because it will not pick up regions where only one or a small number of nonsynonymous mutations have been selectively favored. Rather, it requires a region where a number of such mutations have been favored, thus substantially increasing f_A.

Endo et al. (1996) surveyed DNA sequence databases for homologous sequences having $d_N > d_S$ for the entire coding region. They found 17 cases, corresponding to 0.5% of the groups of homologous genes surveyed. Although this result was interesting in that it gave some indication of the relative rarity of positive selection, there were several aspects of the study that caused it to overlook probable examples of positive selection. First, the authors excluded one gene family, the vertebrate major histocompatibility complex (MHC), in which positive Darwinian selection is well documented (see Chapter 4). Furthermore, including only cases where d_N exceeds d_S for the entire coding region probably led the authors of this study to overlook the vast majority of cases of positive selection already documented. Natural selection rarely favors amino acid change over an entire coding region, but usually is confined to a specific, functionally important domain (see Chapters 4–7). Thus, some knowledge of the protein structure and biology of the protein under study seems essential to detection of positive selection at the molecular level. Indeed, it might be argued that the study of positive selection is of little interest biologically if it is not coupled with specific knowledge of the structure and biology of the molecules involved. As we will see in subsequent chapters, when rates of synonymous and nonsynonymous nucleotide substitution are compared for regions of known function, the results can yield insights into how protein structures adapt to new functions.

There has been a recent tendency in some quarters to use the term "non-neutral evolution" to include both purifying selection and positive selection. These two are grouped because they both represent departures from "strict neutrality," where "strict neutrality" covers only cases in which $f_0 = 1$, in the terms of equation 3.3. This is a rather restrictive condition, which applies to no protein-coding sequences, all of which, being subject to purifying selection, can be said to be evolving "non-neutrally." To me, this is not a very useful term. It is rather like grouping apples and oranges together on the grounds that they are both "non-pears." As we have seen, purifying selection is the norm in the evolution of protein coding genes. Positive selection is a relative rarity—but of great interest, precisely because it represents a departure from the norm.

4

Balancing Selection

The Major Histocompatibility Complex

In this chapter I discuss natural selection on the genes of the major histocompatibility complex (MHC) of vertebrates. The MHC is a multigene family whose products are cell-surface glycoproteins that play a key role in the immune system by presenting peptides to T cells (Klein 1986). The MHC family includes two major subfamilies called class I and class II. In most vertebrates in which these genes have been mapped, the class I and class II families are linked together in a single gene complex. This complex, located on chromosome 6 in humans, is called the HLA complex (for "human leukocyte antigen") (Figure 4.1). In mammals, class I and class II genes are located in different regions of this complex, which are separated by a third region, sometimes called class III, which contains unrelated genes. Interestingly, it has recently been reported that in the zebrafish the MHC class I and class II genes are in different linkage groups (Bingulac-Popovic et al. 1997), but it is unknown whether this condition is widespread in fishes or, indeed, is found in any other vertebrates.

Because many aspects of its biology are quite well understood, the MHC provides an excellent example of adaptive evolution at the molecular level, probably the example that is best understood at present. The type of natural selection that is acting in this case is balancing selection—where natural selection acts to maintain a polymorphism. So that readers unfamiliar with the MHC or with immunology can appreciate the role of natural selection in this case, I begin this chapter with a review of basic MHC biology. Then I review evidence that MHC polymorphism is maintained by balancing selection relating to the peptide-binding function of the MHC molecules and thus, ultimately, to disease resistance.

The MHC complex first became known through experiments in transplantation (Klein 1986). The loci encoding the molecules now known as class Ia molecules, or "classical class I" molecules, were originally named the "major transplantation antigens" because of their role in transplant rejection. These

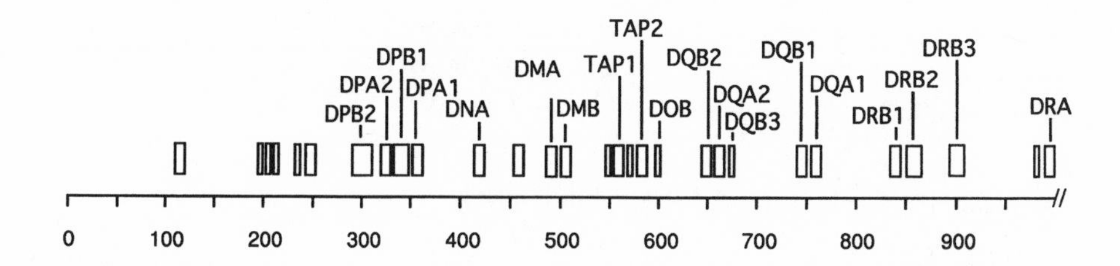

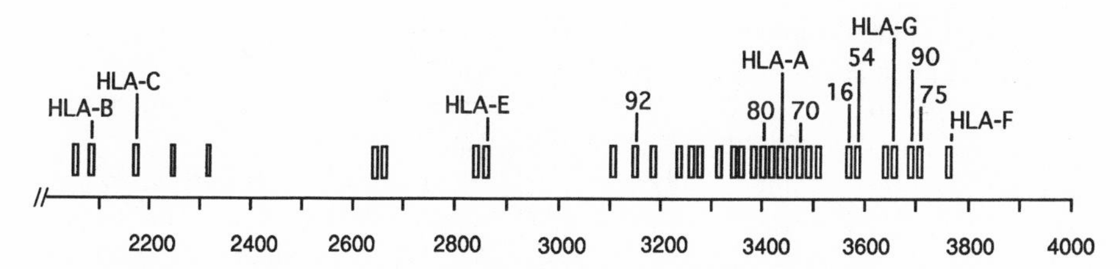

Figure 4.1. Map of the human class I (*bottom*) and class II (*top*) MHC regions. From Hughes (1996a) in M. S. Jackson, T. Strachan, and G. Dover, eds., *Human Genome Evolution,* Bios Scientific Publishers. Reproduced by permission.

loci were found to be highly polymorphic in humans and in the mouse, which was the major experimental organism in transplantation studies. With heterozygosities as high as 80%, these were by far the most polymorphic loci known in vertebrates (Klein 1986). Later, researchers studying genetic differences in immune response between different inbred strains of mice discovered what were called immune response genes. It was a major surprise when the immune response genes were found to map to the MHC region, the very region where the major transplantation antigens were encoded. The immune response genes are now known as class II MHC genes, and several of them are known to be highly polymorphic.

The fact that the polymorphism of MHC loci was known before their function was known has had an important influence on the way in which the field of MHC biology has developed, particularly in the case of evolutionary studies. Before the function of MHC molecules was known, speculative hypotheses were proposed with the goal of simultaneously explaining both the function of MHC molecules and the mechanism of maintenance of polymorphism of the loci. Some of these early hypotheses have taken on a life of their own and continue to persist in the literature, even though they do not take into account recent advances in our knowledge of MHC function.

Intuitively, a hypothesis that accounts for MHC polymorphism in terms of these molecules' function in the immune system seems more plausible than one that ignores that function. Indeed, as we shall see, there is by now fairly convincing evidence that parasites (using the term in the broad sense to include any sort of parasitic organism whether virus, bacterium, fungus, protist, or metazoan) are the main selective agents acting on the MHC of vertebrates and that other selective agents (if present at all) are secondary.

Structure and Function of MHC Molecules

Class I

The polymorphic class I MHC molecules (called the class Ia molecules or class I classical molecules) are glycoproteins expressed on the surface of all nucleated somatic cells; they function to present peptides to cytotoxic T lymphocytes (CTL). The class I molecule is a heterodimer consisting of the following two chains: (1) an α chain or heavy chain, made up of three extracellular domains (designated α_1, α_2, and α_3), a transmembrane region, and a cytoplasmic domain; and (2) a molecule called β2 microglobulin (β2m), which consists of a single domain (Figures 4.2 and 4.3). β2m is noncovalently linked to the α_3 domain. The α chains are encoded within the MHC complex by the polymorphic class Ia loci, of which there are three in humans (*HLA-A*, *HLA-B*, and *HLA-C*) (Figure 4.1). In mammals and probably in other vertebrates, β2m is encoded outside the MHC complex (on chromosome 15 in humans). β2m shows evidence of a distant evolutionary relationship to class I α chains and to class II MHC molecules, but the locus encoding it is not polymorphic.

In all cells, there is a constant turnover of cellular proteins, that are broken down into small peptides by a multimeric proteolytic complex in the cytoplasm known as the proteasome (Orlowski 1990; Rivett 1993). Proteasomes are present in all organisms, but their components are considerably more highly diversified in eukaryotes than in archaebacteria (Rivett 1993). In mammals (and probably

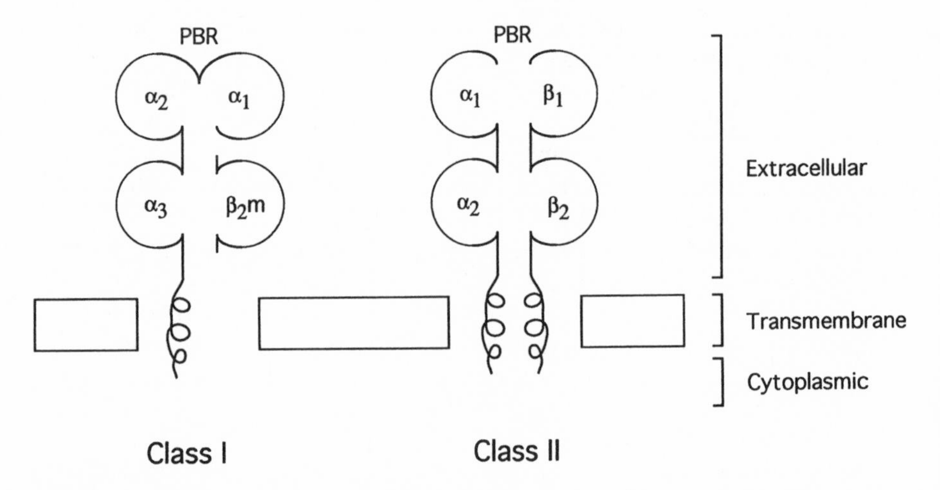

Figure 4.2. Schematic diagrams of MHC class I and class II heterodimers. PBR = peptide-binding region.

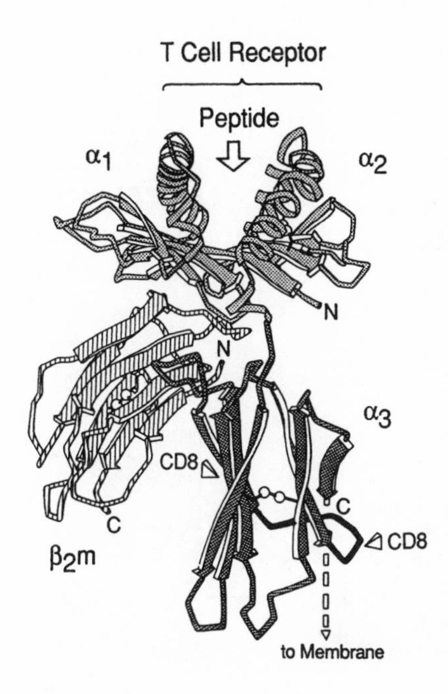

Figure 4.3. The structure of the MHC class I molecule, showing the peptide-binding groove at the top of the molecule. From Lawlor et al. (1990). Reproduced, with permission, from the *Annual Review of Immunology*, volume 9, copyright 1990, by Annual Reviews, Inc.

in most other vertebrates), there are two proteasome components encoded within the MHC class II region, called LMP2 and LMP7. ("LMP" is an abbreviation for "low molecular mass polypeptide.") LMP2 and LMP7 are not expressed under all circumstances. Rather, two constitutively expressed components called X and Y are ordinarily expressed in their place (Belich et al. 1994). The cytokine γ interferon enhances expression of both class I MHC molecules and LMP2 and LMP7. A proteasome containing LMP2 and LMP7 (called an LMP+ proteasome) has an altered specificity with regard to where it cleaves polypeptides; the LMP+ proteasome specifically produces peptides of a sort that are likely to be bound by class I MHC molecules (Driscoll and Finley 1992; Monaco 1992; Germain 1994).

These peptides are transported across the membrane of the endoplasmic reticulum (ER) by a dimeric transporter called TAP. The two subunits of TAP are themselves encoded in the MHC class II region (Figure 4.1). In the ER, a complex is formed involving the class I MHC molecule, the peptide, and β2m, which is then transported to the cell surface. When the first crystal structure of a class I MHC molecule was described, its most striking feature was a groove at the top of the molecule formed by two α helices bordering a β-pleated sheet (Figure 4.4) (Bjorkman et al. 1987a,b). Residues from both the α_1 and α_2 domains contribute to this groove (Figure 4.3). It seemed obvious that this groove was where the peptide is bound, a hypothesis that was confirmed by later crystallographic images of class I molecules complexed with peptides (Guo et al. 1992; Silver et al. 1992). The class I PBR has been shown to consist of five "pockets" (pockets A–F) in which side chains of the peptide residues fit (Saper et al. 1991).

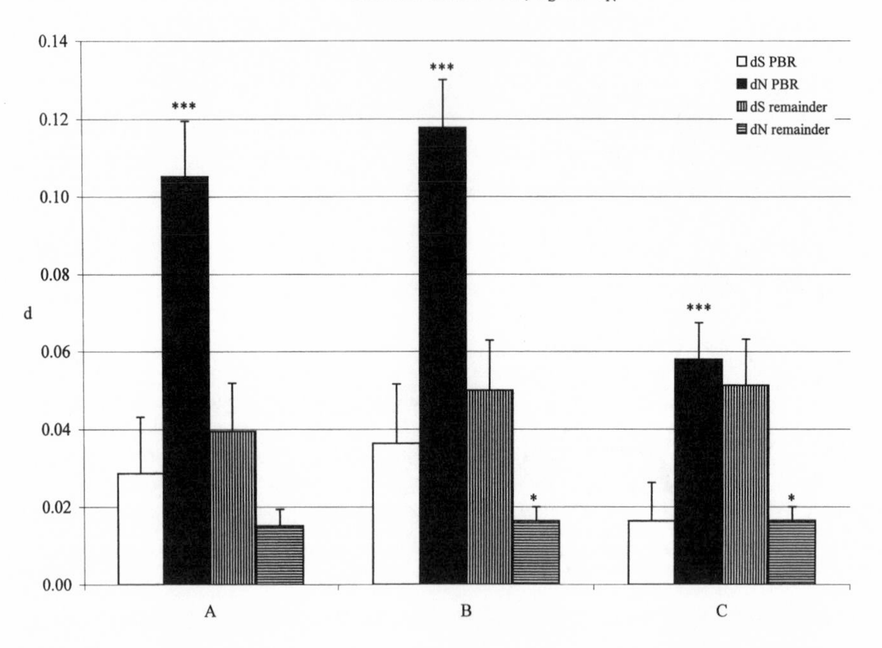

Figure 4.4. Mean numbers of synonymous (d_S) and nonsynonymous (d_N) nucleotide substitutions per site in the peptide binding region (PBR) and remainder of α_1–α_2 domains of human class I MHC loci *HLA-A* (A), *HLA-B* (B), and *HLA-C* (C). Tests of the hypothesis that $d_S = d_N$: $^{*}p < .05$; $^{***}p < .001$. From Hughes and Yeager (1998c). Reproduced, with permission, from the *Annual Review of Genetics*, volume 32, copyright 1998, by Annual Reviews, Inc.

In the case of a normal, healthy cell, the peptides bound by a class I molecule are derived from the cell's own proteins (often called "self-peptides"). CTL exercise a continual surveillance in the body by means of their cell-surface receptors (T cell receptors or TCR). In the development of CTL in the thymus, TCR are selected so that the only CTL permitted to circulate are those that do not attack the complex of self class I MHC and self peptide. However, in the case of infection by a virus or other intracellular parasite, some of the proteins broken down by the proteasome are of parasite origin. Thus, at least some of the class I molecules expressed on the surface of an infected cell will bind "non-self" or "foreign" peptides. When CTL encounter the complex of self class I MHC and foreign peptide, a cytotoxic reaction is initiated that results in killing the infected cell. CTL can only recognize foreign peptides in the context of self class I MHC; this is the phenomenon known as class I MHC restriction of CTL. The CTL and class I MHC together thus provide a drastic solution to the problem of an intracellular parasite: killing all cells that harbor the infection (Berke 1994).

Recently, a strain of laboratory mice has been developed that, because of a mutation, does not express β2m and, thus, does not express class I MHC on its cell surfaces. It has sometimes been said that these mice are "perfectly healthy"—with the implication that the class I MHC is not actually necessary to health. In fact, however, these mice have been shown to suffer severe effects when exposed to intracellular pathogens. For example, these mice showed delayed viral clearance and increase mortality when infected with influenza virus (Bender et al. 1992) and 100% mortality when infected with the intracellular bacterium *Mycobacterium tuberculosis* (Flynn et al. 1992). These results show that the class I MHC plays an essential role in the immune system of vertebrates.

Class II

Class II MHC molecules have a much more restricted expression pattern than do class I molecules, being expressed primarily on antigen-presenting cells of the immune system. The class II molecule presents peptides to helper T cells. In response to a foreign peptide, the helper T cells release cytokines that trigger an appropriate immune response (including the production of antibodies). The class II molecule is similar to the class I molecule in having four extracellular domains, but it achieves this structure in a rather different way (Figure 4.2). The class II molecule is a heterodimer consisting of an α chain and a β chain, each of which in mammals is encoded in the class II region of the MHC complex (Figure 4.1). In placental mammals, the class II region is divided into subregions (designated *DR*, *DP*, and *DQ* in humans), each of which contains a functional α-chain gene and one or more functional β-chain genes (Figure 4.1). The α chain includes two extracellular domains (α_1 and α_2), a transmembrane region, and a cytoplasmic tail.

Like the class I molecule, the class II molecule binds peptides in a groove at the top of the molecule. As with class I, the class II peptide-binding groove consists of two α helices bordering a β-pleated sheet. The difference is that in class II, one of the α helices and about half of the β-pleated sheet are contributed by the α chain, while the other α helix and the other half of the β-pleated sheet come from the β chain. Unlike the peptides presented by class I, which are mainly nine amino acids in length, the peptides presented by class II molecules can vary substantially in length, between about 11 and 17 residues (Rammensee et al. 1995). The reason for this difference is that in the case of class I the ends of the peptide are tucked down into the peptide-binding groove, imposing a limitation of the peptide's length. In class II, the peptide's ends are free, making its length less constrained.

The complex between the class II molecule and its peptide ligand is created by a mechanism quite distinct from that of class I. Before transport to the cell surface, the class II dimer forms a complex with a polypeptide known as the invariant chain (Ii). This complex then travels to an acidic endosome-like compartment (Peters et al. 1991). There Ii is degraded, and the class II molecule binds the peptide, which it transports to the cell surface. A molecule known as

DM serves as a chaperone, facilitating the loading of peptides by class II molecules (Kropshoffer et al. 1997). Interestingly, DM is clearly evolutionarily related to the class II molecule; it consists of an α chain and a β chain, each of which shows clear evidence of homology to the corresponding chains of the class II heterodimer.

Explaining MHC Polymorphism

The Overdominance Hypothesis

Zinkernagel and Doherty (1974) were the first to demonstrate class I MHC restriction of antigen (i.e., peptide) recognition by CTL, work for which they received the Nobel Prize in 1996. Immediately after this discovery, Doherty and Zinkernagel (1975) proposed the first hypothesis to account for MHC polymorphism to take into account the actual biological function of these molecules. Doherty and Zinkernagel presented evidence that different class I MHC gene products differ with respect to the antigens that they can present. In other words, to express the concept in terms of our current knowledge of MHC function, different allelic products bind different arrays of peptides. Thus, they argued, in a population exposed to an array of pathogens, it will be advantageous for an individual to be heterozygous at MHC loci because a heterozygote will be able to present a broader array of antigens and thus resist a broader array of pathogens. Such a mechanism of heterozygote advantage (also known as "overdominant selection") could account for the extraordinary polymorphism found at MHC loci.

Doherty and Zinkernagel's early evidence pointing to a difference between different MHC gene products with respect to the peptides they bind has been spectacularly confirmed in recent years. It is now possible to elute and directly sequence peptides bound by MHC molecules. Comparisons of many such peptides have shown that the peptides bound by a specific MHC allelic product invariably contain one or more characteristic residues—called anchor residues because they anchor the peptide into the binding groove (Table 4.1). In the case of the class I MHC, the anchor residues are the second residue (P2) of the peptide, which fits into the B pocket, and/or the ninth residue (P9) of the peptide, which fits into the F pocket (Table 4.1; Rammensee et al. 1995). In the case of class II, because the ends of the peptide may hang outside the binding groove, the position of the anchor residues relative to the N-terminus of the peptide is not fixed. However, there are usually nine central residues that include the anchors, usually at positions 1, 4, 6, and 9 of these nine residues.

Because positions other than the anchor residues seem to be relatively free to vary, each allelic product can potentially bind thousands of different peptides. Nonetheless, because different allelic products have different anchor motifs, a heterozygote will presumably have a much broader immune surveillance than a homozygote. For example, HLA-B*39011 prefers the positively charged residues R or H in the second residue of the peptide, while HLA-B*4403 prefers

Table 4.1. Examples of anchor residues (in boldface) and auxiliary anchor residues of peptides bound by the human class I loci MHC *HLA-A* and *HLA-B*

	Residue position								
Allele	1	2	3	4	5	6	7	8	9 (10)[a]
A1		**T**	**D**				L		**Y**
		S	E						
A*0201		**L**							**V**
		M							**L**
A68.1		**V**							**R**
		T							**K**
B*39011		**H**				I			
		R				V			L
						L			
B*4403		**E**							Y
									F
B53		**P**							

a. The last residue position of the peptide is usually number 9, but may be number 10. Data from Rammensee et al. (1995).

the negatively charged residue E (Table 4.1). An individual heterozygous for these two alleles will be able to bind both types of peptides.

Overdominant selection was initially discovered as a theoretical possibility by population geneticists, long before any real-world example was known. The best-studied example in a natural population is that of sickle cell anemia in humans. The sickle cell allele, although extremely deleterious in the homozygote, is maintained in populations exposed to malaria because the heterozygote is malaria resistant. Aside from that example, however, others were not easily discovered.

Doherty and Zinkernagel's hypothesis that overdominant selection maintains MHC polymorphism did not initially meet with wide acceptance. Some population geneticists had the mistaken impression that overdominant selection cannot maintain a polymorphism as extensive as that seen at MHC loci. This impression was based on theoretical models that did not take into account the role of mutation in incorporating new alleles. More realistic models that incorporated the role of mutation showed that overdominant selection is indeed capable of maintaining a high level of polymorphism (Maruyama and Nei 1981).

An additional problem was that it was very hard to test the hypothesis of overdominant selection at MHC loci by means of a conventional population study. In an outbred species such as human or mouse, most individuals are heterozygous at most MHC loci. Thus, it would be necessary to survey many thousands of individuals to amass a large enough sample of homozygotes to compare their fitness with that of heterozygotes. Furthermore, if the selective advantage possessed by heterozygotes were small—say one or a few percent—

then the sample size would have to be still larger in order to have the statistical power to test for a difference between homozygotes and heterozygotes.

Pattern of Nucleotide Substitution

Because of these difficulties, Hughes and Nei (1988) decided to take a different approach to testing Doherty and Zinkernagel's hypothesis. At this time a number of nucleotide sequences for MHC class I genes had become available. As mentioned previously (Chapter 3, The Dynamics of Neutral Evolution), in most genes the rate of synonymous nucleotide substitution per site (d_S) exceeds that of nonsynonymous substitution per site (d_N). Because a theoretical study by Maruyama and Nei (1981) had predicted that overdominant selection should enhance the rate of codon substitution, Hughes and Nei reasoned that, if Doherty and Zinkernagel's hypothesis were true, d_N should be enhanced in the case of MHC genes. Furthermore, the first crystal structure of a class I MHC molecule had recently been published (Figure 4.3), revealing the peptide-binding groove. On the hypothesis that MHC polymorphism is maintained by overdominant selection relating to peptide binding, Hughes and Nei (1988) predicted that an enhanced nonsynonymous rate should be seen mainly in the codons encoding the peptide-binding region (PBR) of the molecule.

The results dramatically confirmed this prediction. Figure 4.4 shows the results of recent analyses using far more sequences than were available to Hughes and Nei (1988), but the results are essentially the same as they reported. In the 57 codons encoding the PBR, d_N significantly exceeds d_S (Figure 4.4). By contrast, in the non-PBR portions of the α_1 and α_2 domains, d_S exceeds d_N, as is true of most genes (Figure 4.4). Note that d_S values do not differ greatly from one gene region to another. Because d_S is expected to reflect the mutation rate (the fraction of neutral mutations at synonymous sites being close to 100%), the uniform value of d_S indicates that the enhanced value of d_N in the PBR codons cannot be explained by a higher mutation rate in those codons. Rather, the results strongly support the hypothesis that positive Darwinian selection has acted to enhance the rate of nonsynonymous substitution in the PBR codons and thus to enhance amino acid diversity in the PBR.

In the case of the class II MHC, a hypothetical structure was proposed by analogy with the known class I structure (Brown et al. 1988). Using this hypothetical structure, Hughes and Nei (1989a) found $d_N > d_S$ in the putative PBR. When a class II crystal structure was at last obtained (Brown et al. 1993), this finding was confirmed (Figure 4.5; Hughes et al. 1994).

It is still uncertain whether the selection acting at MHC loci is overdominant, as hypothesized by Doherty and Zinkernagel, or rather represents some other form of balancing selection. One form of balancing selection that has received a great deal of attention in the literature of theoretical population genetics is frequency-dependent selection. Actually, there are several different models of frequency-dependent selection; some of these are theoretically capable of maintaining a high level of polymorphism such as seen at MHC loci (Takahata and Nei 1990). In the case of the MHC, however, there is a clear rationale for

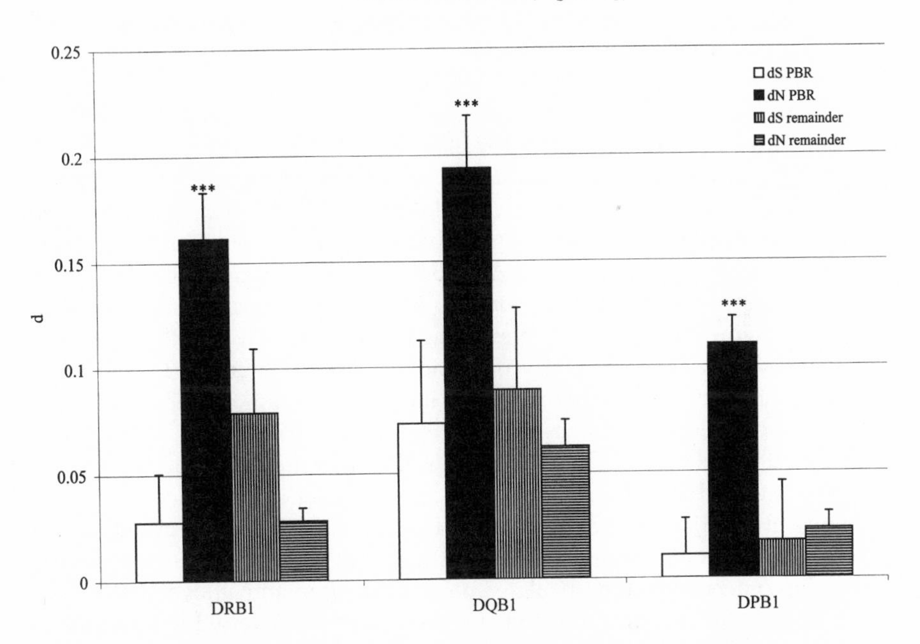

Figure 4.5. Mean numbers of synonymous (d_S) and nonsynonymous (d_N) nucleotide substitutions per site in the peptide-binding region (PBR) and remainder of β_1 domains of human class II MHC loci. Tests of the hypothesis that $d_S = d_N$: $^{***}p < .001$. From Hughes and Yeager (1998c). Reproduced, with permission, from the *Annual Review of Genetics*, volume 32, copyright 1998, by Annual Reviews, Inc.

overdominant selection based on the function of the molecules: namely, that heterozygotes have an advantage derived from a broader immune surveillance because a heterozygote can bind a broader spectrum of peptides than can a homozygote.

Parasites and Overdominant Selection

Hill and colleagues (1991) expressed a common misunderstanding of how overdominant selection might operate at MHC loci. In a study of West African children, they found a particular class I MHC allelic product (HLA-B53) and a particular class II haplotype to be associated with resistance to severe malaria, caused by the protist parasite *Plasmodium falciparum* (see Chapter 5, The Circumsporozoite Protein of *Plasmodium*). Because they did not find heterozygotes to be more resistant to malaria than homozygotes, they claimed to have evidence against the hypothesis of overdominant selection. However, Doherty and Zinkernagel's hypothesis requires that the population be exposed to at least two distinct pathogens. These might simply be two genetically distinct strains of the same parasite species, or they might actually be separate species. If one MHC allele confers resistance to the first parasite, while another allele confers resistance to the second parasite, it is easy to see that heterozygotes bearing both alleles will have an advantage. Because Hill et al. (1991) considered only one parasite, *P. falciparum*, and did not examine genetic polymorphism within the parasite itself, their results are not relevant to the issue of whether selection at MHC loci is overdominant (Hughes and Nei 1992).

In a more recent paper, Hill's research group claimed to have found evidence of overdominant selection on the class II region of West Africans (Thursz et al. 1997). An initial study had shown an association in this population between the presence of the class II allele *HLA-DRB1**1302* and clearance of infection by the hepatitis B virus (HBV) (Thursz et al. 1995). A follow-up study reported that the frequency of heterozygotes in the region of the *HLA-DRB1* and *HLA-DQB1* loci (as assessed by means of restriction enzymes) was significantly lower among individuals with persistent HBV infection than among individuals who had cleared the virus (Thursz et al. 1997). No such difference was seen in the case of the class I loci. In an attempt to explain why they observed an apparent heterozygote advantage in resistance to HBV but not in resistance to malaria, Thursz et al. (1997, p. 12) remarked: "This clearer evidence of overdominant selection with HBV may in part reflect the limited antigenic composition of the virus, encoded by the smallest genome of any human pathogen." However, it is unclear how the size of the viral genome could be relevant to selection on the host's MHC. A more plausible explanation for the difference between the types of selection exerted by HBV and by *P. falciparum* relates to the high mutation rate of the virus and the fact that the malarial peptide presented by the class I MHC is highly conserved.

HLA-B53, the class I molecule associated with resistance to severe malaria, was found to present a peptide (KPIVQYDNF) derived from a protein called liver stage antigen-1 (LSA-1). LSA-1, which is expressed on the surface of the parasite during the stage when it inhabits the host's liver parenchyma cells (see Chapter 5, The Circumsporozoite Protein of *Plasmodium*), is highly conserved; and the peptide presented by HLA-B53 is particularly well conserved. The 3′ region of the LSA-1 gene, which includes this peptide, has now been sequenced from 18 different alleles from around the world (from Africa, Brazil, New Guinea, and Thailand). In 17 of these, the peptide presented by HLA-B53 is totally conserved. In one allele from Brazil, there is a conservative amino acid change; the K at the N-terminus of the peptide is replaced by R. Because both K and R are positively charged, and because this residue is apparently not an important anchor residue for HLA-B53, this variant form of the peptide may be recognized by HLA-B53. In any event, it seems unlikely that West Africans ever encounter allelic forms of LSA-1 that do not include a peptide that HLA-B53 can present.

By contrast, HBV, a virus with a small DNA genome, has a much higher mutation rate than any eukaryote. Mutations in regions of HBV proteins that constitute peptides bound by class I and class II MHC have been detected in patients having chronic HBV infection (Bertoletti et al. 1994; Carman et al. 1997; Tai et al. 1997). Although the contribution of such mutations to persistence of HBV infection is not fully established (Rehermann et al. 1995), it seems a reasonable hypothesis that a heterozygote would be able to defend itself better against such an antigenically variable parasite.

Species with Low MHC Polymorphism

The two best studied MHCs are those of humans and the mouse, both of which have high levels of polymorphism. However, it has been known for some time that there are certain species in which MHC polymorphism is much more limited. For example, in the cottontop tamarin *Saguinus oedipus*, a New World primate, extensive sequencing of class I alleles has revealed only 11 class I classical sequences, which probably represent alleles at three loci (Watkins et al. 1990, 1991). By contrast, over 250 alleles are known from the three human class I classical loci *HLA-A*, *-B*, and *-C* (Bodmer et al. 1997).

It has sometimes been proposed that species having limited MHC polymorphism are ones that lack exposure to many pathogens in their natural environment. An alternative hypothesis is that these species simply have low long-term effective population sizes. Even in the presence of overdominant selection, only a small number of alleles can be maintained at a locus if the effective population size is low. Estimation of rates of synonymous and nonsynonymous nucleotide substitution was used to decide between these hypotheses in the case of the cottontop tamarin. The results of this analysis showed that, in the tamarin just as in humans, d_N significantly exceeds d_S in the PBR of class I molecules (Watkins et al. 1990, 1991). Therefore, a similar type of selection to that seen in humans and the mouse is acting in the case of the cottontop tamarin.

Other Hypotheses for MHC Polymorphism

Mutational Hypotheses

One early hypothesis to explain MHC polymorphism was independent of the molecules' function. This was the hypothesis that MHC loci have an unusually high mutation rate (Bailey and Kohn 1965). DNA sequence data have made it possible to test this hypothesis rigorously. Because d_S is expected to reflect the mutation rate, d_S values for MHC genes can be compared with those of other genes to assess the comparative magnitude of the mutation rate at MHC loci. For example, Figure 4.6 shows d_S between homologous human and murine rodent (mouse or rat) C-type domains of the immunoglobulin superfamily. The conserved domains of MHC molecules (including α_3 of class I and α_2 and β_2 of class II) are immunoglobulin C-type domains. In Figure 4.6, data for class II α_2 and β_2 domains are included because humans and rodents share orthologous class II loci but not class I loci (Hughes and Nei 1989b, 1990). Mean d_S for the class II MHC domains turns out to be slightly lower than the overall mean for 80 domains (Figure 4.6). These and other data (Hughes and Nei 1988, 1989a) show that the mutation rate at MHC loci is not particularly high.

A more recent version of essentially the same hypothesis held that MHC polymorphism was enhanced by interlocus recombination ("gene conversion")

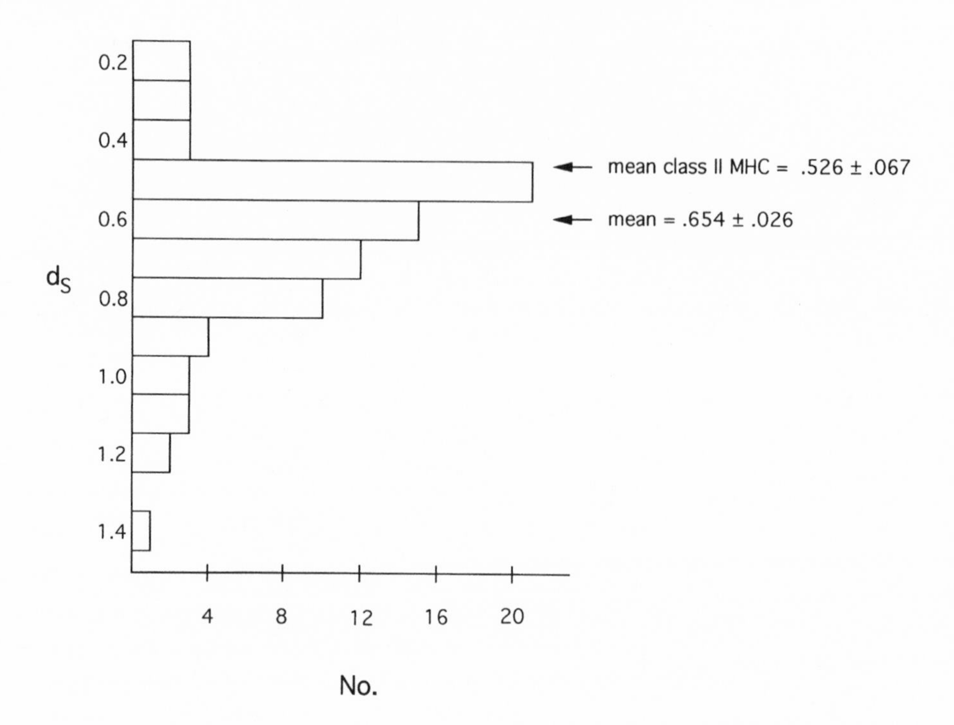

Figure 4.6. Distribution of values of the number of synonymous nucleotide substitutions per site (d_S) in comparison of immunoglobulin superfamily C-type domains between orthologous loci of humans and murine rodents (mouse or rat). The mean for class II MHC and the overall mean are indicated. From Hughes and Hughes (1995a).

(Lopez de Castro et al. 1982; Ohta 1982). Theoretically, it is possible that if members of a gene family have diverged from each other at the sequence level and interlocus recombination subsequently occurs, polymorphism at each locus will be enhanced. However, gene conversion is expected to be an essentially random process; thus, it cannot explain the very specific pattern of $d_N > d_S$ in the PBR codons that characterizes MHC loci.

Maternal–Fetal Interactions

Because the function of MHC molecules was unknown for a long time, the earliest hypotheses to explain MHC polymorphism often attempted to explain both MHC function and polymorphism. This was difficult when the only known function of the MHC was its role in transplant rejection—a function that seemed difficult to correlate with any aspect of a species' natural environment. One researcher tried to make such a connection when he proposed that the function of the MHC was to prevent cancer from being a contagious disease!

Other hypotheses relied on analogies between the MHC and other biological systems. Most influential were hypotheses that saw an analogy between the MHC and the self-incompatibility systems of plants (see Chapter 5, Self-Incompatibility Genes of Plants). The self-incompatibility loci are also extraordinarily polymorphic, and it was tempting to see the MHC as a vertebrate analogue. The result of this analogy was a proliferation of hypotheses relating the MHC to reproduction (e.g., Thomas 1974). Even though the real function of the MHC is now known, some of these hypotheses have assumed a life of their own in the literature, and have continued to attract adherents.

One hypothesis was that MHC polymorphism is maintained by maternal–fetal interactions (Clarke and Kirby 1966). This hypothesis depends on the assumption that the production of maternal antibodies to fetal class I MHC molecules has a beneficial effect on fetal growth and survival. Although some early studies seemed to show such an effect, it has not been supported by subsequent work (James 1965, 1967; Clarke 1971; Wegmann 1984). Furthermore, it is hard to imagine how maternal–fetal interactions would lead to natural selection favoring diversity specifically in the PBR. On this hypothesis, one would predict that selection would enhance the rate of nonsynonymous substitution in epitopes for maternal antibodies. In the case of class I MHC molecules, these epitopes are known to be scattered throughout the α_1 and α_2 domains, rather than being concentrated in the PBR (Nathenson et al. 1986; Bjorkman et al. 1987b). Also, this hypothesis cannot account for the fact that MHC polymorphism is known to be high in fishes, amphibians, and birds, all of which lack maternal–fetal interactions. Finally, it cannot account for class II polymorphism, because class II molecules are only expressed on antigen-presenting cells of the immune system and, thus, are unlikely to be involved in maternal–fetal interactions.

Mate Choice

Another hypothesis is that MHC polymorphism is maintained by disassortative mating on the basis of MHC genotype. This idea was originally proposed by Thomas (1974), and appeared to achieve some support from experiments with mice by Yamazaki and colleagues (1976). However, although these experiments have been widely cited, the results were actually inconclusive. Furthermore, there were a number of serious flaws in the experimental design used by these authors.

Yamazaki and colleagues (1976) used congenic strains of laboratory mice; that is, mice that are identical at all loci except at the MHC. A single male was placed in an enclosure with two females—one of the same strain as himself and the other of a different strain; and the male was said to "prefer" the female with which he mated first. In this design, however, there were several uncontrolled factors, other than choice by the male, that might influence the outcome. One such factor might be a difference between the two females with regard to their motivation or ability to evade the male's pursuit; the "chosen" female might simply be the less successful at evasion. Possible behavioral interactions, including dominance relationships, between the two females represented another uncontrolled factor. Furthermore, both males and females were reused in different tests; individuals' learning regarding the test procedure could, thus, have influenced their behavior in subsequent trials. As a result, the trials were not truly independent, and the reported levels of statistical significance cannot be treated as reliable. From a theoretical point of view these experiments can be criticized for considering male choice of mates rather than female choice of mates, because the latter is believed to be a more important factor in determining mating patterns in natural populations of most animal species, particularly in species like mice, in which males are not involved in rearing the young (Trivers 1972).

Even with inflated significance levels, the results of Yamazaki et al.'s (1976) experiments were hardly conclusive. Only two of six strain combinations tested actually showed "significant" evidence of MHC-associated disassortative mating; three showed no "significant" tendency; and one actually showed a "significant" tendency toward MHC-assortative mating. Other studies using the same experimental design (Yamaguchi et al. 1978; Yamazaki et al. 1978) were equally inconclusive (Table 4.2).

Egid and Brown (1989) introduced a much-improved experimental design into the study of MHC and mate choice in mice. Their apparatus was a Y-maze in with two tethered male mice that could not see each other. The female subject was free to move throughout the apparatus and thus to express a preference by mating with one male or the other. Visual and tactile interactions between the males were thus eliminated, although acoustic and olfactory communication between the two males remained uncontrolled factors. Egid and Brown (1989) used this experimental design with only two strains of inbred mice, one of which showed a significant preference for males of the other strain, while the other showed no preference (Table 4.2). When Eklund et al. (1991) used the same type of design to test male choice of females, they found no significant preference in either of the two strains tested. Manning et al. (1992) used a similar apparatus with females of the inbred strain B10 and with females with one B10 parent and one wild parent; in both cases, there was no significant preference. Overall, then, the results of mate choice experiments do not support the hypothesis that mice mate disassortatively on the basis of MHC genotype (Table 4.2).

Yamazaki and co-workers (1979) showed, using a well-designed experimental procedure, that mice have the ability to differentiate among the urines of certain strains of mice differing only at the MHC. However, this finding does not prove that mice actually use this ability to choose mates (or for any other purpose) in nature. After all, it turns out that humans are also able to distinguish some urines of mice, differing only at the MHC (Gilbert et al. 1986)! Clearly, this is not an ability that humans make use of in daily life. Furthermore, it is interesting to note that not all MHC differences are detectable by mice. Mice lack the ability to discriminate between the urines of certain mutants known to differ only at the class I locus *H-2K* (Yamazaki et al. 1990). Thus, there are limits to the ability of mice to discriminate among MHC genotypes by olfaction.

An additional study was that of Potts and colleagues (1991), which was an observational study of mice released into a large enclosure. The authors described this as a "seminatural" enclosure, but at least with regard to the social environment, the situation was quite unnatural. Nine populations were set up in the

Table 4.2. Studies of MHC and mate choice in mice

Reference	Protocol test subjects	Test stimuli	Genotype combinations	Preference[a]		No reference
				Assortative	Disassortative	
Yamazaki et al. (1976)	Males	Unrestrained females	6	1	2	3
Yamazaki et al. (1978)	Males	Unrestrained females	14	3	4	7
Yamaguchi et al. (1978)	Males	Unrestrained females	6	2	0	4
Eklund et al. (1991)	Males	Tethered females	2	0	0	2
	Male choice totals			6	6	16
Egid and Brown (1989)	Females	Tethered males	2	0	2	0
Manning et al. (1992)	Females	Tethered males	3	0	0	3
	Female choice totals			0	2	3
	All tests			6	8	19

a. Statistically significant preference, as reported by authors of the study. Because of nonindependence of tests, the reported significance levels do not always correspond to true significance levels.

enclosure, each including 24–31 mice of about the same age. Thus, the mice lacked the age structure and kinship patterns that would be found in a natural population. Furthermore, no dispersal outside the enclosure was permitted. There is substantial evidence that in small mammals the major mechanism of inbreeding avoidance is dispersal from the natal area–although it seems to be unclear whether in the house mouse this dispersal is mainly by males (as is true of the majority of mammalian species that have been studied) or by females (Lidicker and Patton 1987; Smith 1993). Especially if female dispersal predominates in the house mouse, the behavior observed in this enclosure was likely to be highly unnatural.

When data from the nine populations were pooled, Potts et al. (1991) found a small but significant excess of heterozygotes in the offspring produced. Because laboratory mating experiments showed no evidence of selective abortion of offspring of MHC-similar parents, they attributed their results to assortative mating rather than to selective abortion. However, because relevant controls were lacking, several alternative explanations could not be ruled out. One simple hypothesis is that mice have an innate tendency to avoid mating with close kin, which are recognized by odors, perhaps including (but not limited to) MHC-associated odors. In the artificial populations of Potts and co-workers, close kin were presumably absent. In this artificial situation the tendency to avoid mating with close kin might lead to a slight tendency to avoid mating with MHC-similar individuals.

It is important to realize that mice—like other vertebrates—recognize individuals, not simply categories of individuals (such as kin versus nonkin or suitable mates versus unsuitable mates) (e.g., Hughes 1989). In the case of mice and most other mammals, olfaction is clearly an important aspect of individual recognition. Because of the high polymorphism of MHC loci, the MHC genotype would be useful in individual recognition if the information it contains could be used by conspecifics. The fact that mice can recognize some MHC-associated differences through urine odors suggests that they are able to make use of some of this information.

The fact that mice recognize individuals suggests a way of discriminating between the hypothesis that mice have a true MHC-based mating preference and the alternative hypothesis that they simply avoid mating with close kin. Each of a number of female subjects would be given a choice between the following: (1) a full brother with which the female has been raised in the same litter but that shares no MHC haplotype with her (as is possible when both parents are heterozygous for MHC haplotype); and (2) an unfamiliar, unrelated male that shares one or both MHC haplotypes with the female. On the hypothesis that mate choice by mice primarily involves avoidance of mating with close kin, females should prefer the latter male. On the hypothesis of MHC-associated mate choice, they should prefer the former male.

One major weakness of any hypothesis that explains MHC polymorphism by mate choice via olfaction is that it seems unlikely to apply to organisms that have poorly developed senses of smell and rely mainly on visual cues in behavioral interaction. Humans and birds are examples; yet in both of these groups MHC polymorphism is high.

Data from the S-leut Hutterite religious isolate have recently been presented as evidence for MHC-associated disassortative mate choice in humans (Ober et al. 1997). It was claimed that spouses in this population shared MHC haplotypes less frequently than would be expected under random mating (Ober et al. 1997). This is an endogamous sect descended from a small number of founders, which has increased rapidly over the past century due to high fertility. Members avoid marriage to first cousins and, of course, to any kin closer than first cousins. In such a population, the avoidance of inbreeding alone will cause spouses to share MHC haplotypes less frequently than would be expected in a random mating population having the same haplotype frequencies. To test the hypothesis that individuals are truly basing marriage choices on MHC type (however, that might be assessed in humans), it is necessary to compare the frequency of MHC haplotype sharing between actual spouses with that between potential spouses (i.e., individuals of the appropriate sex, age, and degree of kinship to have been chosen as spouses). So far, no such comparison has been made. Another recent study in South American Amerindians revealed no evidence of MHC-associated mate choice (Hedrick and Black 1997). The Amerindian populations studied probably have a population structure more typical of the human species throughout most of its history than do the highly inbred Hutterites.

Trans-species Polymorphism at MHC Loci

The comparison of rates of synonymous and nonsynonymous nucleotide substitution is not the only line of evidence that points to balancing selection at MHC loci. There are several independent lines of evidence pointing in the same direction. One is the observation that allelic frequencies at MHC loci do not fit the neutral expectation (Hedrick and Thompson 1983). At polymorphic but neutrally evolving loci, there is usually one common allele and a few other rarer alleles. By contrast, at MHC loci, there are a large number of alleles, most with intermediate frequencies. The latter pattern is suggestive of balancing selection (Hedrick and Thompson 1983).

One especially interesting characteristic of MHC polymorphism that provides additional strong support for the hypothesis of balancing selection is the phenomenon called "trans-species polymorphism." Arden and Klein (1982), on the basis of serological data from mice, suggested that often a given MHC allele of one species might be more closely related to certain alleles of closely related species than it is to any other allele of its own species. Sequencing of class I and class II MHC alleles from humans and chimpanzees supported this idea (Lawlor et al. 1988; Mayer et al. 1988; Gyllensten and Erlich 1989). Figure 4.7 shows an example of trans-species polymorphism in the case of the class II *DQB1* locus in the case of humans and chimpanzees. In the phylogenetic tree of DQB1 alleles, the human alleles *HLA-DQB1*0302* and *HLA-DQB1*03032* cluster with the chimpanzee allele *Patr-DQB1*0302* (Figure 4.7). By contrast, *HLA-DQB1*0601*, *HLA-DQB1*06011*, and related alleles cluster with *Patr-DQB1*O601*, *Patr-DQB1*0602*, and *Patr-DQB1*0604* (Figure 4.8). These clusters of alleles represent allelic lineages that were present in the common ancestor of chimpanzees and humans and have persisted in each population since their divergence 5–7 million years ago.

Neutral polymorphisms are not expected to persist very long in populations. The coalescence theory predicts that, for pairs of neutral alleles selected at random from a locus in a randomly mating population, their mean coalescence time (that is, the time of their last common ancestor) will be $2N_e$ generations (Kingman 1982; Tajima 1983; Takahata and Nei 1985). Assuming a long-term effective population size of 10^4 for humans, the mean coalescence time for neutral alleles would be only 600,000 years. Thus, neutral polymorphism is, with respect to evolutionary time, a relatively transient phenomenon.

Under balancing selection, Takahata and Nei (1990) showed that polymorphisms can persist much longer than in the neutral case. Using computer simulation, these authors studied overdominant selection and several models of frequency-dependent selection. They found that under overdominant selection and one type of frequency-dependent selection (which Takahata and Nei called "minority advantage"), it was possible to maintain polymorphisms for very long

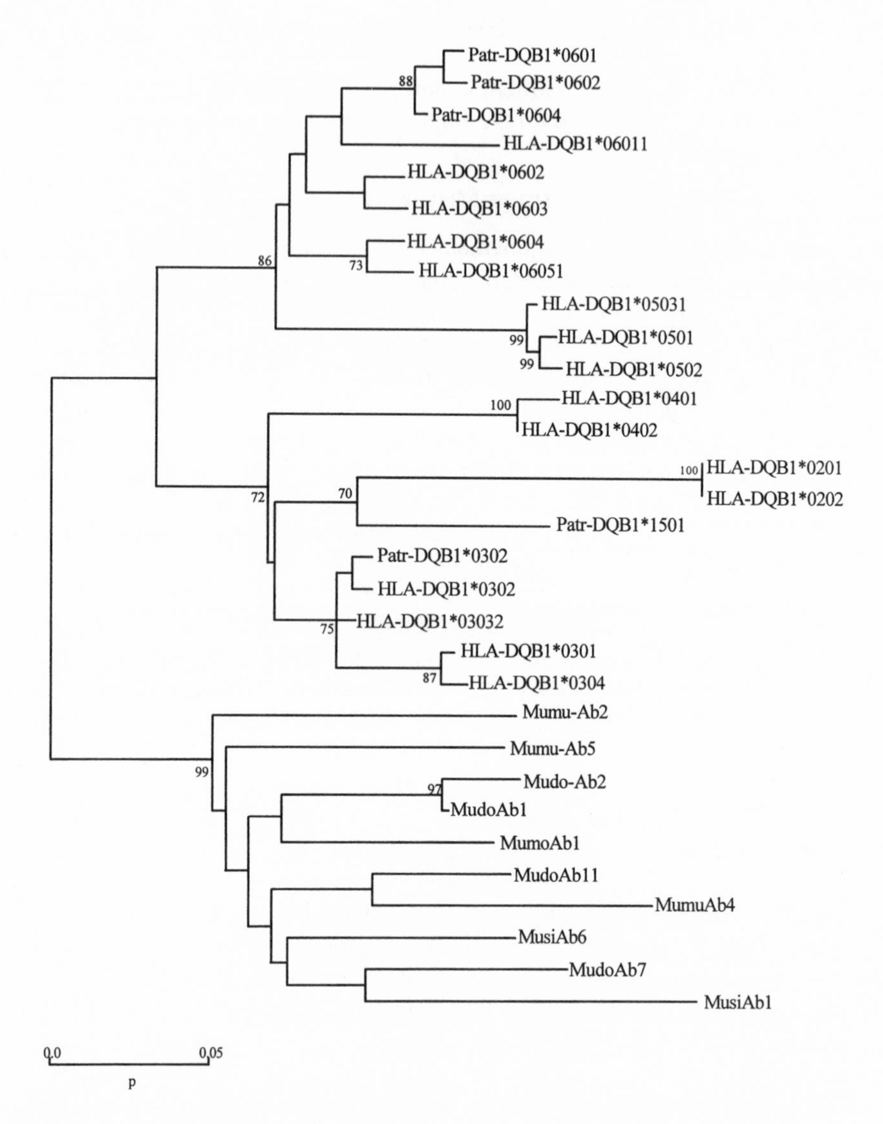

Figure 4.7. Phylogenetic tree of Human (HLA-), chimpanzee (Patr-), *Mus musculus* (Mumu-), *Mus domesticus* (Mudo-), and *Mus spicilegus* (Musi-) class II MHC DQB (Ab) sequences. The tree was constructed by the neighbor-joining method on the basis of the proportion of amino acid difference (p). Numbers on the branches are as in Figure 3.2. From Hughes and Yeager (1998c). Reproduced, with permission, from the *Annual Review of Genetics*, volume 32, copyright 1998, by Annual Reviews, Inc.

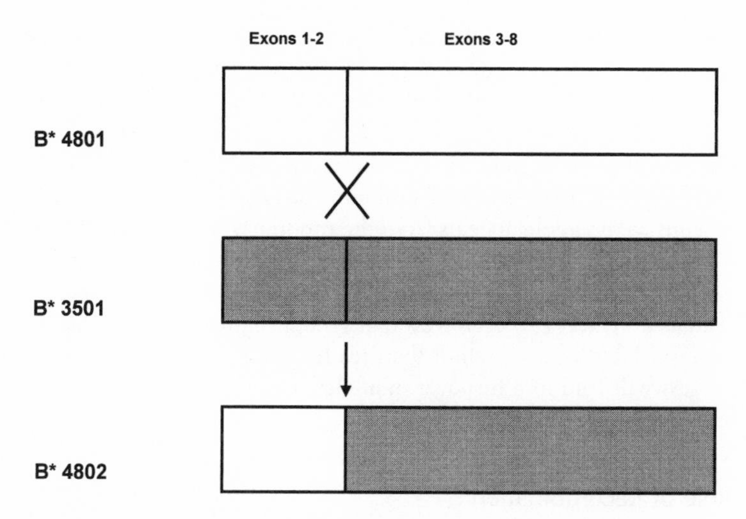

Figure 4.8. Example of a "large-scale" recombination event between two human class I MHC alleles. The *HLA-B*4802* allele, found in the Waorani of South America (Watkins et al. 1992) arose as a recombinant, presumably by crossing over, between the alleles *HLA-B*4801* and *HLA-B*3501*.

times even with relatively modest selection coefficients. Thus, either of these types of balancing selection can account for the long coalescence times of alleles at MHC loci.

In the model of minority advantage, it was assumed that a genotype had a selective advantage whenever it became rare in the population (Takahata and Nei 1990). Mathematically, this model turns out to be essentially the same as that of overdominant selection; however, from a biological point of view it might well be questioned whether it is truly applicable to the MHC (Takahata and Nei 1990).

Consider a parasite species that is easily eliminated by its host because most members of the host species bear an MHC allele (*A1*) whose product can bind and present a peptide from a given protein of the parasite. Then, suppose that a mutation occurs in the parasite so that this MHC allele can no longer bind the peptide, and the parasite is now able to infect most members of the host species with impunity. Suppose, however, that there is a rare MHC allele (A2) in the host species whose product can bind another peptide from this parasite and protect against infection. Clearly, the A2 allele will have a selective advantage and will increase in frequency.

Now consider what will happen to the *A1* allele. The minority advantage model assumes that, once the A2 allele becomes common, the *A1* allele will again have a selective advantage. But how would this happen realistically in the case of the MHC? It might be that the parasite will mutate again so that the A2 allelic product can no longer efficiently bind a peptide from the parasite.

But it seems rather unlikely that this new escape mutant will somehow restore binding by the *A1* allelic product. Yet this is precisely what the minority advantage model requires. Therefore, this model seems inapplicable to the case of the MHC.

On the other hand, after the parasite mutates so that the A2 allelic product no longer binds a peptide from its proteins, it might happen that still another new mutant MHC allele appears (A3), the product of which can efficiently bind a peptide from the parasite. If this happens, we can expect that the A3 allele will increase in frequency. This model is called the *pathogen adaptation model* by Takahata and Nei (1990). However, this model cannot explain what is happening at MHC loci because, rather than leading to a long-lasting polymorphism, this process will lead to a turnover of alleles over time (Hughes and Nei 1988; Takahata and Nei 1990).

The Role of Recombination

In one respect, the observation of trans-species polymorphism at MHC loci is paradoxical. Many authors, in examining MHC sequences, have suggested that recombination frequently occurs among alleles at MHC loci (e.g., Lawlor et al. 1990; She et al. 1991). Some putative recombination events have apparently resulted from ordinary crossing over, leading to hybrid alleles including one or more exons from each parent allele (Figure 4.8). These were termed "large-scale" recombination events by Hughes et al. (1993). In other cases, sequence comparison suggests that a relatively short sequence motif has been transferred from one allele to another (Figure 4.9). These were termed "small-scale" recombination events by Hughes et al. (1993). One problem with reconstructing small-scale recombination events is that sharing of a short sequence motif by two otherwise distantly related alleles may reflect convergent evolution rather than recombination. Certainly, there is evidence that convergent evolution has occurred at certain MHC loci. For example, certain human and bovine class II DRB alleles have serine residues at corresponding sites, but these are encoded by different serine codon sets, suggesting that the residue has evolved independently rather than being inherited from the common ancestor of the two species (Andersson et al. 1991).

The paradox arises from the expectation that, if recombination events have occurred at a high frequency over evolutionary time, allelic lineages will be scrambled; thus, trans-species polymorphism will not be detectable. The resolution of this apparent paradox seems to be that there are differences among loci with respect to the proportion of alleles that are recombinants. At loci with high proportions of recombinant alleles, the identity of allelic lineages is indeed obscured, whereas at loci with lower proportions of recombinants, allelic lineages can maintain distinct identities for long periods.

Such differences among loci are perhaps most striking in the case of the human class I loci *HLA-A*, *-B*, and *-C* (Hughes et al. 1993; McAdam et al. 1994, 1995; Yeager and Hughes 1996). Recent cases of large-scale recombination—cases in which there is no doubt that a recombination has occurred—have all

```
consensus   CAGATCTCCAAGACCAACACACAGACTTACCGAGAGAGCCTGCGGAACCTGCGCGG   320

B*0702      .......A....G..C.GG........G............................
B*0801      .......T...................G............................
B*1301      .....................................A......C.C.GC..T.C.
B*1302      .....................................A......C.C.GC..T.C.
B*1401      .......G...................G............................
B*1402      .......G...................G............................
B*1501      ........................................................
B*1801      ........................................................
B*2702      .......G....G....GG........G.........A........T.GC..T.C.
B*2705      .......G....G....GG........G........GA........C.....T.C.
B*3501      .......T................................................
B*3502      .......T................................................
B*3701      ....................................GA........C.....T.C.
B*4001      ........................................................
B*4002      ........................................................
B*4101      ........................................................
B*4201      .......A....G..C.GG........G............................
B*4401      .....................................A......C.C.GCTGC.C.
B*4402      .....................................A......C.C.GC..T.C.
B*4601      ....AG.A....CG.C.GG........G......T.....................
B*4701      ....................................GA........C.....T.C.
B*4901      .....................................A........T.GC..T.C.
B*5101      .......T.............................A........T.GC..T.C.
B*5201      .....................................A........T.GC..T.C.
B*5301      .......T.............................A........T.GC..T.C.
B*5701      .G..A.A.G...G..TC.G.G................A........T.GC..T.C.
B*5801      .G..A.ATG...G..TC.G.G................A........T.GC..T.C.
B*7801      .......T...................G............................
B*3903      .......G...................G............................
B*4801      ........................................................
B*4802      ........................................................
B*3504      .......T................................................
B*52012     .....................................A........T.GC..T.C.
B*1504      ........................................................
B*3505      .......T................................................
B*3506      .......T................................................
B*4003      ........................................................
B*4004      ........................................................
B*5104      ........................................................
Patr-B1     .G..ATATG...G..TC.G........G.........A........T.GC..T.C.
Patr-B2     ..................G.....................................
Patr-B3     .G..A.GTG...G..TC.G........G.........A........T.GC..T.C.
Patr-B5     ..................G.....................................
```

Figure 4.9. A portion of the DNA sequence of exon of humans and chimpanzees (Patr-) class I MHC *B* locus alleles, showing a sequence motif (underlined) that has apparently been repeatedly exchanged among alleles. From Hughes et al. (1993).

involved recombination between the second and third exons of the gene. Exon 2 encodes the α_1 domain, while exon 3 encodes the α_2 domain; so recombination between these two exons will create a new form of the PBR. Such a recombinant may be selectively favored because it creates a new peptide-binding specificity, and thus may increase in frequency in the population. Because of the frequency of recombinants between exons 2 and 3, one simple way of gauging the frequency of recombination in the evolutionary history of a locus is to compare the sequence divergence of alleles in exons 1–2 with that in exons 3–8.

Figure 4.10 shows plots of the number of nucleotide substitutions per site (*d*) in exons 3–8 versus *d* in exons 1–2 for pairwise comparisons among human and chimpanzee alleles at the A, B, and C loci. In the case of the A locus, there is a strong positive correlation between *d* in exons 1–2 and that in exons 3–8 (Figure 4.10A). This means that, at the A locus, when two alleles are very similar in sequence in exons 1–2, they are also likely to be similar in exons 3–8; and when two alleles are very divergent in exons 1–2, they are also likely to be divergent in exons 3–8. Such a pattern is expected when both portions of the gene are evolving as a unit, with little recombination. By contrast, in the case of the *B* locus, there is no correlation between *d* in exons 3–8 and that in exons 1–2 (Figure 4.10B). This indicates that many *B* locus alleles arose as a result of recombination between these two gene regions. Finally, the C locus resembles the A locus in having a positive correlation between *d* in exons 3–8 and that in exons 1.2 (Figure 4.10C).

Several other lines of evidence have supported the hypothesis that a much higher proportion of *B* locus alleles than of *A* or *C* locus alleles have arisen as a result of recombination events. Stephens (1985) developed a statistical method for detecting putative small-scale recombination events. Application of this method revealed more such putative events in the case of the *B* locus than in the case of the *A* or *C* loci (Hughes et al. 1993; Yeager and Hughes 1996). A phylogenetic tree of *B* alleles based on exon 2 (Figure 4.11A) showed a very different topology from that based on exons 3–8 (Figure 4.11B), providing further support for the hypothesis that different regions of these alleles have different evolutionary histories. Interestingly, only exon 2 shows evidence of trans-species polymorphism, with certain chimpanzee alleles clustering with human alleles rather than with other chimpanzee alleles (Figure 4.11A).

Figure 4.10. (facing page) Plot of the number of nucleotide substitutions per site in exons 3–8 versus that in exons 1–2 for pairwise comparisons among the class I MHC A (**A**), *B* (**B**), and *C* (**C**) loci of humans and chimpanzees. The lines shown are linear regression lines: (**A**) $Y = 0.0176 + 0.4837X$, $r = .526$ ($p < .001$); (**B**) $Y = 0.0705 + 0.0304X$, $r = 0.153$ (n.s.); (**C**) $Y = 0.0108 + 0.7317X$, $r = .532$ ($p < .001$). To provide a conservative test of the correlation coefficient in this case (where the pairwise comparisons are not independent of each other), a two-tailed test with $N - 2$ degrees of freedom (where N = the number of sequences compared) was used. From Yeager and Hughes (1996).

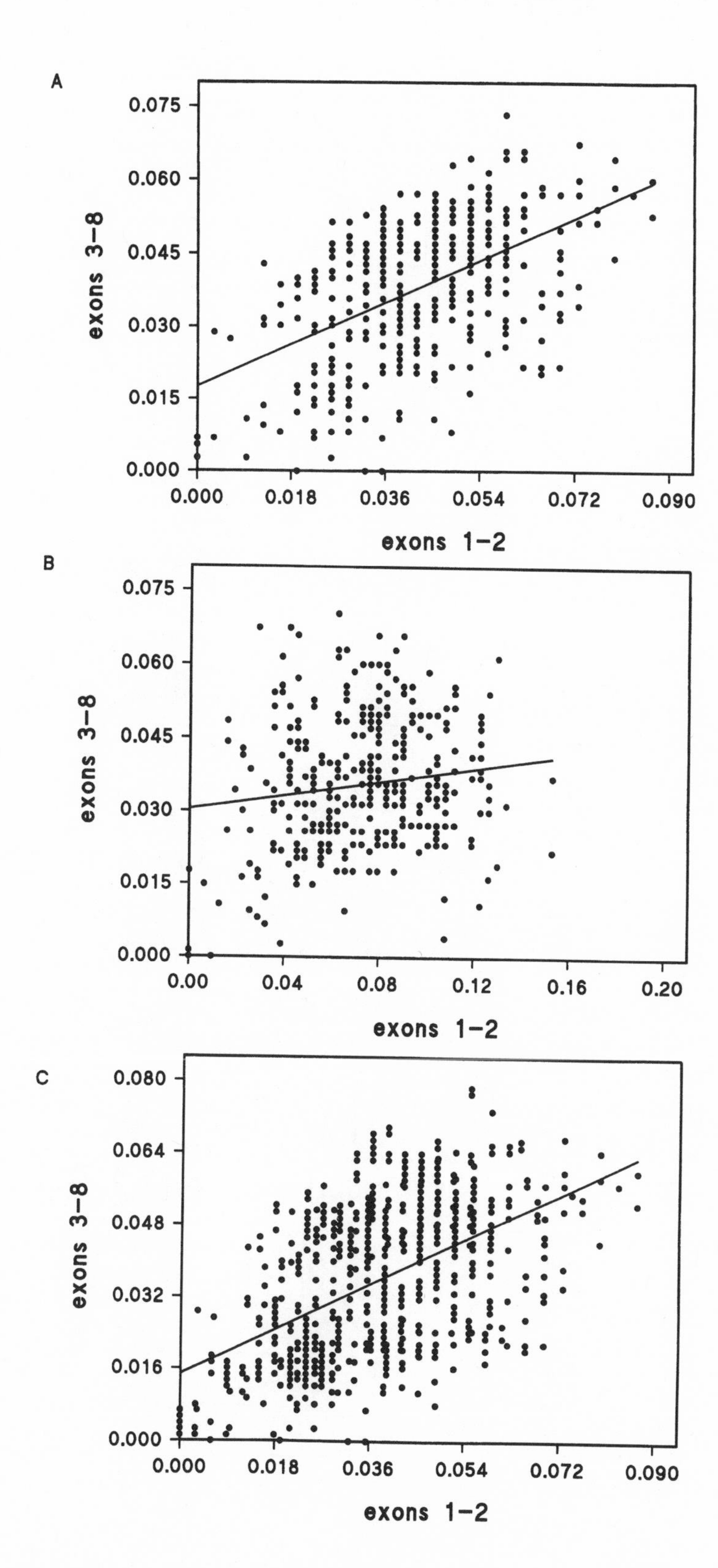
A
0.075
0.060
0.045
0.030
0.015
0.000
exons 3-8
0.000
0.018
0.036
0.054
0.072
0.090
exons 1-2
B
0.075
0.060
0.045
0.030
0.015
0.000
exons 3-8
0.00
0.04
0.08
0.12
0.16
0.20
exons 1-2
C
0.080
0.064
0.048
0.032
0.016
0.000
exons 3-8
0.000
0.018
0.036
0.054
0.072
0.090
exons 1-2

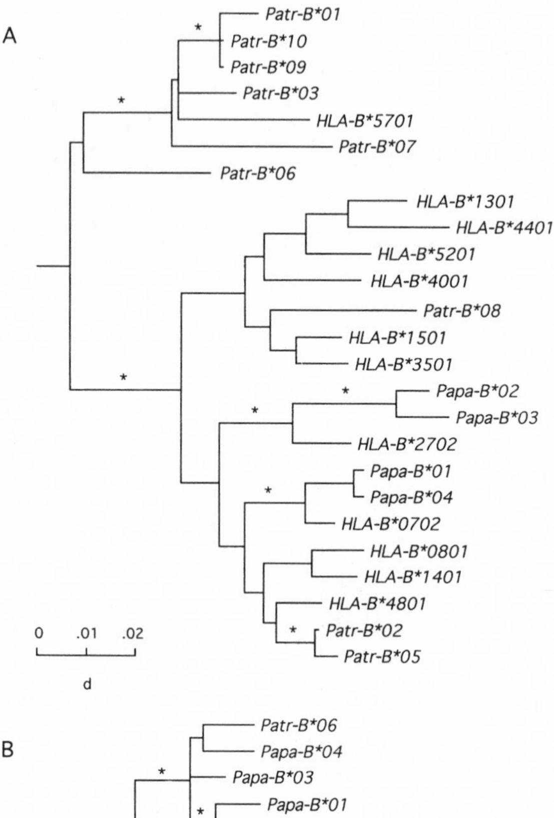

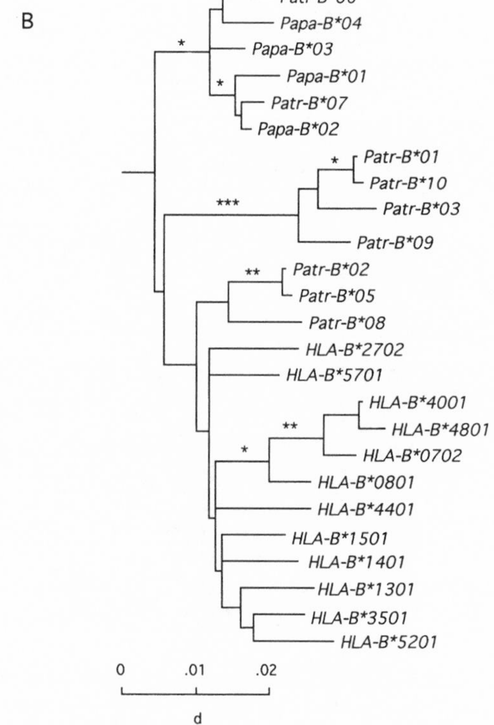

Figure 4.11. Phylogenetic trees of exons 1–2 (**A**) and exons 3–8 (**B**) of human and chimpanzee class I MHC *B* locus alleles. The trees were constructed by the neighbor-joining method based on the number of nucleotide substitutions per site (*d*). Tests of the hypothesis that branch length equals zero: $^{*}p < .05$; $^{**}p < .01$; $^{***}p < .001$.

In the case of class II β-chain loci, several studies have suggested that recombination events have occurred in both primates and rodents (e.g., Gyllensten et al. 1991; She et al. 1991). These events are hypothesized to have taken place within exon 2 of the gene, which encodes the β_1 domain. Because this exon is relatively short (containing about 90 codons), it has been difficult to provide statistical support for the hypothesis of recombination. The β_1 domain consists of one α-helix and a portion of a β-pleated sheet. Gyllensten et al. (1991) showed that MP trees based on these two regions had different topologies; however, because of the small numbers of sites involved, it was unclear whether this difference was due to stochastic error rather than to past recombination events.

Because of the difficulties of testing for recombination at class II loci, Hughes and Yeager (1998b) took a new approach, in which they looked for evidence of nonhomogeneity of sequence similarity in a sliding window of 10 nucleotide positions along the length of class II β-chain exon 2 sequences. They computed the proportion of nucleotide difference (p) in each such window for each pairwise comparison along alleles at a locus. The coefficient of variation (C.V.) of p provides an index of the degree of nonhomogeneity of similarity along the sequence. Nonhomogeneity of similarity along sequences can be caused by other factors besides recombination. The main such factor, of course, in the case of the MHC is that natural selection is acting to favor nonsynonymous substitutions at codons encoding PBR residues but not at other codons. To minimize the influence of positive selection, the codons encoding PBR residues were eliminated from the sequences prior to analysis.

Comparisons of C.V. values for the three human and two mouse polymorphic class II β-chain loci are illustrated in Figure 4.12. A much higher proportion of comparisons had high C.V. values in the case of *HLA-DRB1* and *-DPB1* loci

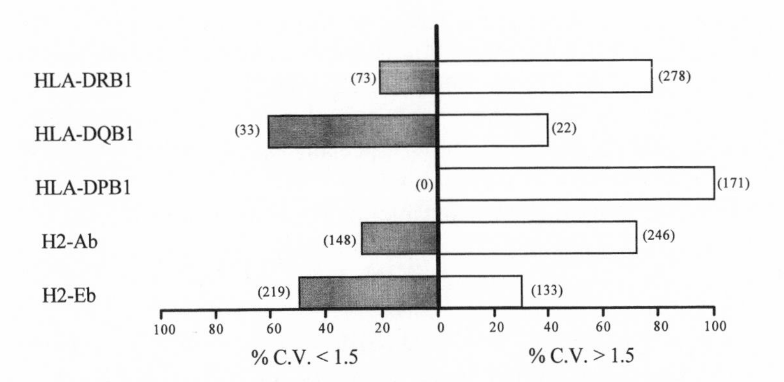

Figure 4.12. Coefficient of variation (C.V.) in a sliding window of nucleotide positions (excluding the peptide binding region) in exon 2 of human (HLA-) and mouse (H2-) MHC class II β-chain genes. The percentage of C.V. values greater than and less than 1.5 is shown. From Hughes and Yeager (1998b).

than in the case of the -*DQB1* locus (Figure 4.12). Thus, the latter locus appears to have a lower proportion of recombinant alleles than do the former two loci. In the mouse, the results suggest that a greater proportion of alleles at *Ab* are recombinants than at *Eb* (Figure 4.12). These results are consistent with the conclusions of previous studies. For example, phylogenetic analyses suggested that recombination has occurred frequently at the human *DRB1* locus (Gyllensten et al. 1991) and at the mouse *Ab* locus (She et al. 1991). In the case of the human *DPB1* locus, an experimental study involving PCR amplification of sequences from a single sperm suggested that there is a high rate of recombination at this locus (Zangenberg et al. 1995).

Statistical analyses of sequence data thus suggest that, in both class I and class II, MHC loci differ with respect to the proportion of alleles that have arisen as a result of intralocus recombination. These analyses cannot directly address the mechanisms responsible for such differences. One hypothesis to explain such differences would be that loci, perhaps as a result of some property of the sequence, differ with respect to the rate at which recombination among alleles occurs. An alternative hypothesis would be that the rate of recombination is the same for all loci; but at certain loci new recombinant alleles are more likely to be selectively favored and thus to increase to appreciable frequencies in the population. So far, available data do not permit us to decide between these two hypotheses. The studies involving PCR of *DPB* alleles from sperm might be taken as evidence that the germline rate of recombination at this locus is high. However, this conclusion is uncertain because the same technique has not been applied to any other locus—whether MHC or non-MHC—and, thus, we have no basis for comparison.

On the other hand, there is some evidence that seems to implicate natural selection in explaining observed patterns. Related loci often differ markedly with respect to the proportion of alleles that are recombinant. For example, *HLA-B* and -*C* diverged 10–20 million years ago, yet the proportion of recombinants is much higher among alleles at the former locus (Yeager and Hughes 1996). In class II, human *DQB* and mouse *Ab* are orthologues, but the proportion of recombinants at the latter locus seems much higher (Figure 4.12). If recombination rate is a function of sequences features, such results might not be expected. However, further comparative studies—especially experimental studies of germline recombination rates—are required before this question can be answered.

Some authors have suggested that interlocus recombination or "gene conversion" has played an important role in MHC evolution. There has been empirical evidence of such a process in a few cases (Nathenson et al. 1986; Hughes 1995a), and several theoretical models have considered the interaction of interlocus recombination with natural selection at MHC loci (e.g., Ohta 1991). However, the overall evidence is that interlocus recombination at MHC loci is relatively rare. For example, phylogenetic analyses of alleles from human class I classical loci revealed locus-specific clusters (Nei et al. 1997). This pattern would not be expected if interlocus recombination is frequent. Some immunologists, while conceding that interlocus recombination is rare in primates, have maintained that it is common in rodents (e.g., Rada et al. 1990). Again, this view is simply not supported by the data. Class I classical alleles of rodents also form locus-specific clusters in phylogenetic analyses (Hughes 1991b).

Class I Introns

Recently, intron sequences have become available for a number of alleles at the human class I loci *HLA-A*, *-B*, and *-C*. The evolutionary dynamics of class I introns seems to differ strikingly from that of exons in ways that may seem surprising to some immunologists. Yet the properties of class I introns are, in fact, exactly what one would predict in the case of balancing selection. Thus, the analysis of class I introns has provided an additional, independent line of evidence that polymorphism at these loci is maintained by balancing selection.

Figure 4.13 shows plots of mean proportion of nucleotide difference (p) for all pairwise comparisons among alleles at the *HLA-A*, *-B*, and *-C* loci. The regions of the gene compared are exons 2–3, which encode the α_1 and α_2 domains, including the PBR codons; exon 4, which encodes the conserved α_3 domain; and the first three introns of the gene (introns 1–3). One striking aspect of these plots is that the mean p is generally lower in the introns than in the exons; this is particularly true of intron 3, which is much longer than either intron 1 or intron 2 (Figure 4.13). In most genes, this pattern would be reversed: p would be higher in introns than in exons because purifying selection eliminates most nonsynonymous mutations in exons (Hughes and Yeager 1997a).

A detailed examination of patterns of nucleotide substitution explains the unusual results of the sliding-window analysis. At each locus, the mean number of nucleotide substitutions per site (d) in introns 1–3 was compared with mean d_S in exons 2–3 (Table 4.3). Mean d_S in the exons was generally higher than mean d in the introns. This was particularly true for intron 3, and was most striking in the case of the *B* locus. At the *B* locus, mean d in intron 3 was less than 1%, whereas mean d_S in exons 2–3 was nearly 5% and seven times higher than mean d in intron 3 (Table 4.3). This result is very unusual because in the case of most genes d in introns and d_S in exons are about equal (see Figure 3.3).

The most reasonable explanation for the fact that d in introns of human class I genes is often lower than d_S in exons is that introns are homogenized by interallelic recombination and subsequent genetic drift (Cereb et al. 1997). The exons of these genes—particularly those encoding the PBR—are quite ancient, having been maintained for millions of years by balancing selection. However, this selection does not apply to introns. Although intron sequences may "hitch-hike" along with exon sequences to some extent, if recombination and drift lead to loss of ancient polymorphism in an intron, this will be selectively neutral. Thus, introns of MHC genes are expected to be evolutionarily younger on average than are the exons encoding the PBR. Both d_S in exons and d in introns are expected to reflect the mutation rate, because most mutations at synonymous sites and at sites in introns are selectively neutral. When d_S in the exons is much higher than d in adjacent introns, the most straightforward interpretation is that the exons are older than the introns.

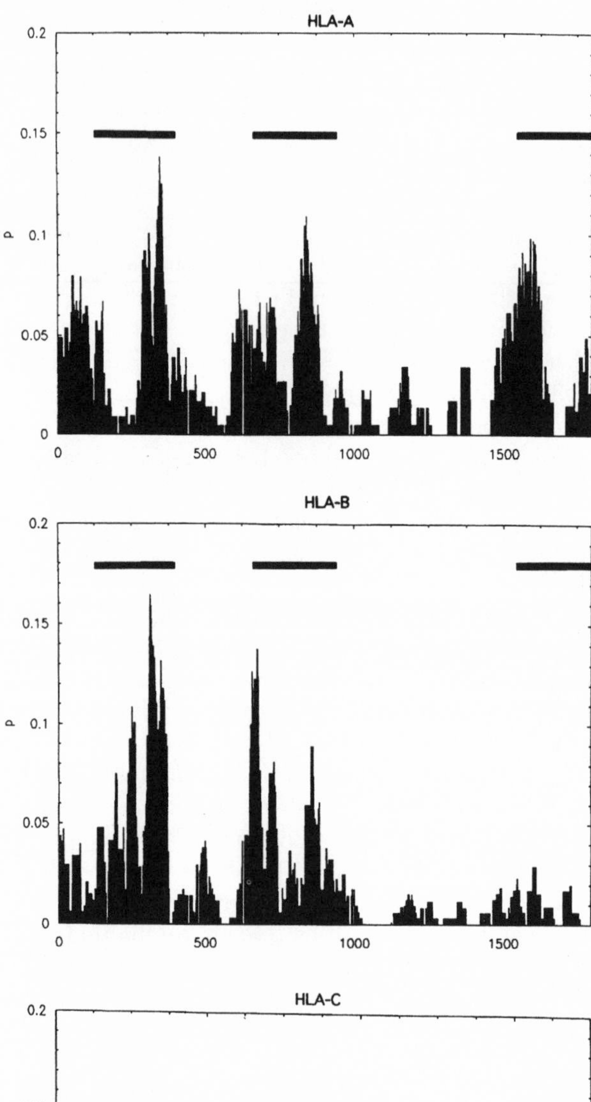
HLA-A
0.2
0.15
0.1
0.05
0
p
0
500
1000
1500
HLA-B
0.2
0.15
0.1
0.05
0
p
0
500
1000
1500

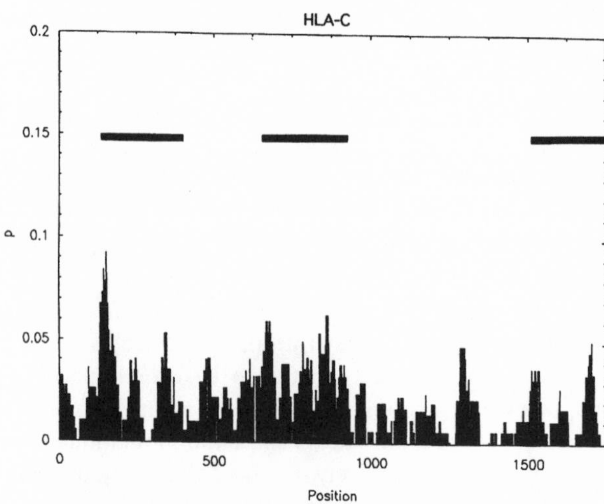
HLA-C
0.2
0.15
0.1
0.05
0
p
0
500
1000
1500
Position

Table 4.3. Number of nucleotide substitutions per 100 sites (d) in introns 1, 2, and 3 and per 100 synonymous sites (d_S) in exons 2–3 for comparisons among HLA class I alleles

Comparison		d			d_S
		Intron 1	Intron 2	Intron 3	Exons 2–3
Means for all pairwise comparisons (intralocus)[a]	*HLA-A* locus	4.2 ± 1.2	2.3 ± 0.6	2.2 ± 0.4	3.5 ± 0.8
	HLA-B locus	2.5 ± 0.9	1.6 ± 0.4***	0.7 ± 0.2***	4.9 ± 0.9
	HLA-C locus	1.6 ± 0.6	1.8 ± 0.5	1.4 ± 0.3**	3.6 ± 0.9
	All intralocus	2.7 ± 0.9	1.8 ± 0.5***	1.1 + 0.3***	4.4 + 0.9
Selected individual comparisons	A*1101 vs. A*3002	0.8 ± 0.8**	2.8 ± 1.0*	0.7 ± 0.4**	9.3 ± 2.8
	A*2501 vs. A*2601	0.0 ± 0.0	0.0 ± 0.0	2.6 ± 0.7***	0.0 ± 0.0
	B*0702 vs. B*4201	0.0 ± 0.0	1.3 ± 0.7	0.7 ± 0.4	2.7 ± 1.4
	B*0702 vs. B*5401	5.0 ± 0.2	0.0 ± 0.0	1.5 ± 0.5	6.8 ± 2.3
	Cw*0602 vs. Cw*1203	1.6 ± 1.2	1.7 ± 0.9	0.4 ± 0.3*	4.6 ± 1.9

a. Numbers of alleles compared for each locus are as follows; *HLA-A*, 15; *HLA-B*, 23; *HLA-C*, 12. Tests of the hypothesis that d in an intron equals d_S in exons 2–3: $^*p < .05$; $^{**}p < .01$; $^{***}p < .001$.

Population geneticists have extensively studied the problem of polymorphism at a locus linked to one under balancing selection (Nei and Li 1980; Strobeck 1983; Kreitman and Hudson 1991). These studies predict that the degree of hitchhiking—and, thus, the extent of polymorphism—at such a locus will be a function of the extent of recombination between that locus and the one under selection. Extending these predictions to the case of polymorphic class I MHC loci, we expect that introns more closely linked to exons 2–3 (encoding the PBR) will show higher levels of polymorphism than those less closely linked to exons 2–3. Introns 2 and 3 are relatively short (130 and 268 aligned nucleotide sites, respectively); and intron 1 is located just 5′ to exon 2, while intron 2 is located between exons 2 and 3. By contrast, intron 3, located 3′ to exon 3, contains 653 aligned sites; so, on average, nucleotide sites in intron 3 will be less closely linked to nonsynonymous PBR sites than will sites in introns 1 and 2. Thus, we might predict that intron 3 will be more likely to be homogenized by recombination and subsequent drift than will introns 1–2; the latter two introns are predicted to hitchhike more closely with the PBR exons, and thus to have higher levels of nucleotide diversity.

These predictions are supported by the data (Table 4.3). Intron 3 sequences show the lowest mean d for all three loci (Table 4.3). The extent of homogeniza-

Figure 4.13. (facing page) The proportion of nucleotide difference (p) is a sliding window of 30 aligned nucleotide positions in comparisons among alleles at human class I MHC loci. The region covered is from the first intron to the fourth exon. Horizontal bars indicate the positions of exons 2, 3, and 4. From Cereb et al. (1997).

tion of intron 3 is particularly striking in the case of *HLA-B*. This is interesting because of evidence (see this Chapter, Trans-species Polymorphism at MHC Loci) that this locus has a higher rate of recombination than *HLA-A* or *-C*, and it suggests that the higher rate of recombination at the *B* locus characterizes both introns and exons. The difference between introns and exons is that in the latter case recombination can enhance polymorphism because natural selection will maintain at least some of the new variants that it creates, whereas in introns, recombination will serve to reduce polymorphism over evolutionary time by breaking up linkage between introns and exons.

Comparisons between individual sequences reveal some apparent recent cases of recombination. For example, the alleles A*2501 and A*2601 are identical in exons 2–3 and in introns 1 and 2, yet differ markedly in intron 3 (Table 4.3), suggesting that this intron has been recently donated to one of these alleles by a more distantly related allele. On the other hand, *B*0702* and *B*5401*, though identical in intron 2, are highly divergent in the remainder of their sequence (Table 4.3). In this case, recombination seems to have caused the homogenization of intron 2 between these two alleles. These examples show that recombination in itself does not cause homogenization of introns among all alleles at a locus. Rather, over evolutionary time, genetic drift will lead to homogenization of introns, given that recombination occurs.

Bergstrom and colleagues (1997) recently sequenced portions of the second intron from the human class II *DRB1* locus. They found that the intron sequences were much more similar to each other than were sequences of exon 2 (which includes the PBR); indeed, intron sequences were even more similar than were synonymous sites in exons. These results are very similar to the class I results, suggesting that introns are younger than exons because introns have been homogenized relative to exons by recombination and subsequent genetic drift. However, Bergstrom et al. (1997) favor a different interpretation of their data. They argue, in fact, that the introns reveal the true age of the MHC alleles. Thus, according to these authors, allelic lineages are not really as old as predicted under the hypothesis of "trans-species" polymorphism. Rather, they argue that *DRB1* alleles are really very recent.

These authors do not address the fact that their conclusions contradict those of previous studies—including studies by the same group—that suggested that MHC alleles are very ancient. Moreover, if they are right, the class II exons have experienced very rapid recent evolution *at synonymous sites*. Positive selection will increase the rate of nucleotide substitution at nonsynonymous sites, but there is no known mechanism that will increase the rate of substitution at synonymous sites. Rather, we expect the rate of substitution at synonymous sites to be very similar to that at sites in introns; and this prediction is supported by comparison of mammalian introns and exons in the case of non-MHC genes (see Chapter 3, The Dynamics of Neutral Evolution). Despite their interpretation, what Bergstrom et al.'s (1997) data clearly suggest is that *DRB1* exon 2 sequences are much older than *DRB1* intron 2 sequences—as seen in class I and as predicted by population genetics theory (Cereb et al. 1997).

Bergstrom and colleagues include in their paper a brief discussion of alternatives to their hypothesis. They discuss "hitch-hiking" but confuse the effects of hitch-hiking when a linked locus is under directional selection (which leads to fixation of a selectively favored allele) with the effects of hitch-hiking with a locus under balancing selection. In the former case, both the locus under selection and a closely linked locus will show reduced polymorphism compared to neutral loci because of the recent fixation of the favorable allele (see Chapter 6, Selective Sweeps). This phenomenon is often referred to as a "selective sweep." By contrast, a locus closely linked to a locus under balancing selection will show higher polymorphism than a neutral locus (Nei and Li 1980; Strobeck 1983). The polymorphism seen at such a linked locus will be a function of how tightly it is linked to the selected locus. In the case of *DRB1*, the selection is acting on nonsynonymous sites in the PBR codons in exon 2. Because sites in intron 2 of *DRB1* are less closely linked to PBR nonsynonymous sites than are synonymous sites in exon 2, the latter are expected to show a higher level of polymorphism because of their hitchhiking with the PBR nonsynonymous sites. And this is, of course, exactly what Bergstrom et al.'s (1997) data show.

Thus, no fewer than four independent lines of evidence support the hypothesis that MHC polymorphisms are selectively maintained: (1) the distribution of allelic frequencies does not fit the neutral expectation (Hedrick and Thomson 1983); (2) the rate of nonsynonymous nucleotide substitution significantly exceeds the rate of synonymous substitution in the PBR codons (Hughes and Nei 1988, 1989a); (3) polymorphisms have been maintained for long periods of time ("trans-species polymorphism"); and (4) introns have been homogenized relative to exons over evolutionary time, suggesting that balancing selection acts to maintain diversity in the latter, not the former (Cereb et al. 1997).

Patterns of Variation in Class I Peptide Ligands

Monos and colleagues (1984) first drew attention to an interesting aspect of amino acid polymorphism at class I MHC loci: many of the amino acid differences among class I allelic products result in residue charge differences. When crystal structures of the proteins encoded by different alleles at the *HLA-A* locus were first compared, it was suggested that differences in the pattern of amino acid residue charge in the PBR might be important for differences in peptide-binding specificity (Garrett et al. 1989). As mentioned previously (Chapter 2, Conservative and Radical Nonsynonymous Differences), the method of Hughes et al. (1990) can be used to test whether amino acid differences occur at random with respect to a residue property such as charge. This method was applied to 36 residues accessible to the six pockets (A–F) of the class I PBR into which side chains of bound peptides fit (Hughes et al. 1990). In the case of *HLA-A* and *HLA-B*, the proportion of nonsynonymous differences in the pockets that cause a charge change was significantly higher than the proportion that conserve charge ($p_{NR} > p_{NC}$) (Figure 4.14). This was not true, however, for *HLA-C* (Figure 4.14). Given the potential importance of the pattern of residue charges (called

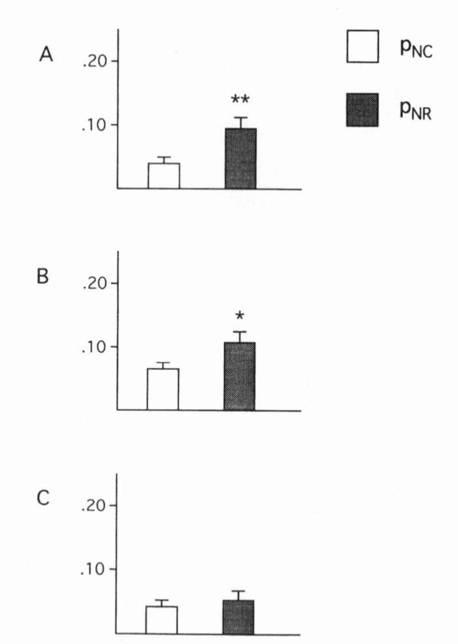

Figure 4.14. The proportion of conservative (p_{NC}) and radical (p_{NR}) nonsynonymous nucleotide difference, with respect to residue charge, in peptide binding groove of alleles at MHC class I *HLA-A* (**A**), *HLA-B* (**B**), and *HLA-C* (**C**) loci.

"charge profile" by Hughes and colleagues, 1990) for peptide binding, the pattern seen at the *HLA-A* and *-B* loci strongly suggests that peptide binding is at the basis of the positive selection acting on these loci. However, it is unclear why natural selection does not focus on the charge profile of *HLA-C* alleles.

Because natural selection seems to act to vary charge profile in pockets A–F of the PBR of *HLA-A* and *-B* molecules, it is worthwhile to ask how the pattern of residue charge in these regions correlates with charge patterns in the actual peptides bound by different MHC allelic products. As mentioned, by now a substantial number of peptides bound in vivo by MHC molecules have been eluted and sequenced. Therefore, Hughes and Hughes (1995a) decided to examine the relationship between charge variation among allelic products and among the peptides they bind.

Figure 4.15. (facing page) (**A**) Mean proportion charge difference at each residue (1–9) in comparisons within the set of nonamer peptides bound by a particular allelic product at the *HLA-B* locus plotted against corresponding values for the *HLA-A* locus. The data set included a total of 61 peptides bound by seven *HLA-A* allelic products and 82 peptides bound by 13 *HLA-B* allelic products. There was a significant positive correlation between PCD for *HLA-A* and PCD for *HLA-B* ($r = .836$; $p < .005$). The linear regression line is shown. (**B**) Mean PCD at each residue position for comparisons among nonamer peptides bound by different *HLA-A* allelic products. (**C**) Mean PCD at each residue position for comparisons among nonamer peptides bound by different *HLA-B* allelic products. From Hughes and Hughes (1995a).

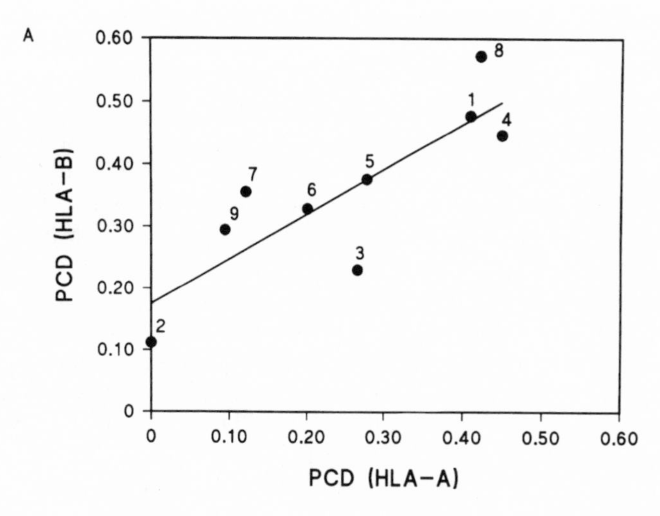
A
0.60
0.50
0.40
0.30
0.20
0.10
0
PCD (HLA-B)
0
0.10
0.20
0.30
0.40
0.50
0.60
PCD (HLA-A)
1
2
3
4
5
6
7
8
9

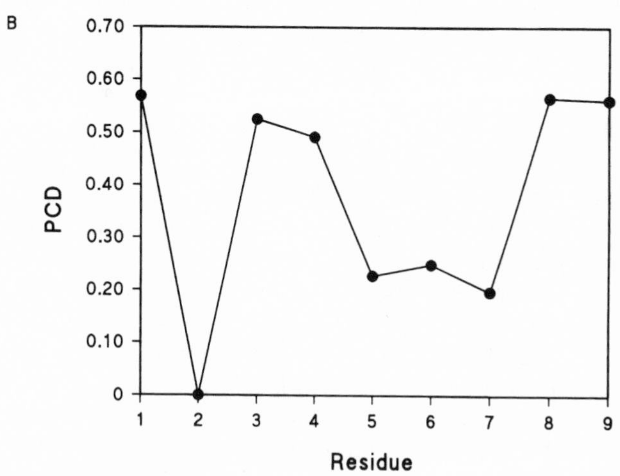
B
0.70
0.60
0.50
0.40
0.30
0.20
0.10
0
PCD
1
2
3
4
5
6
7
8
9
Residue

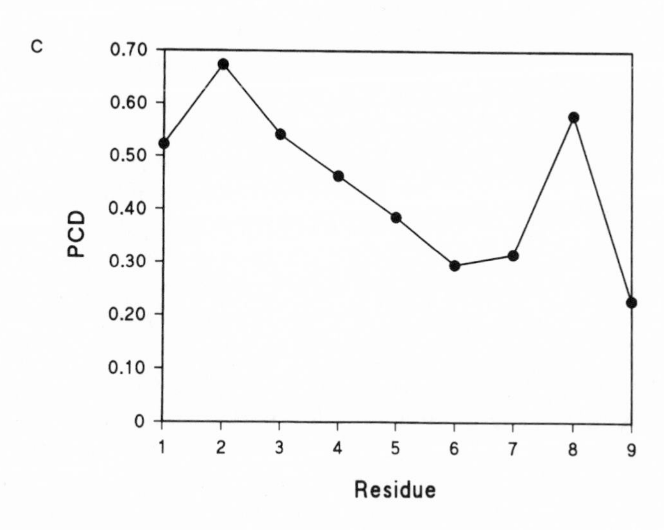
C
0.70
0.60
0.50
0.40
0.30
0.20
0.10
0
PCD
1
2
3
4
5
6
7
8
9
Residue

As a statistical measure of charge diversity at a given residue position within a set of sequences Hughes and Hughes (1995a) used the proportion of comparisons (out of a set of pairwise comparisons) that showed a charge difference; they called this proportion charge difference (PCD). They computed PCD for each of the nine amino acid positions of 61 nonamer peptide ligands of seven *HLA-A* allelic products and 13 *HLA-B* allelic products. Only nonamers (which constitute the vast majority of the reported ligands of class I molecules) were used so that amino acid positions within the peptide could be compared unambiguously.

Initially, PCD was computed for comparisons within the sets of peptides bound by a particular allelic product; in other words, all peptides bound by HLA-B7 were compared with other peptides bound by HLA-B7, and so forth for other alleles. When this was done, mean PCD for *HLA-A* alleles was closely correlated with that for *HLA-B* alleles (Figure 4.15A). This correlation is explained by the fact that, at both of these loci, the same residues tend to be anchor residues characteristic of the allelic product, and thus conserved among all peptides bound by it. Thus, for both *HLA-A* and *HLA-B*, positions 2(P2) and 9(P9) (the main anchor residues) have the lowest PCD in these comparisons (Figure 4.15A). Likewise, the positions in the peptide most free to vary with respect to charge seem to be P1, P4, and P8; these positions had high PCD for comparisons at both loci (Figure 4.15A).

Next, PCD was computed for comparisons among the sets of peptides bound by different allelic products at the same locus. Now the peptides bound by HLA-B7 were compared with those bound by HLA-B8, and so forth. The results showed striking differences between the two loci. In the case of *HLA-A*, mean PCD was zero for P2 but quite high (about 60%) for P9 (Figure 4.15B). By contrast, in the case of *HLA-B*, mean PCD was very high (nearly 70%) for P2 and much lower (about 20%) for P9 (Figure 4.15C). Thus, in comparisons among the sets of peptides bound by different allelic products, the *HLA-A* and *HLA-B* loci are in a sense mirror images of each other. *HLA-A*–bound peptides show high charge variation at P9 and none at P2, while *HLA-B*–bound peptides show high charge variation at P2 and little at P9.

These observations are particularly interesting in light of the pattern of charge variation among the *HLA-A* and *HLA-B* allelic products themselves. Charge variation among these alleles was summarized by computing PCD at residues accessible to pockets A–F; PCD for *HLA-B* alleles was then plotted against that for *HLA-A* alleles (Figure 4.16). These quantities were positively correlated, but certain points were outliers to the general pattern of correlation. Positions 9 and 45 showed no charge variation among the *HLA-A* alleles but relatively high PCD among the *HLA-B* alleles (Figure 4.16). By contrast, positions 66, 70, 77, and 116 showed low PCD among the *HLA-B* alleles but high PCD among the *HLA-A* alleles (Figure 4.16).

The two residues showing high PCD at the *HLA-B* locus and zero PCD at the *HLA-A* locus, 9 and 45, are accessible to pocket B. Pocket B accommodates the side chain of the P2 residue of the bound peptide. P2 is also the position of highest PCD in between-allele comparisons of *HLA-B*–bound peptides (Figure 4.15C) but showed zero PCD in between-allele comparisons of *HLA-A*–bound

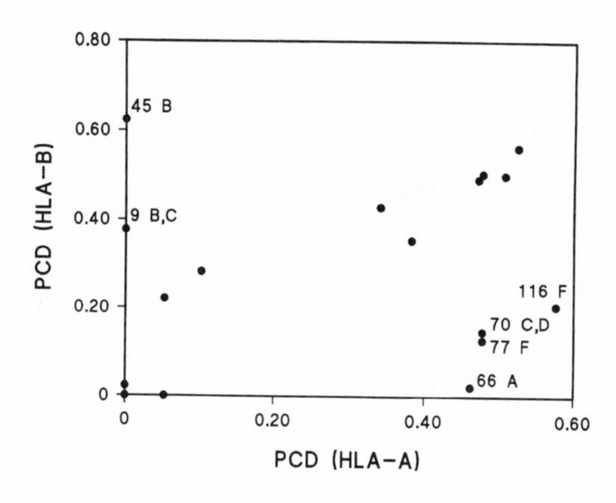

Figure 4.16. Mean PCD at 34 HLA residue positions accessible to pockets A through F. PCD among *HLA-B* alleles is plotted against that for *HLA-A* alleles. From Hughes and Hughes (1995a).

peptides (Figure 4.15B). Thus, in the case of *HLA-B*, there are charge differences in the B pocket among allelic products, corresponding to charge differences in the P2 residue bound there.

By contrast, the positions that showed high PCD among *HLA-A* alleles but low PCD among *HLA-B* alleles include two positions (77 and 116) that are accessible from the F pocket (Figure 4.16). The F pocket is responsible for binding the P9 residue, which shows high PCD among *HLA-A*–bound peptides and low PCD among *HLA-B*–bound peptides (Figure 4.15).

Thus, the differences between *HLA-A*–bound peptides and *HLA-B*–bound peptides are reflected by differences among the allelic products at the two loci. *HLA-B* alleles have high charge variation in the B pocket and correspondingly high variation at P2 among the peptides they bind. *HLA-A* alleles, on the other hand, have high charge variation in the F pocket and correspondingly high variation at P9 among the peptides they bind. This correspondence between variation in MHC molecules and in the peptides they bind is only explainable on the hypothesis that the primary selective force behind the maintenance of MHC polymorphism relates to peptide binding. These results complement other studies showing that the focus of natural selection in these molecules is the PBR and favor the hypothesis that parasites are the major selective agents acting on these loci.

5

Other Examples of Balancing Selection

This chapter surveys additional cases of balancing selection, besides that of the MHC, which have been studied at the DNA sequence level. I begin with cases involving genes that encode immunogenic proteins of two species parasitic on humans: (1) the protist parasite *Plasmodium falciparum*, which causes the most virulent form of human malaria; and (2) the virus human immunodeficiency virus-1 (HIV-1), the more geographically widespread of the two viruses causing acquired immune deficiency syndrome (AIDS). Then I review recent studies of the evolution of self-incompatibility genes in plants. These genes share with the MHC loci the distinction of being the most highly polymorphic loci known in multicellular organisms. Finally, I consider the evidence for balanced polymorphism at the alcohol dehydrogenase locus in the fruitfly *Drosophila melanogaster*; this polymorphism was studied extensively using allozyme data and has more recently been studied at the DNA level.

The Circumsporozoite Protein of *Plasmodium*

The malaria parasites of the genus *Plasmodium* are blood parasites of vertebrates transmitted by a mosquito intermediate host (Figure 5.1). In the life cycle of the parasite, there is a brief diploid phase, but most stages are haploid. Male and female gametocytes are produced within the vertebrate host. When these are taken up by a mosquito in a blood meal, fertilization occurs in the midgut of the mosquito (Figure 5.1). The production of a diploid zygote is immediately followed by meiosis, which produced haploid sporozoites. The sporozoites migrate to the salivary glands of the mosquito and are injected into the vertebrate host with the mosquito's saliva when it feeds. The sporozoite is thus the infective stage of the malaria parasite.

Sporozoites migrate to parenchyma cells of the liver. Now called trophozoites, they divide mitotically, eventually causing rupture of the liver cell and

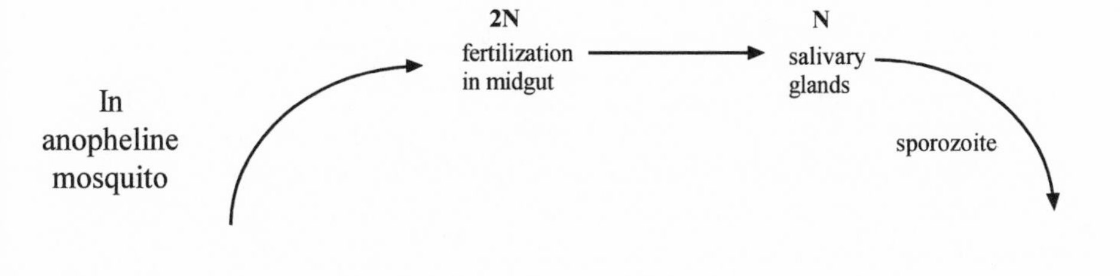

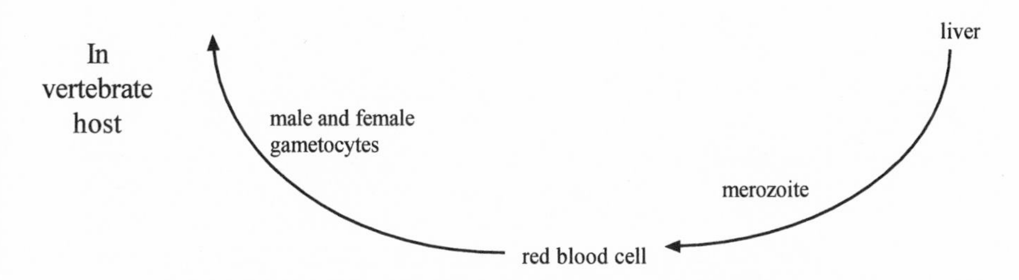

Figure 5.1. Life cycle of malaria parasites (genus *Plasmodium*).

release of a new stage of the parasite, called the merozoite. In some *Plasmodium* species, merozoites may enter the liver cells for additional cycles of replication before entering the red blood cells; in other species, they proceed directly to the red blood cells. In either case, when the merozoite enters the red blood cell, it divides mitotically, leading to rupture of the red blood cell and release of more merozoites. Some of these develop into male or female gametocytes, which can then complete the life cycle if ingested by a mosquito (Figure 5.1).

Surface proteins of both the sporozoite and merozoite stages have attracted the interest of malaria researchers in recent years primarily because of the hope that one or more of these molecules might provide the basis for an antimalarial vaccine. Malaria, in particular that caused by *Plasmodium falciparum,* remains a major public health problem in many parts of the world, accounting for up to 2 million deaths per year (Reeder and Brown 1996). The most intensively studied surface protein of *P. falciparum* has been the circumsporozoite protein (CSP). The CSP covers the surface of the mature sporozoite (Yoshida et al. 1980); in one *Plasmodium* species, it has been estimated that the CSP accounts for 10–20% of the protein synthesized by the sporozoite (Cochrane et al. 1982).

The CSP is encoded by a single polymorphic locus, and the mechanism maintaining this polymorphism has been controversial. In the case of *P. falciparum,* the first available DNA sequences from the CSP locus showed a rather surprising characteristic: there were no synonymous differences among them, but rather all differences were nonsynonymous. This led Arnot (1989) to propose that CSP polymorphism is selectively neutral and that some mechanism—unspecified by Arnot—acts to prevent synonymous substitution at this locus. On the other hand, Good and colleagues (1988) argued that CSP polymorphism is maintained by natural selection exerted by the host immune system, specifically by MHC binding and T cell recognition.

The three-dimensional structure of the CSP is unknown, but CSPs of *Plasmodium* species studied to date share certain features of primary structure. All CSPs have a central region containing amino acid repeats, surrounded by *N*-terminal and *C*-terminal nonrepeat regions. The repeat region consists of numerous tandem repeats (with occasional slight variation) of a short amino acid motif. In *P.falciparum*, for example, the repeat region contains the amino acid motif NANP (or a close variant) repeated 40–45 times; there are variations among alleles with respect to the number of repeats. In 18 CSP alleles from *P. falciparum* analyzed by Jongwutiwes et al. (1994), 92.1% of the repeats were NANP, while 7.9% were NVDP.

The monkey parasite *P. cynomolgi* shows much greater variation in the repeat region than does *P. falciparum*. In this species, the length of the repeated motif differs strikingly among alleles. For example, an allele called "Mulligan" has 52 repeats of the four–amino acid motif NAGG (or occasional variants), while an allele called "London" has 19 six–amino acid repeats (usually ADGRA) followed by six 11–amino acid repeats (usually GNQAGGQAGAG) (Galinski et al. 1987). The repeat regions of different *P. cynomolgi* CSP alleles thus cannot even be aligned with one another. Actually, *P. cynomolgi* may represent a species complex rather than a single species; if so, perhaps the great differences among alleles are less surprising. In the nonrepeat regions, *P. cynomolgi* CSP alleles are somewhat more divergent from each other than are alleles in other species, but not greatly so (Table 5.1). This suggests that the repeat regions may diverge very rapidly and implies that some as yet unknown mechanism is operating to duplicate a new repeat type and replace a previous repeat type.

An analysis of the more homogeneous repeats of *P. falciparum* CSP alleles implicated three different mechanisms in their diversification (Jongwutiwes et al. 1994). Some pairs of alleles were found to differ by a single nucleotide substitution in one of the repeat regions, suggesting point mutation. Sometimes,

Table 5.1. Mean numbers of synonymous (d_S) and nonsynonymous (d_N) nucleotide substitutions per 100 sites (±SE) in *N*-terminal (NTNR) and *C*-terminal (CTNR) nonrepeat regions of *Plasmodium* CSP genes

	NTNR			CTNR		
Species (No. alleles)	Codons	d_S	d_N	Codons	d_S	d_N
P. falciparum (12)	103	1.4 ± 1.1	0.5 ± 0.3	99	0.0 ± 0.1	1.9 ± 0.5**
P. vivax (7/9)	93	0.0 ± 0.0	0.4 ± 0.3	97	0.3 ± 0.3	0.8 ± 0.4
P. berghei (3)	92	6.9 ± 2.9	4.0 ± 1.1	97	3.6 ± 2.1	0.0 ± 0.0
P. knowlesi (3)	96	3.3 ± 1.9	0.9 ± 0.5	98	1.0 ± 1.0	0.6 ± 0.4
P. yoelii (3)	138	4.5 ± 1.9	4.7 ± 1.0	97	4.8 ± 2.4	0.6 ± 0.4
P. cynomolgi (5)	97	10.2 ± 2.9	4.9 ± 1.0	96	5.5 ± 2.0	3.2 ± 0.8

**Test of the hypothesis that $d_S = d_N$, $p < .01$. From Hughes and Hughes (1995b).

Table 5.2. Mean percentage of conservative (p_{NC}) and radical (p_{NR}) nonsynonymous nucleotide differences (±SE), with respect to amino acid residue charge, in NTNR and CTNR of *Plasmodium* CSP genes

	NTNR		CTNR	
Species	p_{NC}	p_{NR}	p_{NC}	p_{NR}
P. falciparum	0.6 ± 0.5	0.4 ± 0.3	0.9 ± 0.4	3.4 ± 1.2*
P. vivax	0.4 ± 0.4	0.5 ± 0.5	0.9 ± 0.5	0.6 ± 0.6
P. berghei	4.3 ± 1.6	3.3 ± 1.4	0.0 ± 0.0	0.0 ± 0.0
P. knowlesi	0.6 ± 0.6	1.2 ± 0.8	1.0 ± 0.7	0.0 ± 0.0
P. yoelii	5.2 ± 1.4	3.9 ± 1.2	0.9 ± 0.7	0.0 ± 0.0
P. cynomolgi	4.2 ± 1.3	5.3 ± 1.4	4.5 ± 1.2	1.1 ± 0.7*

*Tests of the hypothesis that $p_{NC} = p_{NR}$, $p < .05$. From Hughes and Hughes (1995b).

there was evidence that the repeat region had been lengthened as a result of internal duplication. Finally, one allele (K1) seems to have been created as a result of crossing over within the repeat region of alleles similar to MAD20 and T9/94 (Jongwutiwes et al. 1994).

In *P. falciparum,* T cell epitopes—the regions from which peptides bound by class I and/or class II MHC molecules are derived—are known to be located predominantly in the C-terminal nonrepeat region (CTNR) of the CSP (Good et al. 1988). Reviewing reported T cell epitope regions in the CSP, Zevering et al. (1994) list 19 in the CTNR as opposed to only two in the N-terminal nonrepeat region (NTNR). Furthermore, known T cell epitopes essentially cover the entire CTNR (Zevering et al. 1994). Table 5.1 shows mean d_S and d_N in the NTNR and CTNR of CSP alleles from six species of *Plasmodium*. In addition to *P. falciparum,* the following species were analyzed: *P. vivax,* another human parasite; *P. cynomolgi* and *P. knowlesi,* parasites of Old World monkeys; and *P. berghei* and *P. yoelii,* two parasites of rodents. In *P. falciparum,* d_S and d_N do not differ significantly in the NTNR; but in the CTNR, d_N is significantly greater than d_S (Table 5.1). *P. falciparum* is unique in having an enhanced value of d_N in the CTNR; in all other species analyzed, there is no significant difference between d_S and d_N in either repeat region (Table 5.1).

Nonsynonymous differences were categorized as conservative or radical with respect to amino acid residue charge, and the proportions of conservative (p_{NC}) and radical (p_{NR}) nonsynonymous differences were estimated by the method of Hughes et al. (1990) (Chapter 2, Conservative and Radical Nonsynonymous Differences). In the CTNR of *P. falciparum,* p_{NR} was significantly greater than p_{NC} (Table 5.2). These results indicate that, in this region, nonsynonymous differences among alleles occur in such a way as to change residue charge to a greater extent than expected under random substitution. No other region shows $p_{NR} > p_{NC}$. In fact, in the CTNR of *P. cynomolgi,* p_{NC} is significantly greater than p_{NR}, indicating that in this species nonsynonymous differences in the CTNR occur in such a way as to conserve residue charge to a greater extent than expected under random substitution.

The analyses summarized in Tables 5.1 and 5.2 indicate that positive selection is acting to diversify the CTNR of CSP alleles in *P. falciparum* but not in other *Plasmodium* species and that residue charge changes are selectively favored. Because MHC-bound peptides are derived from this region, and because the pattern of residue charges is known to be important for binding of peptides by MHC molecules, this suggests that the underlying basis of this selection is evasion of immune recognition by the host. Because residue charge is particularly important in the case of MHC class I-bound peptides (Chapter 4, Patterns of Variation in Class I Peptide Ligands), evasion of recognition by cytotoxic T cells (Chapter 4, Structure and Function of MHC Molecules) may be particularly important. Red blood cells of mammals do not express class I MHC molecules; therefore, cytotoxic T cell control of malaria parasites must focus on the stage when they are present in liver cells. There is substantial evidence that cytotoxic T cells play a critical role in resistance to *Plasmodium* infection. For example, mice that do not express class I MHC molecules are unable to resist infection even with attenuated sprozoites of *P. berghei*, which invade liver cells but not red blood cells (White et al. 1996).

As mentioned previously, Arnot (1989) hypothesized that some mechanism might eliminate synonymous substitution at the CSP locus but did not specify a possible mechanism. One factor that might reduce the observed level of synonymous substitution is a bias in nucleotide content. Such a bias is likely to express itself most strongly at synonymous sites because the amino acid composition of the protein will not be affected by a nucleotide content bias at these sites. It is known that the genomes of *Plasmodium* species are extraordinarily AT-rich (Webber 1988; Hughes and Verra 1998). However, the pattern of nucleotide substitution seen at the CSP locus of *P. falciparum* cannot be explained on the basis of nucleotide content alone. As seen in Table 5.1, there are some synonymous substitutions among CSP alleles. Furthermore, d_N in the CTNR is over three times that in the NTNR (Table 5.1), and this difference is statistically significant ($p < .05$). This difference obviously cannot be explained by nucleotide content bias.

The CSP provides evidence of ancient polymorphism in *P. falciparum*, as expected under balancing selection (see Chapter 4, Trans-species Polymorphism at MHC Loci). By comparing *P. falciparum* CSP genes with those of *P. reichenowi*, a parasite of chimpanzees, we can calibrate a molecular clock for the CSP on the assumption that *P. falciparum* and *P. reichenowi* diverged when humans and chimpanzees did (about 5 million years ago). Mean d_S in the NTNR and CTNR of CSP between 13 *P. falciparum* alleles and the one available *P. reichenowi* allele is 0.035 ± 0.016, whereas d_S between the most divergent pair of *P. falciparum* CSP alleles is 0.015 ± 0.011. The divergence time between these two alleles is thus estimated at 2.1 ± 1.5 million years (Hughes and Verra 1998). In spite of the relatively large standard error of this estimate, it is clear that this polymorphism is quite old, most likely between 0.6 and 3.6 million years old.

It might be supposed that the type of selection maintaining the CSP polymorphism in *P. falciparum* is frequency-dependent selection, which has often been proposed as the mechanism maintaining polymorphisms at loci encoding parasite antigens (May and Anderson 1990). However, there is evidence that the selection at the CSP locus is at least in part overdominant. Certain epitopes for cytotoxic T cells (CTL) derived from the CTNR of CSP have the property of antagonizing CTL specific for each other (Gilbert et al. 1998). This phenomenon is known as altered peptide ligand (APL) antagonism. In APL antagonism, by an unknown mechanism, a small amount of mutational difference in a peptide bound by class I MHC molecules and presented to CTL will actually downregulate the CTL response against the original peptide. Thus, it will be advantageous from the parasite's point of view if two mutually antagonistic peptides are present in the same host at the same time. This can happen if, by chance, the host is bitten within a short period of time by two different mosquitos, each of which injects sporozoites expressing one of the two mutually antagonistic CSP alleles. However, coinfection by the two antagonistic alleles will be more likely if the parasite is a heterozygote. Recall that the malaria parasite undergoes a brief diploid stage in the midgut of the mosquito (Figure 5.1). If this diploid is a heterozygote for the antagonistic CSP alleles, both alleles will be present in sporozoites injected by the same mosquito bite, giving rise to heterozygote advantage.

Regarding the repeat regions of the CSP of *P. falciparum*, as mentioned, it is possible to implicate processes of point mutation, internal duplication, and interallelic recombination in their diversification. However, there is no evidence that positive selection plays a role. This is not the case with *P. cynomolgi*, however. As mentioned, the repeats are very different among different alleles in this species. However, the repeat motifs share features suggesting that there is an evolutionary relationship among them (Galinski et al. 1987). Comparison of these motifs suggests a model for allelic diversification: one or more nonsynonymous mutations occurs in the motif, and when such a new motif is selectively favored it is replicated rapidly, thus replacing the ancestral motif (Hughes 1991c).

Other *Plasmodium* Surface Proteins

The merozoite is the stage of the malaria parasite that invades the host's red blood cells (Figure 5.1). One of its major surface proteins is merozoite surface antigen-1 (MSA-1). This same protein has sometimes been referred to in the literature as merozoite surface protein-1 (MSP-1), major merozoite protein, p190, and p195. Like the CSP, it is encoded by a single polymorphic locus. The primary translation product is a polypeptide of about 1,600 amino acids (with some length variation among allelic products) and a molecular mass of about 195 kDa. This polypeptide is processed proteolytically to yield a number of fragments present on the surface of the merozoite (Lyon et al. 1986; Holder et al. 1987; McBride and Heidrich 1987). Only one of these, a 19-kDa fragment, is carried by the merozoite into the red blood cell (Blackman et al. 1990). Regarding interaction with the host immune system, MSA-1 is known to contain epitopes for both antibodies and helper T cells (Holder and Riley 1996).

Analysis of DNA sequences clearly showed that different regions of MSA-1 genes have quite different evolutionary histories, as the result of recombination among alleles (Hughes 1992a). These recombinational events can be divided into two classes with respect to their effect on the relationships among alleles: reciprocal and nonreciprocal. In a reciprocal event, two alleles have exchanged portions of sequence, as expected in the case of crossing over (Figure 5.2). In a nonreciprocal event, two alleles have become homogenized in one region relative to other regions. In the latter case, the recombination causes the two alleles to share a sequence in a particular region; the two may then begin to diverge in this region as mutations accumulate. In categorizing recombinations in this way, I am not implying that the two types of recombination necessarily occur by different mechanisms. It is possible that nonreciprocal recombination can occur as a result of gene conversion. However, it can occur as a result of ordinary crossing over as well, if one of the nonrecombinant chromosomes and one of the recombinant chromosomes resulting from the crossover event is subsequently lost from the population (Figure 5.2).

Figure 5.3 shows the amino acid sequence of one *P. falciparum* MSA-1 allelic product with different regions differentiated on the basis of evolutionary history by Hughes (1992a). Tanabe et al. (1987) divided *P. falciparum* MSA-1

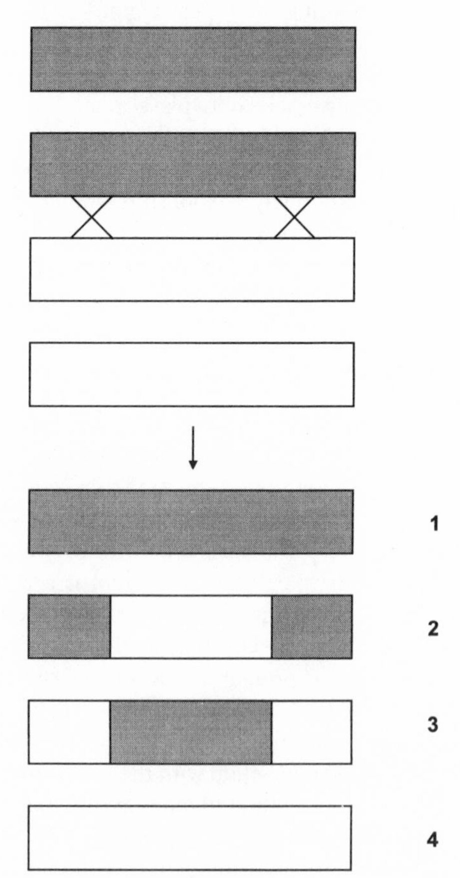

Figure 5.2. Crossover products (2 and 3). If both parental alleles (1 and 4) and at least one recombinant allele (2 or 3) are present in the population, it is easy to see that reciprocal recombination has been involved. In other cases, however, the same crossover event can give rise to apparently non-reciprocal recombinants. For example, if at some subsequent time only products 2 and 4 are found in the population (products 1 and 3 having been lost as a result of either selection or drift), the crossover has had the effect of nonreciprocal recombination, homogenizing the alleles in the central (unshaded) region.

```
                   1 2                                             2 3
MKIIFFLCSFLFFIINTQC|VTHESYQELVKKLEALEDAVLTGYSLFQKEKMVL|NEGTSGTAVTTSTPGSSGSVT
                                                      3 2
SGGSVASVASVASGGSGGSVASGGSGNSRRTNPSDNS|SDSNTKTYADLKHRVQNYLFTIKELKYPELFDLTNHML
                                                                                2
TLSKNVDGFKYLIDGYEEINELLYKLNFYYDLLRAKLNDACANSYCQIPFNLKIRANELDVLKKIVFGYRKPLDN|
4
IKDNVGKMEDYIKKNKTTIANINELIEGSKKTIDQNKNADNEEGKKKLYQAQYNLFIYNKQLQEAHNLISVLEKRI
       4 5                                          5 6
DTLKKNEN|IKKLLEDIDKIKTDAENPTTGSKPNPLPENKKKEVEGHE|EKIKEIAKTIKFNIDSLFTDPLELEYY

LREKNKKVDVTPKSQDPTKSVQIPKVPYPNGIVYPLPLTDIHNSLAADNDKNSYGDLMNPDTKEKINEKIITDNKE

RKIFINNIKKQIDLEEKNINHTKEQNKKLLEDYEKSKKDYEELLEKFYEMKFNNNFDKDVVDKIFSARYTYNVEKQ

RYNNKFSSSNNSVYNVQKLKKALSYLEDYSLRKGISEKDFNHYYTLKTGLEADIKKLTEEIKSSENKILEKNFKGL

THSANASLEVSDIVKLQVQKVLLIKKIEDLRKIELFLKNAQLKDSIHVPNIYKPQNKPEPYYLIVLKKEVDKLKEF

IPKVKDMLKKEQAVLSSITQPLVAASETTEDGGHSTHTLSQSGETEVTEETEVTEETVGHTTTVTITLPPKEESAP

KEVKVVENSIEHKSNDNSQALTKTVYLKKLDEFLTKSYICHKYILVSNSSMDQKLLEVYNLTPEEEKELKSCDPLD

LLFNIQNNIPAMYSLYDSMNNDLQHLFFELYQKEMIYYLHKLKEENHIKKLLEEQKQITGTSSTSSPGNTTVNTAQ

SATHSNSQNQQSNASSTNTQNGVAVSSGPAVVEESHDPLTVLSISNDLKGIVSLLNLGNKTKVPNPLTISTTEMEK
                                      6 7
FYENILKNNDTYFNDDIKQFVKSNSKVITGLTETQKNA|LNDEIKKLKDTLQLSFDLYNKYKLKLDRLFNKKKELG
                   7 8
QDKMQIKKLTLLKEQLESKLN|SLNNPHNVLQNFSVFFNKKKEAEIAETENTLENTKILLKHYKGLVKYYNGESSP
                                                                          8 9
LKTLSEVSIQTEDNYANLEKFRALSKIDGKLNDNLHLGKKKLSFLSSGLHHLITELKEVIKNKNYTGNSPS|ENNK

KVNEALKSYENFFPEAKVTTVVTPPQPDVTPSPLSVRVSGSSGSTKEETQIPTSGSLLTELQQVVQLQNYDEEDDS

LVVLPIFGESEDNDEYLDQVVTGEAISVTMDNILSGFENEYDVIYLKPLAGVYRSLKKQIEKNIITFNLNLNDILN

SRLKKRKYFLDVLESDLMQFKHISSNEYIIEDSFKLLNSEQKNILLKSYKYIKESVENDIKFAQEGISYYEKVLAK

YKDDLESIKKVIKEEKEKFPSSPPTTPPSPAKTDEQKKESKFLPFLTNIETLYNNLVNKIDDYLINLKAKINDCNV
                                                           9 10
EKDEAHVKITKLSDLKAIDDKIDLFKNTNDFEAIKKLINDDTKKDMLGKLLSTGLVQ|IFPNTIISKLIEGKFQDM
                                                 10 11
LNISQHQCVKKQCPENSGCFRHLDEREECKCLLNYKQEGDKCEENPNP|TCNENNGGCDADATCTEEDSGSSRKKI

TCECTKPDSYPLFDGIFCSSSNFLGISFLLILMLILYSFI
```

Figure 5.3. Amino acid sequence of MSA-1 (allele MAD) from *Plasmodium falciparum*, showing regions identified by distinct patterns of nucleotide substitution. From Hughes (1992a).

alleles into two groups (I and II) on the basis of sequence similarity. The distinction between the two groups is clearest in the extensive region designated region 6 in Figure 5.3. Table 5.3 shows mean d_S and d_N within and between the two groups in the first seven regions. Each group is represented by three alleles. These values highlight the differences between the regions. For example, in region 1, all sequences analyzed are identical at both synonymous and nonsynonymous sites (Table 5.3); thus, in this region, there is no distinction between groups I and II. By contrast, in region 6, although the sequences are very

Table 5.3. Mean numbers of synonymous (d_S) and nonsynonymous (d_N) nucleotide substitutions per 100 sites (±SE) in comparisons of *MSA-1* alleles of *Plasmodium falciparum*

	Group I		Group II		Group I versus Group II	
Region	d_S	d_N	d_S	d_N	d_S	d_N
1	0.0 ± 0.0	0.0 ± 0.0	0.0 ± 0.0	0.0 ± 0.0	0.0 ± 0.0	0.0 ± 0.0
2	1.5 ± 1.9	0.8 ± 0.4	6.1 ± 2.1	3.7 ± 0.9	7.5 ± 2.1	2.8 ± 0.6
3	12.8 ± 5.1	50.9 ± 8.9***	30.4 ± 8.6	83.5 ± 10.7***	10.8 ± 3.9	56.5 ± 7.2***
4	0.0 ± 0.0	0.0 ± 0.0	0.0 ± 0.0	0.5 ± 0.5	0.0 ± 0.0	6.5 ± 2.1**
5	0.0 ± 0.0	0.8 ± 0.6	12.8 ± 5.1	21.6 ± 3.3	8.5 ± 3.0	16.5 ± 2.7*
6	0.0 ± 0.0	01 ± 0.1	0.3 ± 0.2	0.8 ± 0.2	63.6 ± 6.0	46.0 ± 2.0*
7	0.0 ± 0.0	0.0 ± 0.0	0.0 ± 0.0	0.0 ± 0.0	14.2 ± 7.3	10.5 ± 2.9

Regions are defined as in Figure 5.3. Tests of the hypothesis that $d_S = d_N$: $^{*}p < .05$; $^{**}p < .01$; $^{***}p < .001$. From Hughes (1992a).

similar within the two groups, the divergence between the two groups at both synonymous and nonsynonymous sites is extraordinarily high for alleles at a locus in a eukaryotic species (Table 5.3).

Despite the high level of divergence between the two groups of alleles in region 6, d_S between the groups is significantly greater than d_N (Table 5.3). Thus, this region is subject to purifying selection. In region 4, the sequences are also very similar within the two groups. There are no differences among group I alleles (Table 5.3). Group II alleles are also very similar to one another, showing no synonymous differences and a very low mean rate (<1%) of nonsynonymous substitution among them (Table 5.3). Between the groups, although d_S is zero, d_N is significantly greater than zero (Table 5.3). Thus, although the two groups of alleles are less than one-seventh as divergent at nonsynonymous sites in region 4 ($d_N = 6.5\%$) than they are in region 6 ($d_N = 46.0\%$), positive selection appears to have promoted diversification of the two groups in the former region but not in the latter.

How can these remarkable differences between regions of the *P. falciparum* MSA-1 gene be explained? The simplest explanation seems to be one involving different histories of recombination and different patterns in different regions (Hughes 1992a). Region 1, for example, seems to have been homogenized by nonreciprocal recombination between the two allelic groups. The recombination has apparently occurred quite recently, because the two groups have not had time to accumulate any differences. Clearly, no such recent homogenization has occurred in region 6, where the two allelic groups are very divergent from each other. In region 4, the two allelic groups are much more similar to each other than they are in region 6, but not completely identical as in region 1. Here, it seems likely that the two groups were homogenized by nonreciprocal recombination sometime in the past and they have subsequently diverged. Furthermore, the fact that d_N significantly exceeds d_S in region 4 argues that differentiation of the amino acid sequence in this region between the two groups has been favored by positive Darwinian selection.

The reason homogenizing recombination must be invoked to explain the fact that the two allelic groups are much more similar to each other in certain other regions than they are in region 6 is that this similarity is seen at both synonymous and nonsynonymous sites. Functional constraint at the amino acid level would explain similarity at nonsynonymous sites but not at synonymous sites. The rate of synonymous substitution could be elevated in one gene region if that region had a higher mutation rate than other parts of the gene. However, this explanation does not seem applicable to region 6 of MSA-1, because the within-group comparisons show no evidence of an enhanced nonsynonymous rate.

In addition to non-reciprocal recombination that homogenizes alleles between groups, there is evidence that *P. falciparum* MSA-1 alleles have sometimes been involved in reciprocal recombination events that have exchanged sequence between the groups of alleles. Figure 5.4B shows the phylogeny of six MSA-1 alleles based on region 6. The tree shows that in this region there is very little difference within group I (M, PA, and K1) or within group II (CAM, RO, and MAD), but that the groups are very divergent from each other (Figure 5.4B).

Figure 5.4A shows the phylogeny of the same alleles in region 3, and the pattern is very different from that in region 6. From Table 5.3 it can be seen

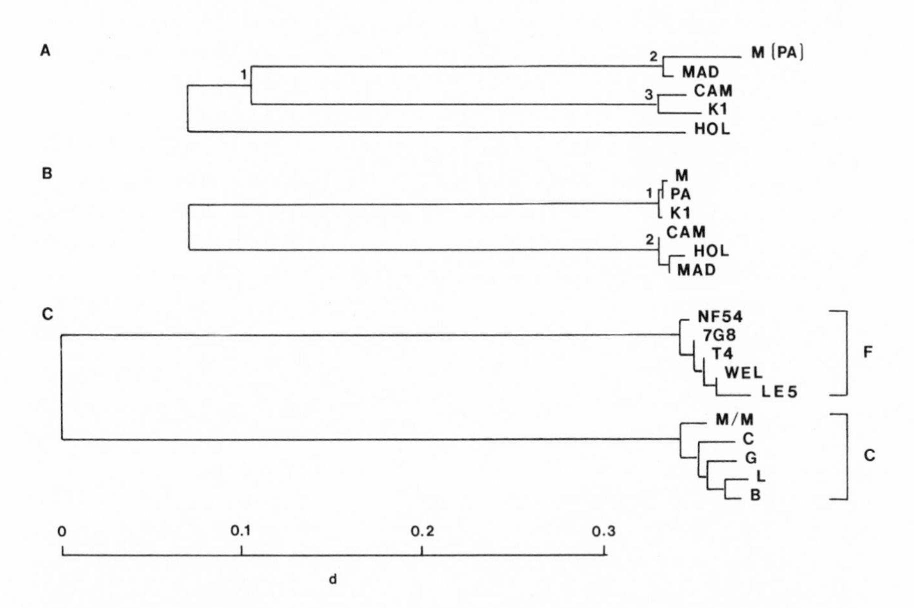

Figure 5.4. Phylogenetic trees constructed by the neighbor-joining method based on the number of nucleotide substitutions per site (*d*). (**A**) Region 3 (Figure 5.3) of *MSA-1* alleles of *Plasmodium falciparum*. (**B**) Region 6 of *MSA-1* alleles of *P. falciparum*. (**C**)) Nonrepeat regions of *CSP* alleles from *P. falciparum* (**F**) and *P. cynomolgi* (**C**).

that region 3 is unusual in that there is a great deal of sequence divergence within the two allele groups as well as between them—a pattern that makes this region unique among the regions analyzed in Table 5.3. The phylogenetic tree shows the reason for this difference. In region 3, the group II allele MAD clusters with the group I alleles M and PA; (Figure 5.4). Thus, region 3 of MAD seems to have been donated by recombination with a group I allele. Conversely, the group I allele Kl clusters with the group II allele CAM in region 3 (Figure 5.4). Thus, region 3 of Kl has apparently been donated by recombination with an allele related to CAM (Figure 5.4A). Finally, in region 3, the group II allele RA is not closely related to either group I or group II alleles (Figure 5.4). Evidently region 3 of RO was donated by recombination with a third, otherwise unknown allelic family.

Regions 3, 4, and 5 all show d_N significantly higher than d_S in the comparison between the two allelic groups (Table 5.3). Only in region 3 is d_N significantly greater than d_S in comparisons with the two allelic groups (Table 5.3). Of course, the reason for this difference is the history of reciprocal recombination in region 3. Because of past recombination, certain pairs of alleles that belong to the same allelic group in region 6 represent different allelic groups in region 3.

Although relatively little is known about the interactions of MSA-1 with the human immune system, several studies have mapped helper T cell epitopes to different protein regions (e.g., Crisanti et al. 1988; Rzepcyk et al. 1989; Quakyi et al. 1994). These epitopes are particularly densely concentrated in region 4. When the proportion of conserved (p_{NC}) and radical (p_{NR}) nonsynonymous differences were calculated, with respect to residue charge, region 4 showed a unique characteristic: p_{NR} is significantly greater than p_{NC} in the comparison between allelic groups (Table 5.4). Thus, in *P. falciparum*, region 4 of MSA-1 shares the characteristics of the T cell epitope-rich C-terminal region of CSP: positive selection favoring diversity at the amino acid level and, in particular, favoring residue charge changes. This most plausible basis for such selection is escape from binding by the vertebrate host's MHC molecules.

The fact that the two MSA-1 allelic groups are very divergent at both synonymous and nonsynonymous sites in region 6 (Table 5.3) suggests that, at

Table 5.4. Mean proportion of conservative (p_{NC}) and radical (p_{NR}) nonsynonymous nucleotide differences (±SE), with respect to amino acid residue charge, in *MSA-1* genes of *Plasmodium falciparum*

	Group I		Group II		Group I versus Group II	
Region	p_{NC}	p_{NR}	p_{NC}	p_{NR}	p_{NC}	p_{NR}
3	32.2 ± 4.0	31.5 ± 5.6	54.9 ± 41.8	41.8 ± 7.3	37.2 ± 3.8	29.5 ± 5.0
4	0.0 ± 0.0	0.0 ± 0.0	0.0 ± 0.0	0.0 ± 0.0	1.7 ± 1.7	9.8 ± 3.4*
5	0.8 ± 0.8	0.9 ± 0.9	18.7 ± 3.6	18.4 ± 3.7	14.3 ± 2.9	13.7 ± 2.8

Regions are defined as in Figure 5.3. Tests of the hypothesis that $p_{NC} = p_{NR}$: *$p < .05$. From Hughes (1992a).

least in this region, the MSA-1 polymorphism may be quite ancient. A conserved portion of the MSA-1 region 6 can be aligned between *P. falciparum* and MSA-1 genes of two parasites of rodents, *P. yoelii* and *P. chabaudi*. Assuming that *P. falciparum* diverged from the ancestor of these two rodent malarias when primates diverged from rodents (about 110 million years ago; Hedges et al. 1996), comparison of MSA-1 sequences from the rodent parasites with the two allelic groups of *P. falciparum* can be used to estimate the divergence time of the two allelic groups in the latter species. Based on d_N values in Table 5.5, the divergence time of the two allelic groups is estimated at about 48 million years ago. If this estimation is accurate, the MSA-1 locus of *P. falciparum* has the oldest polymorphism known in eukaryotes, much older than that at the CSP locus. MSA-1 polymorphism is probably even older than MHC polymorphisms (Chapter 4, Trans-Species Polymorphism at MHC Loci); class II MHC allelic lineages of primates may have been maintained for 20–40 million years (Gyllensten and Erlich 1989; Gyllensten et al. 1990).

Note that this is a conservative estimate for the age of the MSA-1 polymorphism. If d_S values were used instead of d_N, the divergence time estimate would be far higher (about 90 million years). It might be argued that d_S is preferable to d_N for such estimation because synonymous substitution is not expected to be subject to positive selection, which make nonsynonymous substitution a poor clock. However, in this case, because the d_S values between human and rodent malarias are quite high (around 80%; Table 5.5), they have relatively large standard errors; thus, an estimate based on d_S may be unreliable. In any event, there is no evidence of positive selection on region 6 (Table 5.3).

Tibayrenc and colleagues (1991) claimed that *P. falciparum* has a clonal population structure. A clonal population structure is characteristic of haploid organisms such as the bacterium *Escherichia coli*, in which the species consists of a number of separate clones that do not recombine freely. This seems somewhat surprising in a diploid, sexually reproducing organism like *P. falciparum*; and Tibayrenc et al. (1991) went so far as to state that in this species "the possibility

Table 5.5. Mean numbers of synonymous (above diagonal) and nonsynonymous (below diagonal) nucleotide substitutions per 100 sites (±SE) in comparisons between region 6 of two groups of *P. falciparum* *MSA-1* alleles and *MSA-1* from rodent malarias

			P. falciparum	
	P. yoelii	*P. chabaudi*	Group 1	Group 2
P. yoelii		20.2 + 4.6	76.7 + 13.6	63.7 + 11.2
P. chabaudi	5.3 + 1.2		85.8 + 15.4	75.1 + 13.1
P. falciparum				
Group 1	44.1 + 4.1	45.7 + 4.2		68.1 + 12.2
Group 2	36.8 + 3.6	39.2 + 3.7	18.1 + 2.2	

From Hughes (1992a).

of an unknown uniparental cycle cannot be discarded." However, the case of MSA-1 clearly shows that recombination has played a major role in the evolution of *P. falciparum* genes, strongly contradicting the highly improbable conjecture that this extensively studied species has a previously undiscovered asexual mode of reproduction. As mentioned (this Chapter, The Circumsporozoite Protein of *Plasmodium*), the repeat regions of the CSP of *P. falciparum* also provide evidence of recent recombinational events (Jongwutiwes et al. 1994).

Rich and colleagues (1998) advanced the surprising hypothesis that all *P. falciparum* in the world today are descended from a single haploid genotype (which they called a "strain") living between 24,500 and 57,500 years ago. Perhaps we might call this alleged genotype "the malarial Eve." According to these authors, all of the extraordinary polymorphism in this species has accumulated in just under 60,000 years! The MSA-1 data refute this hypothesis easily (Hughes and Verra 1998). Whatever one thinks of the divergence time estimate provided here for the major lineages of MSA-1 alleles, all biologists would agree that it is impossible to accumulate 68% substitution at synonymous sites in 60,000 years. This would imply a mutation rate over 1,000 times as great as those of other eukaryotes, one that would produce, on average, one new substitution per 180,000 base pairs each year.

The divergence of CSP alleles at synonymous sites (this Chapter, The Circumsporozoite Protein of *Plasmodium*) is also too great to be reconciled with a divergence time of 60,000 years (Hughes and Verra 1998). As mentioned, d_S between the most divergent *P. falciparum* alleles is 0.0150 ± 0.011. To accumulate this much difference in 70,000 years would require a mutation rate 100 times as great as those of other eukaryotes.

Rates of synonymous and nonsynonymous nucleotide substitution have been compared for a number of polymorphic loci of *P. falciparum* besides CSP and MSA-1 (Table 5.6). It is interesting that in all surface proteins analyzed there

Table 5.6. Results of tests for positive selection at polymorphic protein coding loci of *Plasmodium falciparum*

Molecule	$d_N > d_S$ [a]	$p_{NR} > p_{NC}$ [a,b]
Sporozoite and merozoite surface proteins:		
Circumsporozoite protein (CSP)	+	+
Merozoite surface antigen-1 (MSA-1)	+	+
Merozoite surface antigen-2 (MSA-2)	+	−
Sporozoite surface protein 2 (TRAP)	+	+
Apical membrane antigen 1 (PF83)	+	+
Other proteins		
Liver stage antigen-1 (LSA-1)	−	−
Knob-associated histidine-rich protein (KAHRP)	−	−
Ring-infected erythrocyte surface antigen (RESA)	−	−
S-antigen	−	−

a. + Indicates a significant difference in one or more domain.
b. With respect to amino acid residue charge. Data in Hughes and Hughes (1995b).

is at least one region where d_N significantly exceeds d_S, suggesting positive selection. However, this is not true of any nonsurface protein, even though there may be sequence data for a number of alleles. One particularly interesting case of a nonsurface protein is liver stage antigen-1 (LSA-1). As discussed (Chapter 4, Structure and Function of MHC Molecules), a peptide derived from LSA-1 is presented by HLA-B53, a class I molecule that confers resistance to severe malaria in West Africa (Hill et al. 1991, 1992). The fact that a peptide from this molecule can serve as a protective epitope is consistent with its not having previously been subject to diversifying selection exerted by the host's immune system.

Selection on HIV-1

Since its discovery as the most geographically widespread of the viruses causing human acquired immune deficiency syndrome (AIDS), human immunodeficiency virus-1 (HIV-1) has been one of the most intensely studied of all viruses. However, many aspects of its basic biology, including the mechanism of pathogenesis, remain mysterious. Unlike most common viruses infecting vertebrates, HIV-1 in humans is not defeated by the ordinary arsenal of antiviral defenses; namely, the immobilization of extracellular virus particles by antibody and the killing of infected host cells by cytotoxic T cells (CTL).

HIV-1 has an RNA genome; it is a retrovirus, meaning that it reverse-transcribes a DNA copy of its genome (the provirus), which then integrates into the host's nuclear genome. Like other RNA viruses, HIV-1 has a very high mutation rate (Holland et al. 1982; Hahn et al. 1986; Fisher et al. 1988). The rate of synonymous substitution in HIV-1 coding regions, which should reflect the mutation rate, has been estimated to be about 1×10^{-2} substitutions per site per year (Li et al. 1988). This rate is about seven orders of magnitude faster than the mammalian mutation rate. Its high mutation rate alone guarantees that HIV-1 will be highly polymorphic at the DNA level. Sequence comparisons have revealed that a substantial amount of the virus' polymorphism is nonsynonymous, and this raises the question whether natural selection has promoted amino acid diversity in HIV-1. In particular, it would be interesting to know whether selection exerted by the host's immune system has played a role in favoring such diversity. Moreover, if viral polymorphism is a mechanism for evading immune recognition by the host, does evasion of immune recognition play a central role in the disease process and in the host's eventual inability to clear this virus?

Theoretically, selection exerted by the host's immune system on viral protein-coding genes might act in a variety of ways. First, the different components of the host's immune system might be expected to have different selective effects on the virus. Obviously, selection favoring evasion of recognition by host antibodies would act on those protein domains that form epitopes for antibodies, while selection favoring evasion of binding by MHC molecules and recognition by T cells would act on the appropriate T-cell epitopes. Usually, these two types of epitopes do not overlap. In addition, the effect of selection acting within a single host may differ from that over the whole of viral population inhabiting a number of different hosts. For example, selection within a single host may be directional, leading to fixation of an allele that successfully evades immune recognition by that host (see Chapter 6, Adaptive Divergence between Species). However, because hosts may differ genetically—particularly at their MHC loci—a viral genotype that evades immune recognition in one host may not do so in another. Therefore, directional selection may favor different parasite genotypes in different hosts. At the level of the population as a whole, this type of directional selection within hosts will lead to polymorphism, through a type of balancing selection based on heterogeneity of the environment. On the other hand, there may be balancing selection within a single host, whereby viral polymorphism within the host is favored as a strategy for defeating the host's immune defenses. In the case of HIV-1, several authors have suggested that such within-host diversifying selection is acting on the virus.

Figure 5.5 shows a schematic diagram of the structure of the HIV-1 virus. The genomes of HIV-1 and related retroviruses contain three major genes called *gag*, *pol*, and *env*; proteins encoded by these genes make up the bulk of the infective virion (Arnold and Arnold 1991). There are additional genes, such as *nef*, which are believed to play an important role in pathogenesis of the virus

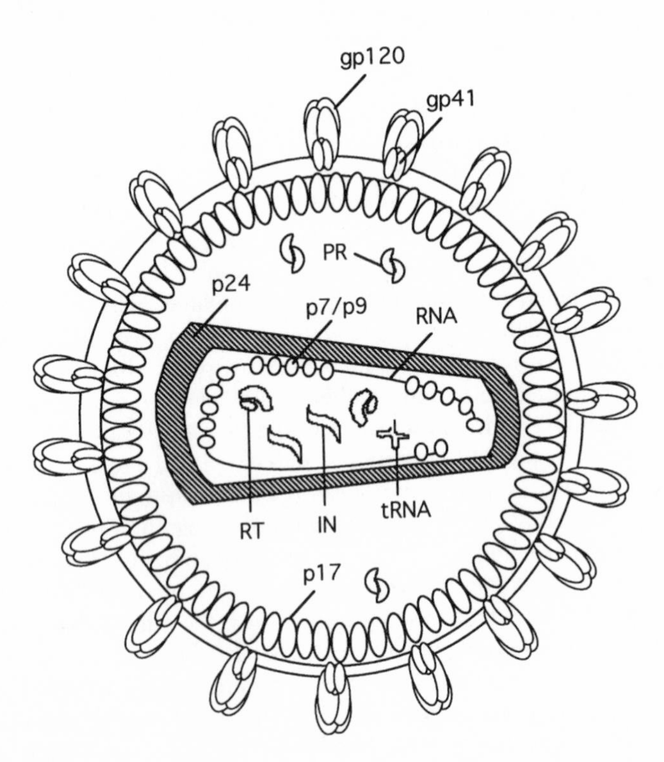

Figure 5.5. HIV-1 virion. IN = integrase; PR = protease; RNA = genomic RNA; RT = reverse transcriptase. From Seibert et al. (1995).

(Collette 1997). Each of the *gag, pol,* and *env* genes encodes a polyprotein, which is subsequently cleaved into separate proteins by the viral protease. The *gag* gene encodes the virion structural proteins p17 (matrix), p24 (capsid), and p7/p9 (nucleocapsid) (Figure 5.5). The *pol* gene encodes the viral protease, the component proteins of reverse transcriptase (p66, p51, and RNAse H), and integrase. The *env* gene encodes the envelope glycoproteins gp120 and gp41 (Figure 5.5).

Sequence comparisons led to the identification of five hypervariable regions (V1–V5) in gp120 (Modrow et al. 1987). These regions, particularly V3, have been identified as epitopes for antibodies against the virus (Modrow et al. 1987; Rusche et al. 1988). Because gp120 is located on the surface of the virion, it is exposed to the host's antibodies. Thus, it has been suggested that natural selection has favored amino acid variation in V1–V5 as a mechanism for avoiding neutralization by host antibodies. There is evidence that host antibodies can select for amino acid replacements in viral proteins; for example, in the hemagglutinin protein of the influenza virus (Wiley et al. 1981). Simmonds and colleagues (1990) tested this hypothesis in the case of V3, V4, and V5 of HIV-1 gp120 by amplifying DNA sequences encoding this domain from proviral sequences of 11 AIDS patients. These researchers estimated rates of synonymous and nonsynonymous substitution per site in the V3 domain and sequences flanking it. They reported the ratio of d_S/d_N to be 0.67. This certainly suggests that amino acid diversity in the V3 region is favored by positive selection; however, the authors did not report any statistical test of the equality of synonymous and nonsynonymous rates (Simmonds et al. 1990).

In a subsequent study, Holmes and colleagues (1992) analyzed 89 sequences for V3 obtained over a seven-year period from a single hemophiliac patient. These included 24 different sequences for the V3 loop. In the first year, all V3 sequences were of a single type, which was evidently the type that caused the initial infection. A phylogenetic tree revealed two major lineages of viral sequences, which were well differentiated from each other by the third year of the infection (Holmes et al. 1992). Interestingly, certain amino acid sequence motifs at V3 appear to have evolved independently within the two separate lineages. Such independent or "convergent" evolution of the same sequence motif suggests that positive selection may be acting on this region (see Chapter 6, Convergent Sequence Evolution); however, especially when the mutation rate is high, independent evolution of the same sequence motif is not proof of positive selection.

To further test the hypothesis that positive selection acted to diversify the V3 region of gp120 within this patient, I estimated d_S and d_N for each year (years 3–7) for which data were available both within and between the two lineages of sequences (called C/D and E) (Table 5.7). Within each lineage, d_S and d_N were not significantly different in any one year, nor were the overall means significantly different (not shown). In the C/D lineage, d_S exceeded d_N in the first two years, while d_N exceeded d_S in the final two years. In the E lineage, d_S exceeded d_N for all years for which comparisons could be made. In the comparison between lineages, on the other hand, d_N was higher than d_S in every year

Table 5.7. Mean numbers of synonymous (d_S) and nonsynonymous (d_N) nucleotide substitutions per 100 sites (±SE) in comparisons of V3 regions of HIV-1 gp120 between two viral lineages (C/D and E) in a single patient

Year	d_S	d_N
3	1.3 ± 0.7	8.9 ± 3.5*
4	7.7 ± 3.3	13.5 ± 4.8
5	3.3 ± 3.3	15.6 ± 5.2*
6	11.6 ± 6.6	19.1 ± 5.4
7	11.2 ± 6.1	21.1 ± 5.9
Mean	7.0 ± 2.1	15.6 ± 2.1**

The two lineages are defined by Holmes et al. (1992). Tests of the hypothesis that $d_S = d_N$: *$p < .05$; **$p < .01$.

(Table 5.7). Although these differences were only significant in years 3 and 5, the overall mean d_N was significantly higher than d_S (Table 5.7). Thus, natural selection acts to diversify V3 between the two lineages of gp120 genes within a single patient.

The basis of this selection is not clear. One obvious hypothesis is that the major selecting agent is the host antibody, which exerts frequency-dependent selection on the V3 region. On this hypothesis, because new virus variants can initially escape neutralization by antibodies, they enable virions to infect new cells. Thus, when the host's major antibody response is directed against a common viral form of the V3 epitope, a mutant V3 may be favored because it is not recognized by the prevalent host antibody type, and thus it will be able to infect cells until the host begins to mount an antibody response against it. However, it is not obvious why this type of selection would lead to the maintenance of two major allelic lineages in the virus or why it would promote amino acid diversification between these lineages but not within them. Perhaps such diversification occurs because, when the dominant antibody response is directed against an abundant allele belonging to one lineage, a mutant allele of the other lineage is favored because it is more divergent from the abundant allele than any mutant of that allele's own lineage could be, and thus less likely to be recognized by antibodies against that allele.

Seibert and colleagues (1995) studied the evolution of the *env* gene at the level of the HIV-1 population as a whole, comparing sequences that represent isolates from different patients. Figure 5.6 shows a phylogeny of the gp120

Figure 5.6. (facing page) Phylogenetic tree of gp120 sequences, constructed by the neighbor-joining method based on proportion of amino acid difference (p). Tests of the significance of internal branches: *$p < .05$; **$p < .01$; ***$p < .001$. From Seibert et al. (1995).

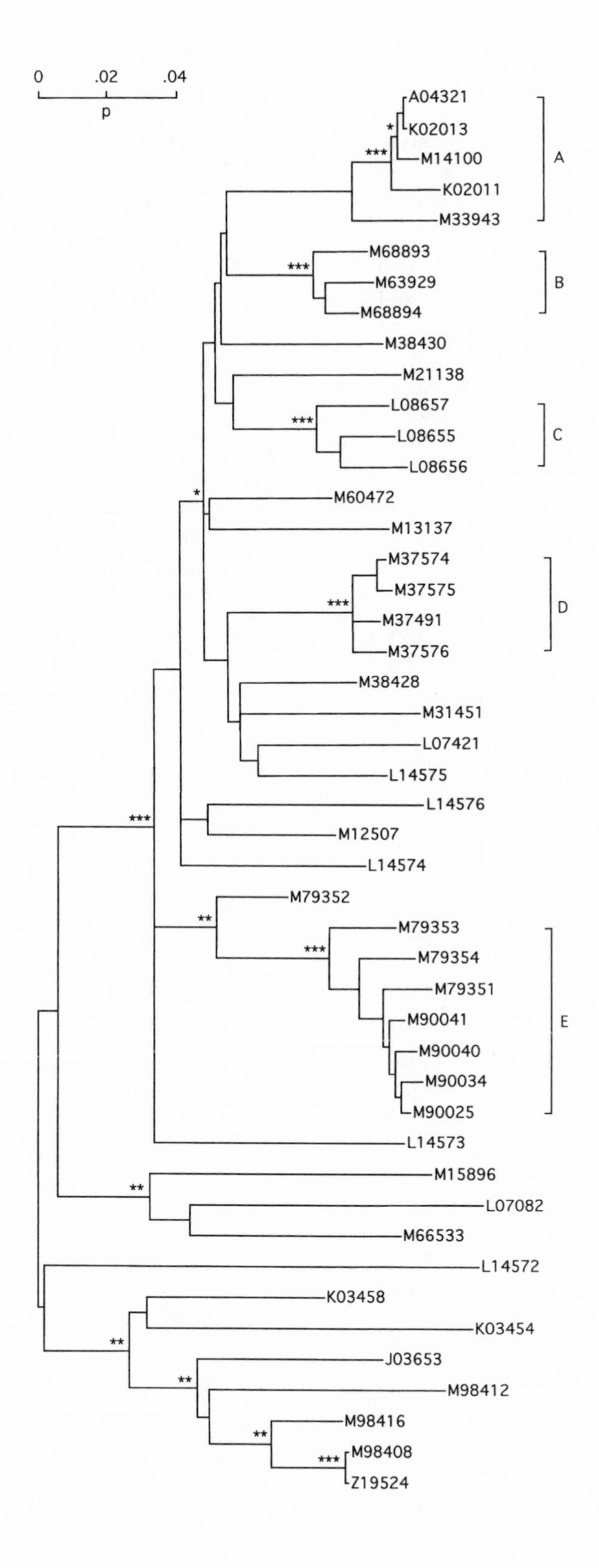

0
.02
.04
p
A04321
K02013
M14100
K02011
M33943
A
M68893
M63929
M68894
B
M38430
M21138
L08657
L08655
L08656
C
M60472
M13137
M37574
M37575
M37491
M37576
D
M38428
M31451
L07421
L14575
L14576
M12507
L14574
M79352
M79353
M79354
M79351
M90041
M90040
M90034
M90025
E
L14573
M15896
L07082
M66533
L14572
K03458
K03454
J03653
M98412
M98416
M98408
Z19524

sequences that they analyzed; a similar phylogeny was observed in the case of gp41, the other protein encoded by the *env* gene (Seibert et al. 1995). In the phylogeny they identified a number of families of closely related sequences (A–E in Figure 5.6). Within each of these families, mean numbers of synonymous and nonsynonymous substitutions per site were estimated for the five variable regions (V1–V5) of gp120, for known T-cell epitopes in gp120 and gp41, and for the remainder of gp120 and gp41 (Table 5.8).

In three of the five families (B, C, and E), mean d_N in V2 was significantly greater than mean d_S (Table 5.8). In two of these families (B and C), mean d_N in V3 was significantly greater than mean d_S; and in one (C), mean d_N in V4 was significantly greater than mean d_S (Table 5.8). In V1 and V5, although in some families mean d_N exceeded mean d_S, the difference was not significant in any family (Table 5.8). When overall mean d_S and d_N were computed for all five families, mean d_N was significantly greater than mean d_S only in V2 and V3 (Table 5.8). In both the T-cell epitopes and the remainder of both gp120 and gp41, d_S was greater than d_N in most comparisons; and, for all five families, overall mean d_S was significantly greater than mean d_N (Table 5.8).

Overall, these results imply that the V regions are subject to positive selection favoring diversity at the amino acid level. However, they suggest that different

Table 5.8. Mean numbers of synonymous (d_S) and nonsynonymous (d_N) nucleotide substitutions per site (±SE) in HIV-1 gp120 sequences

	Family A		Family B		Family C	
Region	d_S	d_N	d_S	d_N	d_S	d_N
V1	24.9 ± 15.4	18.1 ± 5.1	6.7 ± 8.4	14.2 ± 7.0	0.0 ± 0.0	0.0 ± 0.0
V2	1.9 ± 2.0	3.8 ± 1.5	0.0 ± 0.0	3.6 ± 1.8*	1.4 ± 2.0	8.0 ± 2.6*
V3	0.0 ± 0.0	2.5 ± 1.5	0.0 ± 0.0	4.5 ± 2.2	0.0 ± 0.0	4.4 ± 2.2*
V4	7.8 ± 8.2	1.8 ± 2.3	9.1 ± 9.4	8.5 ± 4.4	0.0 ± 0.0	7.4 ± 3.7*
V5	16.5 ± 16.3	21.6 ± 8.3	18.4 ± 15.7	10.9 ± 6.0	31.8 ± 35.8	9.7 ± 7.0
TCE	1.1 ± 0.8	1.3 ± 0.4	2.3 ± 1.3	1.7 ± 0.6	0.8 ± 0.8	1.0 ± 0.5
Remainder	1.6 ± 0.7	0.8 ± 0.3	2.0 ± 1.0	1.0 ± 0.4	3.3 ± 1.2	2.3 ± 0.5

	Family D		Family E		Overall means	
Region	d_S	d_N	d_S	d_N	d_S	d_N
V1	13.5 ± 9.9	17.3 ± 4.9	17.2 ± 12.2	11.7 ± 4.0	12.5 ± 4.7	12.3 ± 2.1
V2	0.0 ± 0.0	2.5 ± 1.4	0.0 ± 0.0	4.6 ± 1.3	0.7 ± 0.6	4.5 ± 0.8***
V3	5.3 ± 4.4	3.3 ± 1.8	1.3 ± 1.3	5.2 ± 2.0	1.3 ± 0.9	3.9 ± 0.9*
V4	25.1 ± 18.0	14.9 ± 6.3	7.4 ± 6.0	1.3 ± 0.9	9.9 ± 4.5	6.8 ± 1.8
V5	7.7 ± 8.0	2.0 ± 2.0	9.1 ± 9.2	20.5 ± 7.4	19.3 ± 8.5	13.3 ± 2.9
TCE	0.0 ± 0.0	0.3 ± 0.2	3.2 ± 1.1	1.3 ± 0.4	2.5 ± 0.6	3.2 ± 0.4
Remainder	0.3 ± 0.3	0.6 ± 0.3	3.6 ± 1.0	2.1 ± 0.4	3.5 ± 0.5	1.9 ± 0.2**

V1–V5 are variable regions 1–5; TCE are T-cell epitopes. Tests of the hypothesis that $d_S = d_N$: *$p < .05$; **$p < .01$; ***$p < .001$. Data from Seibert et al. (1995).

HIV-1 lineages may be subject to quite different modes of selection. For example, the selection favoring amino acid diversity in V4 of family C is clearly not present in families A and E, in both of which mean d_S in V4 substantially exceeds mean d_N (Table 5.8). Similarly, the strong positive selection on V3 in families B and C seems absent in family D (Table 5.8). These differences suggest that some aspects of the phylogenetic history or the environment encountered by these separate lineages may influence the way natural selection acts on them. In the case of a virus like HIV-1, environmental variables influencing natural selection might include factors such as the genetic constitution of the host as well as possible other factors such as concurrent infection by other pathogens that may impinge on the host's immune response.

Because of the key role of CTL in clearing viral infections, most studies of natural selection on HIV-1 have concentrated on the role of CTL escape mutants; that is, mutant alleles of viral genes that alter CTL epitopes, thereby preventing binding by the host's MHC or recognition by CTL clones that are widespread in the host (reviewed by McMichael and Phillips 1997). One of the first studies to demonstrate the widespread occurrence of CTL escape mutants was a study of HIV-seropositive hemophiliacs by Phillips et al. (1991). Although these authors proposed that CTL escape mutants are positively selected, the mere occurrence of such mutants does not prove that they are selected.

Some studies have provided evidence for directional selection on CTL epitopes (see Chapter 6, Selective Sweeps). One interesting, although artificial example involved a radical AIDS therapy whereby a large number of HLA-A3–restricted CTL were injected into a patient (Koenig et al. 1995). These CTL recognized an HLA-A3–bound peptide from the nef protein. Rather than ameliorating the disease, the effect of this treatment was to lead to a rapid increase of the frequency in this patient of a new mutant form of nef from which this epitope was deleted. A more natural example of directional selection by CTL involved a group of hemophiliac patients, all of whom expressed HLA-B27. Each patient was shown to mount a strong CTL response against a peptide derived from gag p24 (KRWIILGLNK). Arginine in the second position of the peptide is particularly important for binding by B27. In one patient, after 12 years of infection, a mutant occurred that replaced this arginine with lysine, and the mutant allele increased to 100% frequency (Goulder et al. 1997).

It remains unclear whether CTL can actually maintain a polymorphism in the HIV-1 virus. One study that appears to suggest such an effect involved an HLA-B8 epitope in the nef protein (Price et al. 1997). Nef sequences were monitored in a single patient over a six-month period directly after initial infection. In the region around the HLA-B8 epitope, there were initially no nonsynonymous differences. However, at the end of six months, nonsynonymous substitutions exceeded synonymous substitutions in this region.

There is evidence that certain CTL epitopes of HIV-1 cause APL antagonism (see this Chapter, The Circumsporozoite Protein of *Plasmodium*) (Klenermann et al. 1994). Certainly, APL antagonism is expected to select for polymorphism in the virus. Furthermore, it has been suggested that in the case of HIV-1 the presence of numerous CTL epitopes may swamp immune defenses, contributing to disease progression (Nowak et al. 1995). Whether such a mechanism contributes to the progression to AIDS remains uncertain.

Self-Incompatibility Genes of Plants

Like the MHC genes of vertebrates, the self-incompatibility or self-sterility genes of flowering plants represent an extraordinarily polymorphic system first discovered by classical genetic methods. However, unlike the MHC, the function of self-incompatibility genes was obvious from the moment of their discovery: these genes serve to prevent selfing in many species of flowering plants. In plants with bisexual flowers, the anther (producing pollen) and pistil (containing ovules) are in close physical proximity, and sexual reproduction might involve frequent selfing unless mechanisms existed to prevent it. Self-incompatibility is one such mechanism, and it has been estimated to occur in between one-third and one-half of all angiosperm species (Franklin et al. 1995).

Sewall Wright (1939) modeled the population genetics of self-incompatibility in a species (*Oenothera organensis*) that is remarkable for maintaining some 37 *S* alleles at a single self-incompatibility locus in a population of only about 1000 individuals. He showed that it is possible to maintain such a degree of polymorphism if the population is subdivided. Formally, the type of selection occurring in the case of a self-incompatibility locus is equivalent to a symmetric overdominance in which homozygotes have a fitness of zero and all heterozygotes have equivalent fitness. Because any combination of pollen and pistil that would lead to an offspring homozygous at the *S* locus leads to rejection of the pollen, the results are equivalent to the case of a locus at which homozygosity is a lethal condition (Yokoyama and Hetherington 1982).

There are two major strategies of self-incompatibility in flowering plants, called homomorphic and heteromorphic (Kao and McCubbin 1996). In the former, there is a single morphological type of flower, whereas in the latter there are two or three flower morphs, each of which can fertilize flowers of a different morph but not of the same morph. Homomorphic self-incompatibility systems are, in turn, classified as either sporophytic or gametophytic. In gametophytic self-incompatibility, the haploid genotype of the pollen itself determines its incompatibility type. In sporophytic self-incompatibility, the diploid genotype of the pollen's parent plant determines its incompatibility type. In either case, pollen cannot fertilize a pistil of a plant with which it shares an incompatibility type.

Gametophytic self-incompatibility has been most extensively studied in the Solanaceae, the family that includes potato, tomato, and tobacco. In this family, classical genetic studies attributed self-incompatibility to the action of a single polymorphic locus (the *S* locus). Later, Lewis (1949, 1960) proposed that the *S* locus encodes three factors: the allelic specificity factor that determines the specificity of pollen and pistil, the pollen activity factor, and the pistil activity factor. The activity factors are assumed to activate the allelic specificities of pollen and pistil, respectively. Their existence was proposed because mutants were discovered that affected the activity of pollen but not of the pistil of the same plant from which it is derived, as well as mutants affecting the response of the pistil but not activity of the pollen from the same plant. However, this model does not clarify whether the *S* locus actually consists of three separate, but very closely linked genetic loci, each encoding a separate protein, or whether the three factors might represent different domains of a single protein.

S allele sequences have been obtained from a number of solanaceous plant species by identifying genes expressed in the pistil that cosegregate with *S*-allele specificity (Kao and McCubbin 1996). The proteins these genes encode are RNases, and their RNase activity in the pistil is necessary for the rejection of pollen. It is as yet uncertain whether these proteins are expressed in the pollen itself or how the *S*-allele specificity of the pollen is recognized by the pistil.

Ioerger et al. (1990) and Clark and Kao (1991) were the first to study the molecular evolution of *S* allele sequences from Solanaceae. They observed that sequences of alleles in these species are sometimes highly divergent from one another. The *S* locus is expected to be subject to balancing selection; and, as we have seen in the case of the MHC (Chapter 4, Trans-species Polymorphism at MHC Loci) and surface proteins of malaria (this Chapter, second and third sections), such selection can maintain polymorphisms for long periods of time. Thus, Ioerger et al. (1990) tested for transspecies polymorphism by a phylogenetic analysis. They found that an allele of one species is often more closely related to an allele of another species than to other alleles of the same species.

Figure 5.7 shows a phylogeny of selected *S* alleles from solanaceous species. The tree is rooted with plant RNases that are not involved in self-incompatibility and with an RNase from *Drosophila* (Figure 5.7). In several cases, there are strongly supported clusters including alleles from different species. For example, S11 and S12 from *Lycopersicon peruvianum* cluster with H-SC from *Solanum carolinense*, while *L. peruvianum* S3 clusters with *S. carolinense* D-SC (Figure 5.7). The existence of such long-lasting polymorphisms supports the hypothesis that balancing selection is operating at the *S* locus of Solanaceae.

The sequences available to Clark and Kao (1991) were so divergent that comparison of the rates of synonymous and nonsynonymous substitution could not be used to test for positive selection. They computed that a greater proportion of nonsynonymous sites than synonymous sites were involved in polymorphisms shared between species, consistent with the hypothesis that selection is favoring polymorphism at the amino acid level.

More recently, Richman et al. (1996) sequenced 17 closely related *S* alleles from the ground cherry (*Physalis crassifolia*). A phylogenetic analysis showed that these formed two major families (data not shown). Representatives of both families are among the sequences included in the phylogeny of Figure 5.7. *P. crassifolia* S1, S8, and S10 represent family 1, while *P. crassifolia* S16, S17, and S28 represent family 2. When mean d_S and mean d_N were estimated within and between these families, mean d_N exceeded mean d_S (Table 5.9). The difference was highly significant in the case of family 2, whose members were more closely related to each other than those of family 1 (Table 5.9). These results strongly support the hypothesis that selection at this locus favors heterozygosity of the *S* proteins at the amino acid level. They also suggest that evidence for such

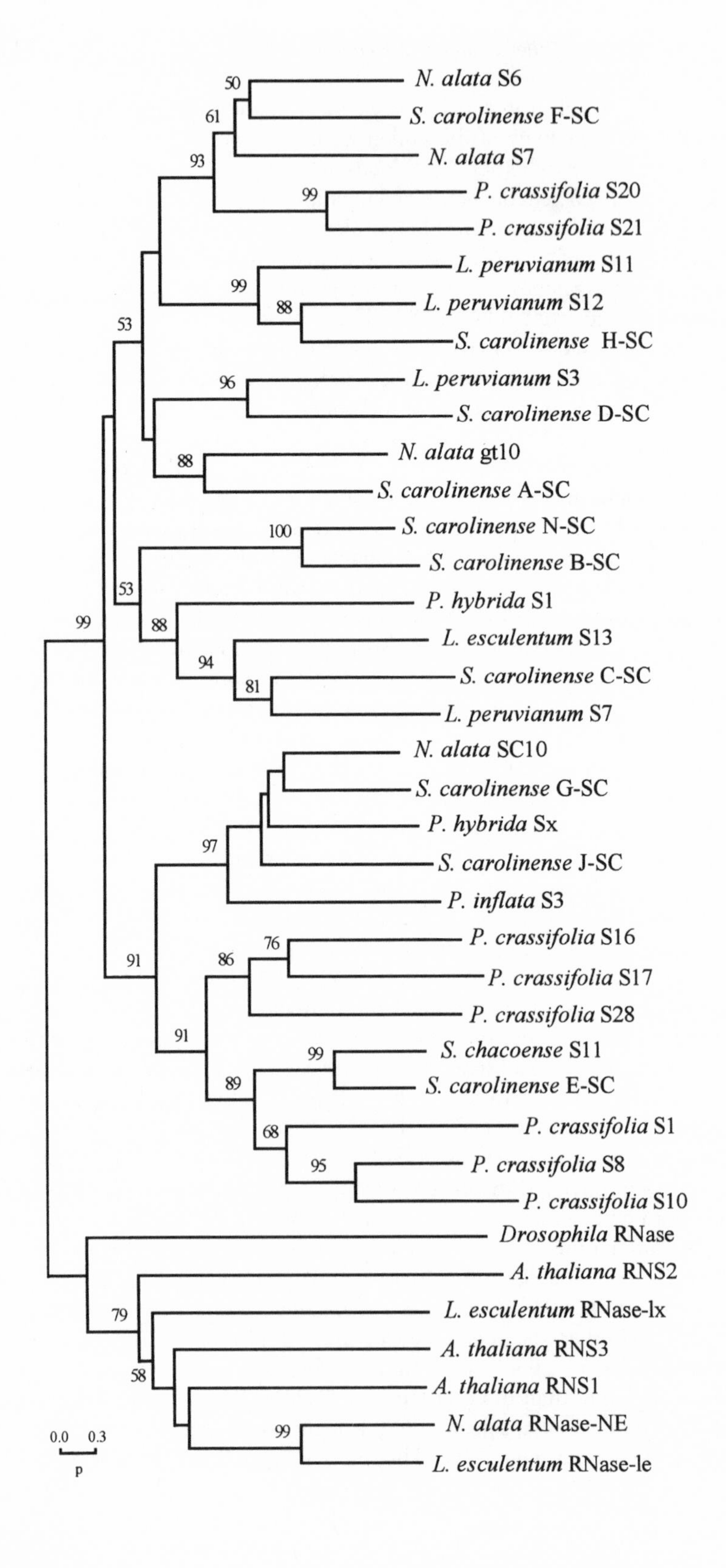
N. alata S6
S. carolinense F-SC
N. alata S7
P. crassifolia S20
P. crassifolia S21
L. peruvianum S11
L. peruvianum S12
S. carolinense H-SC
L. peruvianum S3
S. carolinense D-SC
N. alata gt10
S. carolinense A-SC
S. carolinense N-SC
S. carolinense B-SC
P. hybrida S1
L. esculentum S13
S. carolinense C-SC
L. peruvianum S7
N. alata SC10
S. carolinense G-SC
P. hybrida Sx
S. carolinense J-SC
P. inflata S3
P. crassifolia S16
P. crassifolia S17
P. crassifolia S28
S. chacoense S11
S. carolinense E-SC
P. crassifolia S1
P. crassifolia S8
P. crassifolia S10
Drosophila RNase
A. thaliana RNS2
L. esculentum RNase-lx
A. thaliana RNS3
A. thaliana RNS1
N. alata RNase-NE
L. esculentum RNase-le
50
61
93
99
99
88
53
96
88
100
53
99
88
94
81
97
91
76
86
91
99
89
68
95
79
58
99
0.0 0.3
p

Table 5.9. Mean numbers of synonymous (d_S) and nonsynonymous (d_N) nucleotides per 100 sites (±SE) in comparisons within and between two families of S alleles of *Petunia crassifolia*

Comparison (No. of sequences)	d_S	d_N
Family 1 (10)	17.0 ± 3.2	17.3 ± 1.6
Family 2 (7)	10.8 ± 2.7	23.3 ± 2.2***
Family 1 versus Family 2	26.3 ± 5.0	38.1 ± 3.3
All comparisons	20.8 ± 3.4	29.0 ± 2.1*

Tests of the hypothesis that $d_S = d_N$: *$p < .05$; ***$p < .001$.

selection is most easily seen in comparisons of relatively closely related S alleles. Once S alleles have reached a certain level of divergence at nonsynonymous sites, selection no longer seems to favor further amino acid level divergence. Thus, in comparisons between more distantly related sequences, the number of synonymous substitutions per site eventually seems to catch up with and surpass the number of nonsynonymous substitutions per site. Interestingly, such a phenomenon has not been observed in the case of alleles at the MHC loci, suggesting a difference between the type of selection acting in these two cases.

In the case of sporophytic incompatibility, most molecular study has concentrated on the genus *Brassica*. In this genus, two different genes linked to the classical S-locus have been found: these encode the S-linked glycoprotein (SLG) and a serine-threonine receptor protein kinase (SRK) (Franklin et al. 1995). These genes are closely linked, so that in *Brassica*, it is considered more accurate to speak of "S haplotypes" rather than "S alleles." The other surprising thing about the SLG and SRK genes is that they are clearly related evolutionarily, with the soluble SLG being homologous to the extracellular portion of SRK. It has been suggested that the sequence similarity between SLG and SRK is maintained by gene conversion between them (Kusaba et al. 1997), but there is really no evidence of this. It is at least equally plausible that new SLG genes arise repeatedly by duplication of the SRK gene and occurrence of a mutation causing a premature stop codon.

Figure 5.8 shows a phylogenetic tree of SLGs and the homologous region of SRKs from *Brassica* species. The tree is rooted with other serine-threonine kinases from plants. The tree shows a trans-species pattern of evolution, with

---

Figure 5.7. (facing page) Phylogenetic tree of S alleles from plants of the family Solanaceae, constructed by the neighbor-joining method based on the proportion of amino difference (p). The tree is rooted with non-S RNases from plants and *Drosophila*. Species are as follows: *Arabidopsis thaliana; Lycopersicon esculentum; Lycopersicon peruvianum; Nicotiana alata; Petunia hybrida; Petunia inflata; Solanum carolinense; Solanum chacoense; Solanum crassifolia*. Numbers on branches are as in Figure 3.2.

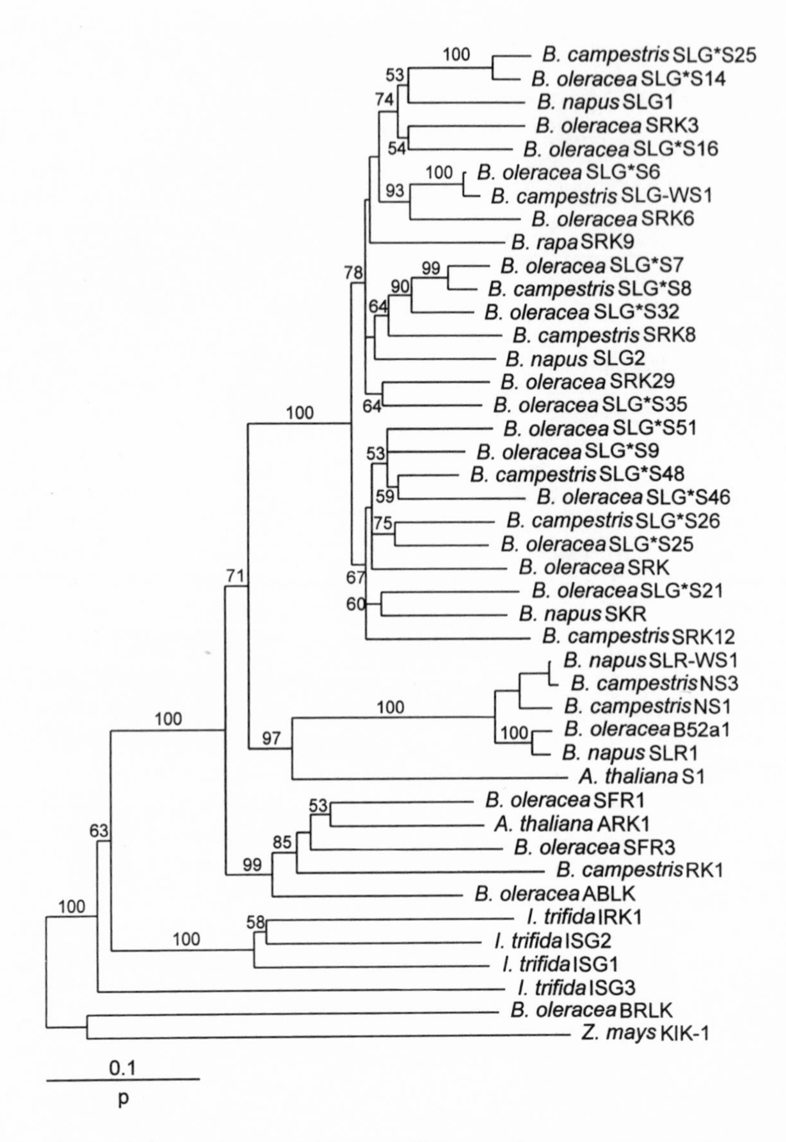

Figure 5.8. Phylogenetic tree of SLG and SRK from the genus *Brassica* (*B. campestris, B. napus, B. oleracea*, and *B. rapa*), constructed by the neighbor-joining method based on the proportion of amino acid difference (p). The tree is rooted with related serine-threonine kinases from *Brasccica* and other plants (*Arabidopsis thaliana, Ipomoea trifida*, and *Zea mays*). Numbers on branches are as in Figure 3.2.

SLGs of *B. oleracea* frequently clustering with those of *B. campestris* (Figure 5.8). Examination of rates of synonymous and nonsynonymous substitution showed $d_N > d_S$ in the hypervariable regions of the molecules (Kusaba et al. 1997) in certain comparisons between closely related genes (data not shown). Thus, the SLGs of *Brassica* may be subject to a type of selection similar to that acting on the *S* alleles of Solanaceae.

Alcohol Dehydrogenase in *Drosophila melanogaster*

Alcohol dehydrogenase (Adh) is an enzyme that catalyzes the detoxification of alcohols by converting them to aldehydes and ketones. In the fruitfly, *Drosophila melanogaster*, the locus (*Adh*) encoding this enzyme is polymorphic, and this polymorphism has been among the most intensively studied of all naturally occurring polymorphisms. Using a variety of different approaches, researchers have obtained evidence that polymorphism at this locus is selectively maintained. However, both biochemically and ecologically, the basis of this selection remains obscure.

The *Adh* gene encodes a 255–amino acid polypeptide (Chambers 1988). Protein electrophoresis was able to detect two major allelic classes called "fast" (*Adh-F*) and "slow" (*Adh-S*). At the amino acid level, the two allelic classes differ by a single amino acid substitution at position 192, where *Adh-F* has threonine and *Adh-S* has lysine (Kreitman 1983). *Adh-F* homozygotes gave a two- to threefold higher enzyme activity than do *Adh-S* homozygotes (Laurie et al. 1991). This difference occurs both because *Adh-F* homozygotes produce greater quantities of the protein than do *Adh-S* homozygotes and because *Adh-F* encodes a protein with higher catalytic efficiency. In vitro site-directed mutagenesis studies determined that the single amino acid difference between *Adh-F* and *Adh-S* gene products is responsible for the difference in catalytic efficiency but has no effect on concentration of the enzyme (Choudhary and Laurie 1991). There is a sequence-length polymorphism in the first intron of the gene that is in linkage disequilibrium with the fast–slow polymorphism. There is an insertion in this intron in most *Adh-F* alleles, which seems to be associated with their higher expression levels, presumably regulating expression by an unknown mechanism (Laurie et al. 1991).

Drosophila melanogaster originated in tropical Africa and subsequently migrated to the rest of the world (David and Capy 1988). It is believed to have spread to Europe and Asia 10,000 to 15,000 years ago, probably without human assistance. Much more recently, humans have spread it to the rest of the world. It was presumably brought to South America several hundred years ago during the post-Columbian colonization, and to Australia in the past 200 years. It reached North America only in the middle of the last century.

Despite the widely differing times of the species' appearance in different parts of the world, corresponding latitudinal clines in *Adh* allele frequency are seen in different parts of the world (Oakeshott et al. 1982). In Africa, *Adh-S* is the more frequent allele, with the frequency of *Adh-F* increasing in a clinal fashion as one moves northward in Europe. Similar clines are observed in both North America and Australia: as one moves away from the equator, the frequency of *Adh-F* increases. It has been proposed that *D. melanogaster* was brought to South America primarily from Africa at the time of the slave trade. If *D. melanogaster* colonized southern North America from South America but northern North America in a second wave of importation from Northern Europe, the observed cline in *Adh* frequencies in North America might be explained as a result of colonization history without the involvement of natural selection. It should be noted, however, that it would be very difficult to devise a similar historical scenario to account for the Australian cline.

Berry and Kreitman (1993) studied this question by means of a population study using restriction enzymes. They examined 44 polymorphic restriction sites in the *Adh* gene region, one of which was associated with the nonsynonymous difference that causes the *Adh-F*/*Adh-S* electrophoretic difference. Surveying over 1,500 individuals from 24 localities in eastern North America (from Maine to Florida), they found that the north–south cline existed only at the site associated with the fast–slow polymorphism, not at other polymorphisms in the *Adh* gene region. Because this implies that restriction sites in this region have recombined freely, it supports the hypothesis that the north–south cline in *Adh* allele frequencies is selectively maintained and not a by-product of colonization history.

The strongest evidence that the polymorphism at the *Adh* locus is selectively maintained comes from comparisons of complete genomic sequences of a number of *Adh-F* and *Adh-S* alleles (Kreitman 1983). When the proportion of nucleotide differences among allelic sequences is computed within a sliding window along the gene, a striking pattern is revealed: there is a sharp peak of nucleotide diversity right around the site where a nonsynonymous difference gives rise to the fast–slow polymorphism (Figure 5.9). This suggests that the polymorphism closely linked to the fast–slow polymorphism is much older than the polymorphism in the rest of the gene, and is thus hitchhiking along with the fast–slow polymorphism. Such a pattern would only be observed if the fast–slow polymorphism is being maintained by balancing selection.

Thus, the comparison of *Adh* alleles reveals a phenomenon very similar to the comparison of class I MHC exons and introns discussed in Chapter 4, Patterns of Variation in Class I Peptide Ligands. In the class I MHC, balancing selection operates on certain nonsynonymous sites in exons 2–3; these, and other sites in the exons that hitchhike along with them, are more divergent between alleles than are the introns. Not being closely linked to the exons, the introns are homogenized over evolutionary time by recombination and subsequent genetic drift. At the *Adh* locus, the regions not closely linked to the single selected site are likewise homogenized by recombination and genetic drift.

As mentioned, it is known that *Adh-F* encodes an enzyme that is produced in higher quantities than that encoded by *Adh-S*, and that the fast enzyme is more catalytically efficient. In the artificial environment of wine cellars in Spain, where ethanol concentration is high, the *Adh-F* allele is near fixation; and experimental exposure of flies to varying ethanol concentrations supports the

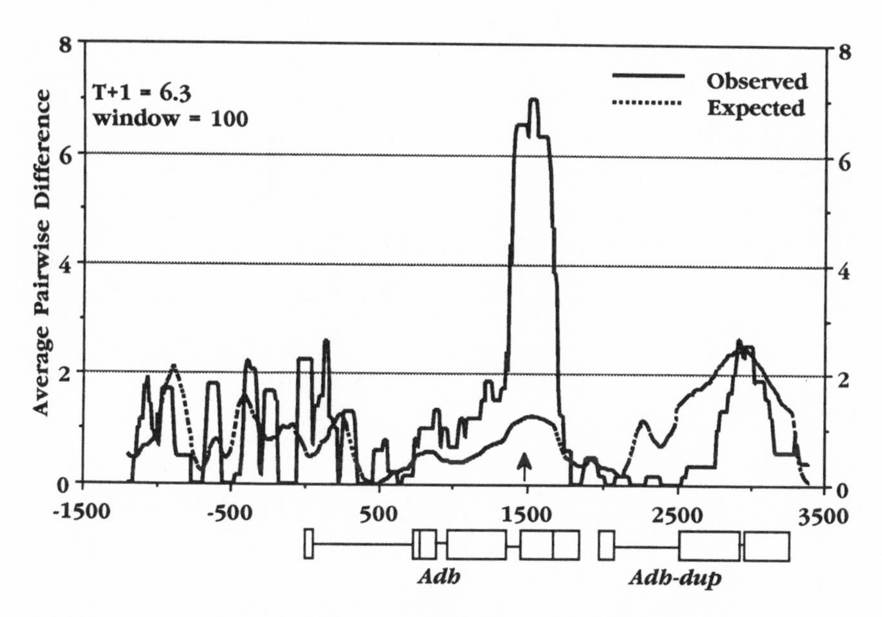

Figure 5.9. Sliding window of nucleotide diversity in a sliding window of 100 nucleotides along the *Adh* gene and adjoining regions in *Drosophila melanogaster*. There is a peak of nucleotide diversity around the amino acid responsible for the fast–slow protein polymorphism (arrow). T + 1 is the coalescent time (in 2*N* generations) estimated between one *D. melanogaster* allele and one *D. simulans* allele. From Kreitman and Hudson (1991). Reproduced by permission.

hypothesis that mortality due to ethanol is the major factor driving this selection (Briscoe et al. 1975).

However, it is much less clear why the frequency of *Adh-F* should increase at high latitudes in free-living populations. One possibility is that in cooler climates, limited quantities of rotting fruit are generally available, forcing flies to choose breeding habitat that is less than optimal, including fruit with a high ethanol concentration. An alternative possibility is that in warmer climates detoxification of ethanol by microorganisms occurs more rapidly, leading to a lower ethanol concentration in fruits colonized by *D. melanogaster*. But these hypotheses are purely speculative at present. In addition, it is unclear to what extent natural selection acts to maintain the polymorphism at this locus within a given population.

To understand the basis of the fast–slow polymorphism at the *Adh* locus in *D. melanogaster*, further study of this species' ecology is necessary. Because there is no evidence of similar selection in other species of the genus *Drosophila* (e.g., *D. pseudoobscura*; Schaeffer and Miller 1992), the environmental factors responsible for this selection must be unique to *D. melanogaster*. Surprisingly, although in terms of its genetics and developmental biology *D. melanogaster* is easily the best known of all multicellular eukaryotes, our knowledge of this species' natural history is remarkably scanty.

There are other enzyme loci at which electrophoretic studies have revealed evidence of balancing selection. One is the lactate dehydrogenase-B locus in the killifish *Fundulus heteroclitus* (Mitton and Koehn 1975). Another is phososphoglucoisomerase in butterflies of the genus *Colias* (Watt 1992). It would be of considerable interest to study these loci at the DNA level, because in these cases, in contrast to *Adh* of *Drosophila*, the biochemical and ecological bases of the balancing selection are relatively well understood (Mitton 1997).

6

Directional Selection

As shown in Chapters 4 and 5, by now a number of examples of balancing selection have been studied at the DNA sequence level. Probably 30–40 million years represents the upper limit for maintenance of polymorphism by balancing selection in species with long-term effective population sizes such as those found in most multicellular organisms (Takahata and Nei 1990). No polymorphisms older than this have been documented in any multicellular eukaryote. In the protist parasite *Plasmodium falciparum*, which may have an effective population size somewhat larger than that of its human host, the MSA-1 polymorphism may have persisted nearly 50 million years (see Chapter 5, Other *Plasmodium* Surface Proteins). Because even balanced polymorphisms are young relative to this history of life, they are particularly amenable to study by comparing rates of synonymous and nonsynonymous nucleotide substitution. Even after 40 or 50 millions years, synonymous sites are far from saturation, and this technique can reveal positive selection. Furthermore, under certain types of overdominant selection, new codon substitutions are continually favored (Maruyama and Nei 1981), leading to repeated nonsynonymous substitution between alleles over time.

However, when most of us think of adaptive evolution, we do not think mainly of balancing selection, but of directional selection. *Directional selection* is a term for cases in which an advantageous mutation occurs in the population and then becomes fixed as a result of positive selection. The hypothesis of evolution by natural selection holds that the numerous exquisite adaptations seen in organisms—from the cilia of protists to the wings of birds—arose by such a process, repeated many times over long periods, leading to the efficiently functioning structures seen in organisms today. However, it has been difficult to study the molecular basis for such adaptations or to reconstruct the sequences of evolutionary events by which they arose.

In the case of distinctive morphological or physiological adaptations, we are usually dealing with evolutionary events that took place in the very distant past. For example, if we were to discover some of the genes responsible for the process by which birds' forelimbs develop into wings and could compare them with homologous genes in other vertebrate classes such as reptiles or mammals, it might still prove difficult to find evidence of positive selection through sequence comparisons alone. First of all, because birds diverged from mammals over 300 years ago and from their closest living reptilian relatives (alligators and crocodiles) soon after that, synonymous sites are completely saturated in any comparison between birds and other vertebrate classes. Second, the adaptive changes in a given protein-coding gene may involve only one or a few nonsynonymous substitutions. However, in 300 million years, a great many neutral nonsynonymous substitutions may also occur in the same gene. In practice, it may be difficult to distinguish a small number of adaptive substitutions in a background of neutral substitution.

Therefore, the cases of directional selection that are most amenable to molecular study are those that have happened relatively recently. In this chapter I consider three types of study of directional selection: (1) cases of directional selection studied within a single species; (2) directional selection in the divergence between related species; and (3) convergent or parallel evolution, whereby species presumably have evolved adaptive traits independently. Another major topic under the general heading of directional selection is that of the functional diversification of genes within multigene families; this latter topic will be covered in Chapter 7.

Selective Sweeps

Consider a new favorable allele resulting from a single point mutation; when it first occurs in a population, it is found on only a single chromosome. Recombination plays an important role in the increase in frequency of the new mutant allele because the allele is recombined out of the genetic background in which it first occurred and placed in new genetic backgrounds as it goes to fixation. (Thus, in a species like an asexual vertebrate that reproduces clonally without recombination between clones, a favorable mutant cannot go to fixation unless all clones not bearing the mutant become extinct.) However, although recombination places the new favorable mutant in new genetic background, the single favorable nucleotide will generally carry with it some of the sequence that surrounded it on the chromosome where it initially occurred. This happens because recombination is not as precise as to excise the single beneficial nucleotide from the nucleotides around it. Thus, a process of "hitchhiking" occurs, whereby a certain portion of the chromosome on which the mutation originated is carried to fixation along with it (Maynard Smith and Haigh 1974; Kaplan et al. 1989; Stephan et al. 1992).

The process whereby linked nucleotide sites are "swept" to fixation along with a linked favorable mutation is called a "selective sweep" (Charlesworth 1992). At the population level, this process will have the effect of reducing variation at sites linked to the locus under selection. Note that "hitchhiking" in the case of directional selection has an opposite effect to that seen in the case of balancing selection. Neutrally evolving sites closely linked to a locus under balancing selection are expected to show higher levels of polymorphism than are less closely linked sites. As we have seen, this prediction is supported in the case of the *Adh* locus of *Drosophila melanogaster* (Chapter 5, Alcohol Dehydrogenase in *Drosophila melanogaster*) and the introns of class I MHC genes (Chapter 4, Class I Introns).

A number of authors have reported evidence for past selective sweeps in *Drosophila melanogaster* from surveys of nucleotide sequence diversity in different genomic regions using restriction enzymes. These studies have found that in regions known to have low recombination rates there are reduced levels of DNA polymorphism within *D. melanogaster*, yet the same regions show no reduction in the extent of divergence between *D. melanogaster* and the closely related species *D. simulans* (Figure 6.1; Begun and Aquadro 1991, 1992; Berry et al. 1991; Martin-Campos et al. 1992; Langley et al. (1993). One possible interpretation of reduced polymorphism in a low-recombination region is that it reflects hitchhiking with a recently fixed advantageous mutation. However, more detailed analyses of the same sort have cast doubt on this interpretation.

In the case of hitchhiking, there should be a relatively high frequency of rare polymorphisms in comparison with the pure neutrality; this should occur because most polymorphisms should be relatively new, having evolved subsequent to the selective sweep (Hudson 1990). In fact, this is not always the case in certain low-recombination regions of the *D. melanogaster* genome (Begun and Aquadro 1995). An alternative hypothesis to that of hitchhiking that might explain the association between low polymorphisms and low recombination is the so-called background selection model (Charlesworth et al. 1993). This model predicts a reduction in polymorphism due to removal of deleterious mutants from the population. However, it does not predict anywhere near as great a reduction as seen in some low-recombination regions of the *D. melanogaster* genome (Begun and Aquadro 1995).

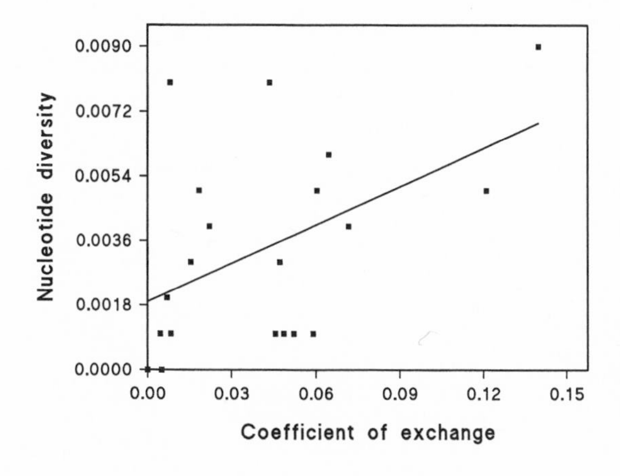

Figure 6.1. Nucleotide diversity plotted against the coefficient of exchange for 20 regions of the *Drosophila melanogaster* genome. The data are from Begun and Aquadro (1992). The line is the linear regression line; $r = .487$, $p < .05$.

Thus, the association between low polymorphism and low recombination in *D. melanogaster* is not fully explained at present. It is possible that the observed pattern has resulted from selective sweeps in some regions as well as other factors. However, because the loci involved in selective sweeps have not been identified, the occurrence of such events remains speculative.

The phenomenon of selective sweeps can also be studied in cases where there is independent evidence that a locus has been subject to directional selection. Interactions of parasites with their hosts provide numerous examples of selective fixation events; for example, the cases in which it has been documented that a particular CTL escape mutant has been fixed within a particular HIV-1 patient (Chapter 5, Selection on HIV-1). One study of a viral CTL escape mutant in a virus that provides a good illustration of a selective sweep involved a mutation in the hepatitis C virus (HCV) in an experimental infection of chimpanzee (Weiner et al. 1995).

Initially, a chimpanzee named Rodney was infected with the virus. Then an inoculum taken from Rodney was used to infect a second chimpanzee called Ross. A short stretch of sequence encoding 77 residues of the HCV protein NS3 and including a known CTL epitope was amplified from RNA extracted from the chimps' plasma. This region was sequenced from Rodney prior to the inoculation of Ross, and then periodically from Ross over a period of two years. After 16 weeks of infection, a new form of the CTL epitope had appeared in Ross, involving a replacement of aspartic acid by glutamic acid (Figure 6.2). Unlike other previously and subsequently observed point mutations, this mutation both occurred and was fixed between the 4th and 16th weeks of infection. That this was a CTL escape mutation is suggested by the fact that Ross was unable to clear the viral infection.

Figure 6.3 shows a maximum parsimony network of the DNA sequences from Rodney and Ross. The sequences from Rodney were all very similar, being within one nucleotide substitution of a common type (corresponding to the sequences designated Rodney 5, 7, and 10; Figure 6.3). The sequences taken from Ross after four weeks were of this common type or (in one case) differed from it by one nonsynonymous substitution (Figure 6.3). Between weeks 4 and 16, the nonsynonymous substitution in the CTL epitope occurred and went to fixation in the HCV population infecting Ross. Thereafter, all sequences found in Ross were within one or two nucleotide substitutions of the new type.

These results illustrate in a simple system several of the features we expect to see in the case of a selective sweep. Genetic diversity in the virus was reduced after the appearance of the CTL escape mutant, even outside the CTL epitope,

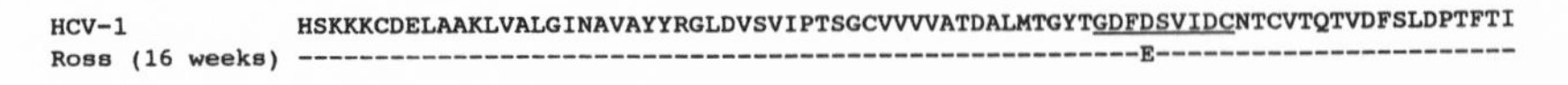

Figure 6.2. Amino acid sequence of a CTL epitope (*underlined*) and flanking regions of the NS3 protein of hepatitis C virus (HCV). The bottom sequence shows the amino acid replacement in the epitope that occurred after 16 weeks in the infected chimpanzee Ross (Weiner et al. 1995).

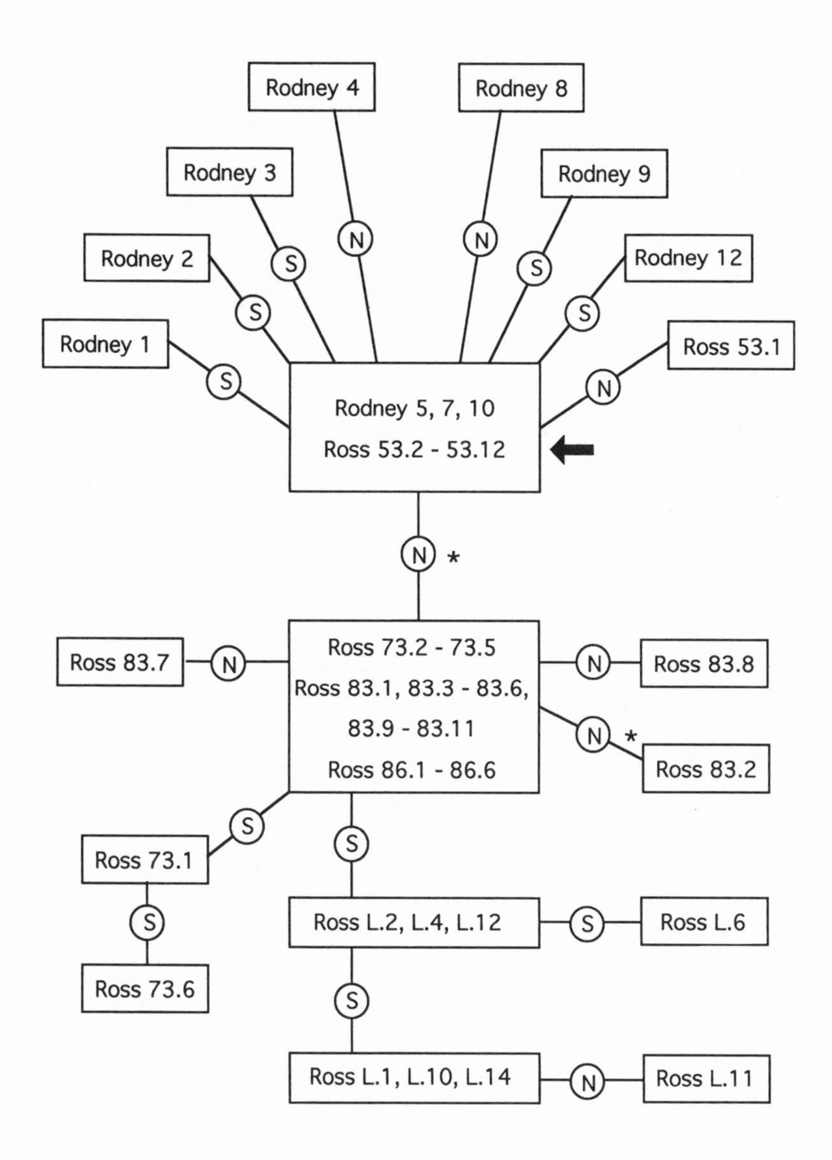

Figure 6.3. Phylogenetic tree constructed by maximum parsimony method for HCV sequences from the chimpanzees Rodney and Ross. By comparison with the published H strain, the tree was rooted at the node indicated by the arrow. Reconstructed nucleotide substitutions are designated S (synonymous) and N (nonsynonymous). Asterisks indicate substitutions in the CTL epitope. From Weiner et al. (1995). Reproduced by permission. Copyright (1995) National Academy of Sciences USA.

showing the effect of hitchhiking with the favorable mutation. As diversity was reestablished afterward, as the result of new mutations, most sequences were rare variants of the ancestral type bearing the CTL escape mutant. There are statistical methods to test whether a phylogeny is indicative of a selective sweep. A pattern indicative of a sweep is one where most substitutions are on terminal branches, as would occur as diversity is reestablished after a sweep. One such method (Fu and Li 1993) was applied to the phylogeny of Figure 6.3; the results indicated that the pattern of substitution is inconsistent with selective neutrality, supporting the selective sweep hypothesis.

Adaptive Divergence between Species

A major question confronting evolutionary biology is the question of the adaptive differentiation of species. By this I mean not simply the question of speciation—that is, how two populations of the same species become reproductively isolated from each other and thus no longer capable of crossfertilization. Rather, I am referring to the phenomenon of adaptive radiation; that is, the process by which two or more species descended from a common ancestor adapt to different habitats or lifestyles by evolving distinct morphological and/or physiological characteristics. A number of recent studies have examined evidence for adaptive divergence between species at the molecular level. I discuss two of these cases in this section. Neither of these studies is a complete study of adaptive divergence, largely because in each case the nature of the selective agents driving adaptive evolution remains speculative. However, these studies point the way to future research that may eventually address questions regarding the selective forces at work.

Divergence of Adh *between* Drosophila *species*

As mentioned (Chapter 5, Alcohol Dehydrogenase in *Drosophila melanogaster*), a polymorphism at the *Adh* locus, encoding the enzyme alcohol dehydrogenase, in the fruitfly *Drosophila melanogaster* is a well-documented example of balancing selection. McDonald and Kreitman (1991) compared sequences from this locus between *D. melanogaster* and the two closely related species, *D. simulans* and *D. yakuba*. They observed a pattern of nucleotide substitution that, they argued, is indicative of adaptive divergence among these species at the *Adh* locus.

The rationale behind McDonald and Kreitman's analysis was based on the expectation that, under the neutral theory, the ratio of nonsynonymous to synonymous substitutions between species should be equal to the ratio of nonsynonymous to synonymous substitutions between alleles within species. McDonald and Kreitman tested this hypothesis by comparing polymorphisms at the *Adh* locus within the species with differences fixed between the species (Table 6.1). The ratio of synonymous to nonsynonymous differences within species (21:1) was much greater than that between species (4.4:1) (Table 6.1). Thus, McDonald and Kreitman concluded that the *Adh* is not evolving neutrally.

Table 6.1. Numbers of synonymous and nonsynonymous differences at the *Adh* locus fixed between *Drosophila* species or polymorphic within species

	Fixed	Polymorphic
Synonymous	17	42
Nonsynonymous	7	2

$p < .01$ (Fisher's exact test). Data from McDonald and Kreitman (1991).

This analysis has been criticized by Graur and Li (1991) and by Whittam and Nei (1991). One problem is that differences between a pair of species were designated as "fixed" only if no polymorphism was detected in the small numbers of alleles sampled (12 for *D. melanogaster*, 6 for *D. simulans*, and 12 for *D. yakuba*). Inclusion of more sequences might lead to reclassification of many of these sites as polymorphic. Even if the sites do truly represent fixed differences, "fixed" sites were defined in such a way as to exclude sites that were polymorphic in one species and fixed in one or both of the others. This definition seems arbitrary and loses information. Both Graur and Li (1991) and Whittam and Nei (1991) suggest that a better approach to this question might be to estimate mean d_S and d_N within and between species, which would make use of the information at all sites. When Whittam and Nei (1991) applied this approach to the comparison between *D. melanogaster* and *D. yakuba*, McDonald and Kreitman's data, they found that the ratio of synonymous to nonsynonymous substitutions between the two species was nearly three times as high as that within species, although the difference was not statistically significant.

I further analyzed *Adh* sequences from these species to understand the factors that might explain the results of previous analyses. I analyzed 14 *Adh* alleles from *D. melanogaster*, 5 from *D. simulans*, and 6 from *D. yakuba*. (Of the sequences used by McDonald and Kreitman, I excluded a partial *D. simulans* sequence and six sequences from *D. yakuba*, including undetermined nucleotides, because for this sort of analysis it is preferable to compare only complete sequences.) One possible complicating factor in this type of study might be that the polymorphism may predate speciation, particularly in the case of very closely related species like *D. melanogaster* and *D. simulans* (Whittam and Nei 1991). However, a phylogenetic analysis suggested that this is not a problem with the *Adh* locus; the alleles from each species form well-differentiated species-specific clusters (Figure 6.4).

In this data set, the ratio of mean d_N to mean d_S for within-species comparisons was 0.06, whereas that for between-species comparisons was 0.08. When d_N was plotted against d_S for the within-species comparisons, the two values were uncorrelated (Figure 6.5). By contrast, in the case of between-species comparisons, d_S and d_N were strongly positively correlated (Figure 6.5). One

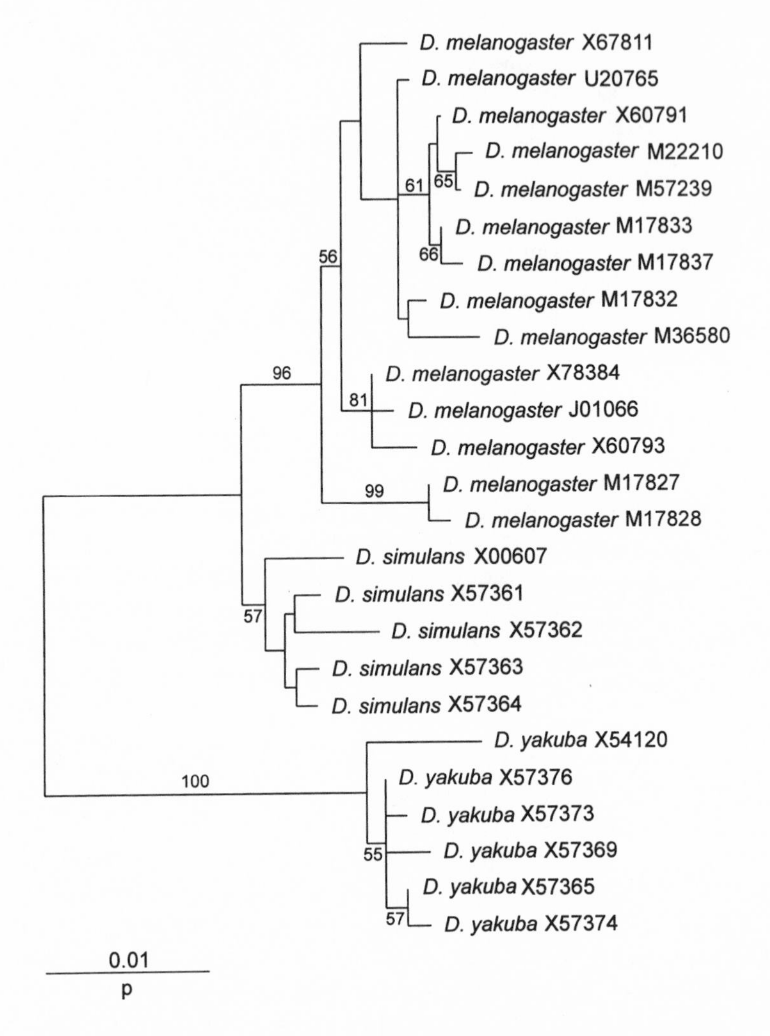

Figure 6.4. Phylogenetic tree of *Adh* genes from *Drosophila* species, constructed by the neighbor-joining method based on the number of nucleotide substitutions per site (*d*). Numbers on branches are as in Figure 3.2.

reason for this difference is that within species, sequences are subject to recombination, which reshuffles polymorphic sites. There is abundant evidence for interallelic recombination at the *Adh* locus in *D. melanogaster* (see Chapter 5, Alcohol Dehydrogenase in *Drosophila melanogaster*). In the language of population genetics, there is no linkage disequilibrium between polymorphic sites in within-species comparisons. This is not true, however, in between-species comparisons. Two sequences from different species that are divergent at synonymous sites will generally also be divergent at nonsynonymous sites;

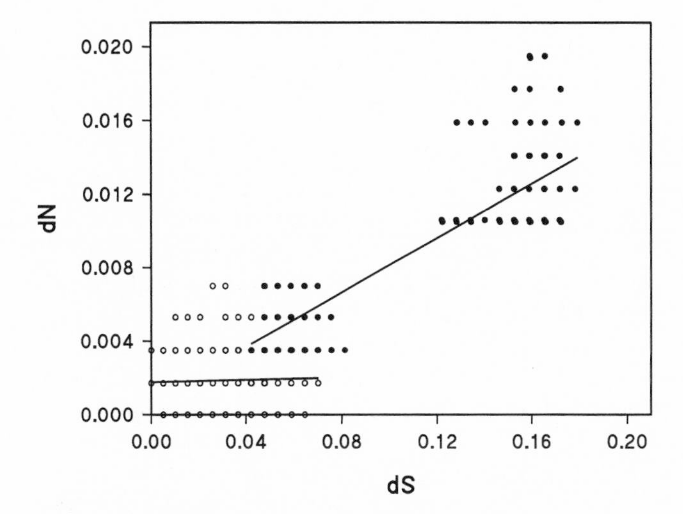

Figure 6.5. Number of nonsynonymous substitutions per site (d_N) plotted versus number of synonynmous substitutions per site (d_S) for pairwise comparisons within (*open circles*) and between (*solid circles*) *Drosophila* species. In each case the linear regression line is drawn. For within-species comparisons, $Y = 0.0031 + 0.0018X$ ($r = .028$, n.s.). For between-species comparisons, $Y = 0.0007 + 0.0740X$ ($r = .857$, $p < .01$). Because pairwise comparisons are not independent, as a conservative test of the correlation coefficient, $N - 2$ degrees of freedom was used where N is the number of sequences compared (25).

likewise, two sequences from different species that are similar at synonymous sites will tend to be similar at nonsynonymous sites.

In the present data set, the more divergent between-species comparisons are mainly those between *D. yakuba* and either *D. melanogaster* or *D. simulans*, whereas the less divergent between-species comparisons are those between *D. melanogaster* and *D. simulans*. The existence of phylogenetic structure to the data (Figure 6.4) explains the correlation between d_S and d_N in between-species comparisons (Figure 6.5). This correlation, in turn, is sufficient to explain why the d_N/d_S ratio is slightly higher in the between-species comparisons than in the within-species comparisons.

Thus, more detailed analyses do not provide unequivocal support for the hypothesis that *Adh* genes have diverged adaptively between these *Drosophila* species. In any event, if such adaptive divergence has occurred, its biochemical and ecological basis remains unknown, as, of course, is also true of the well-documented balanced polymorphism at this locus in *D. melanogaster* (Chapter 5, Alcohol Dehydrogenase in *Drosophila melanogaster*). Nonetheless, McDonald and Kreitman's (1991) paper is important in calling attention to the potential usefulness of within- and between-species ratios of synonymous and nonsynonymous substitution as a means of studying adaptive evolution (see Chapter 7, Diversifying Selection in Multigene Families).

Abalone Sperm Lysins

Abalone (genus *Haliotis*) are marine gastropods that reproduce by releasing gametes into the seawater. In this type of reproduction, it is advantageous for sperm to be able to recognize eggs of the same species so that cross-species fertilization is avoided. Presumably, fertilization of an egg by sperm of another species would result in nonviable offspring and thus a waste of gametes (Palumbi and Metz 1991; Palumbi 1994). However, in marine species that breed by release of gametes, no behavioral or morphological premating isolating mechanisms such as are found in other animal species are available to prevent cross-species mating. Thus, these species appear to have evolved complex biochemical mechanisms for sperm–egg recognition (Vacquier et al. 1997).

In abalone, the egg is encased in a vitelline envelope that the sperm must penetrate before fertilization can occur. In the acrosome of the sperm, there is a vesicle containing two proteins involved in penetrating the vitelline envelope, the better known of which is called lysin. The crystal structure of lysin, which burrows through the viteline envelope by a nonenzymatic method, is known in the case of the red abalone (*Haliotis rufescens*) (Shaw et al. 1993, 1995). Initially, a homodimer is formed, but the subunits dissociate to create the active form. Each subunit is highly basic, with two approximately parallel tracks of positively charged residues running the length of the molecule. There is a patch of hydrophobic residues involved in the interaction between the subunits.

Figure 6.6 shows a phylogenetic tree of lysins from 20 abalone species. Within a group of nine closely related species, seven of which are from California (Figure 6.6), there is strong evidence of positive selection favoring diversification among species at the amino acid level.

When rates of synonymous and nonsynonymous substitution were estimated for pairwise comparisons among these genes, mean d_S exceeded mean d_N in the regions of the gene encoding the signal peptide and the hydrophobic patch, although the standard errors were high because the numbers of codons in these regions are small (Table 6.2). However, d_N significantly exceeded d_S in the remainder of the gene (Table 6.2). So far, the basis for this selection is not known. Selection favoring reproductive isolation among species is only one of several possibilities (Vacquier et al. 1997).

Convergent Sequence Evolution

Convergent evolution is a phenomenon that was initially described at the morphological level. Comparative morphology shows many examples of species that are not closely related but have independently evolved similar structures in response to similar environments. For example, the limbs of whales and dolphins (Cetacea) have been modified to form fin-like appendages in response to their aquatic environment, thus giving these mammals a superficial resemblance to

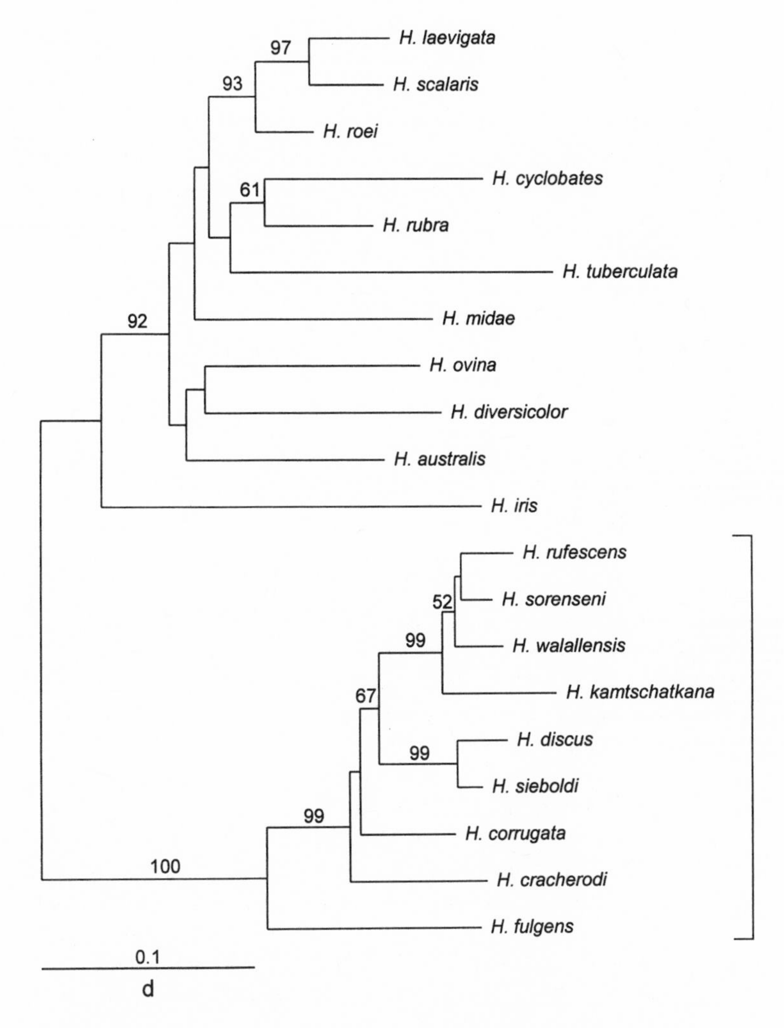

Figure 6.6. Phylogenetic tree of abalone (*Haliotis*) lysin sequences, based on the number of nucleotide substitutions per site (*d*). Numbers on the branches are as in Figure 3.2. The bracket indicates species analyzed in Table 6.2.

fishes. In recent years, there was been frequent reference to the possibility of convergent evolution of proteins. However, Doolittle (1994) has argued that this term is often applied very loosely, without good evidence.

Doolittle (1994) distinguishes four types of convergent evolution at the molecular level:

1. Functional convergence occurs when two unrelated molecules evolve a similar function. Many examples are known among en-

Table 6.2. Mean numbers of synonymous (d_S) and nonsynonymous (d_N) nucleotide substitutions per 100 sites (±SE) in comparisons among sperm lysins of nine *Haliotis* species from California

Region (No. of aligned codons)	d_S	d_N
Signal peptide (17)	9.6 ± 9.5	4.7 ± 2.1[a]
Dimer interface (11)	15.9 ± 12.8	2.3 ± 1.7[a]
Remainder (111)	10.1 ± 2.0	18.7 ± 1.5***

For species analyzed, see Figure 6.6.
[a]Significantly different from d_N in remainder ($p < .001$).
***d_S and d_N significantly different ($p < .001$).

zymes. For example, the alcohol dehydrogenase of vertebrates are unrelated to those of insects (Ikuta et al. 1986).

2. Mechanical convergence occurs when mechanistically similar structures—such as similar active sites—evolve in unrelated sequences having different three-dimensional structures. An example involves the catalytic residues of serine proteases (Kraut 1977). Chymotrypsins of vertebrates have as active site residues histidine at position 57, aspartic acid at position 102, and serine at position 195. The same three residues are present in the catalytic site of bacterial subtilisin, but in this case aspartic acid is at position 32, histidine at position 64, and serine at position 221. The three-dimensional structures bringing the three catalytic residues together are different in the two cases.
3. Structural convergence involves the evolution of similar three-dimensional structures by unrelated proteins. For example, fibronectin type III and immunoglobulin domains have similar structures but lack detectable sequence homology (Leahy et al. 1992). This is most likely due to convergent evolution, although it is possible that the two domain types have a distant common ancestor.
4. Sequence convergence occurs when proteins independently evolve similar primary sequences. If sufficiently extensive, sequence convergence might lead to errors in phylogenetic reconstruction. However, Doolittle (1994) argues that demonstrated cases of sequence convergence are few, and that the number of residues involved in convergent evolution is small. Therefore, the probability that convergent sequence evolution will lead to incorrect phylogenetic reconstruction is slight.

Doolittle (1994) proposes a strict definition of convergent evolution; namely, that convergent evolution involves adaptive amino acid replacements fixed as a result of positive selection. Thus, it excludes chance resemblances. It also excludes what Doolittle (1994) calls "conservative variability." Conservative variability can occur at a site where the amino acid is free to vary within certain limits, but not entirely. For example, a certain protein's function may require a hydrophobic residue at a given site, but it may not matter which hydrophobic residue. At such a site, the same conservative amino acid replacements may independently occur a number of times in different lineages.

In this section I discuss three cases where there is evidence for convergent evolution at the sequence level. In each case, the hypothesis of chance resemblance seems implausible because we have some knowledge of the selective forces that might be expected to favor independent evolution of similar sequence motifs.

Lysozymes of Foregut Fermenters

Lysozymes are antibacterial proteins that are apparently universal in vertebrates and perhaps appear in many other animal taxa (Jolles and Jolles 1984). They are found in a variety of locations in the vertebrate body, including tears, saliva, avian egg white, and mammalian milk. In ruminants, lysozymes have been recruited to play a key role in digestion. Digestion in these animals involves fermentation of plant matter in the foregut by bacteria capable of digesting cellulose. Lysozymes are involved in lysis of these bacteria, which enables the animal to absorb nutrients from them. Interestingly, lysozymes have been recruited for a similar digestive role independently in at least two other groups of leaf-eating vertebrates: the colobine monkeys or leaf-monkeys (family Cercopithecidae, subfamily Colobinae), and the hoatzin (*Opisthocomus hoazin*), a leaf-eating bird native to South America and related to cuckoos.

Because of the similarity in function between lysozymes of ruminants and colobine monkeys, Stewart et al. (1987) sequenced the lysozyme of one of these monkeys, the hanuman langur (*Presbytis entellus*), and compared it to that of bovine. Langur and bovine lysozymes were known to have similar physiological properties, both being active at low pH such as is found in the stomach, and both being resistant to breakdown by pepsin (Dobson et al. 1984). A phylogenetic analysis using the maximum parsimony method actually led to clustering of the langur lysozyme with that of the cow (Figure 6.7B). However, Stewart et al. (1987) considered what they called the biological tree to be a more plausible indication of true relationships (Figure 6.7A). In the biological tree, the langur lysozyme clustered with that of the baboon (*Papio cynocephalus*), an Old World monkey belonging to the subfamily Cercopithecinae and lacking foregut fermentation (Figure 6.7A).

Assuming the biological tree, Stewart et al. (1987) reconstructed amino acid replacements in the evolution of the langur lysozyme by the maximum parsimony method (Figure 6.7C). The reconstruction suggested that more amino acid replacements occurred in the lineage leading to the langur than in the lineage leading to the baboon, another Old World monkey that is not a foregut fermenter (Figure 6.7C). Furthermore, seven amino acid replacements on the lineage leading to the langur paralleled those seen in bovine (Figure 6.7C). Two of these bovine-like substitutions were reconstructed as occurring in the Old World monkey lineages after its divergence from the human lineage but before the

A)

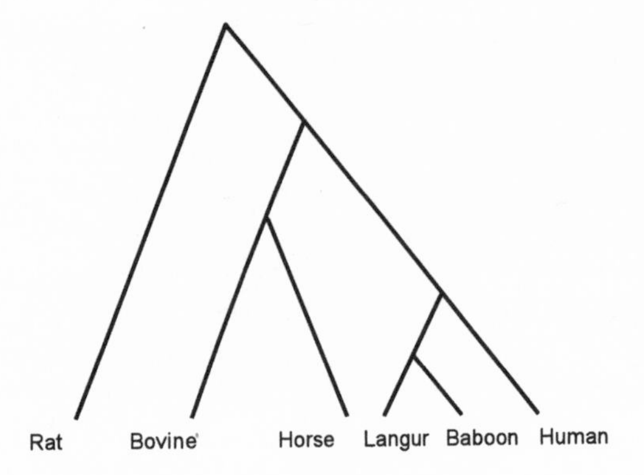

B)

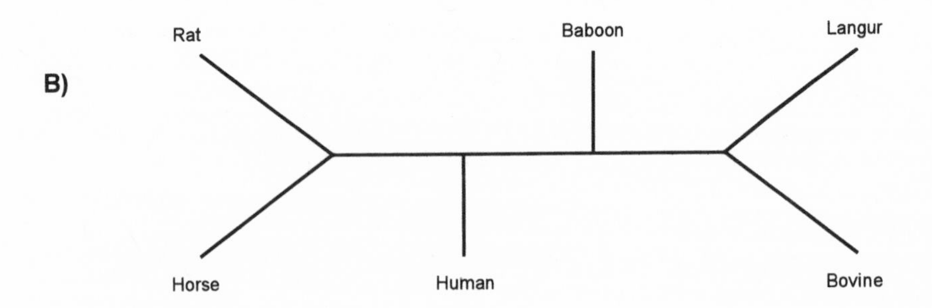

C)

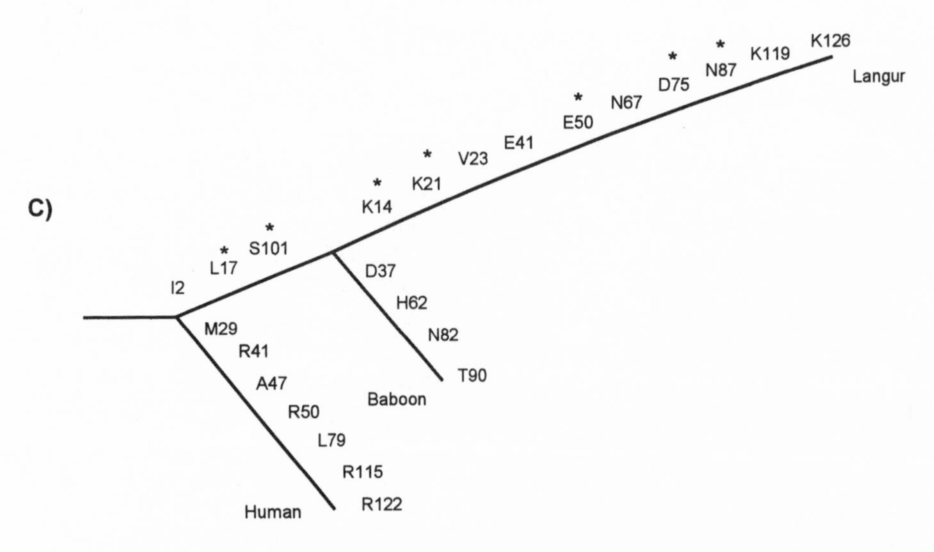

Figure 6.7. Mammalian lysozyme evolution: (**A**) tree showing assumed species relationships; (**B**) maximum parsimony tree obtained by Stewart et al. (1987); (**C**) reconstructed amino acid replacements in primate lysozymes. Asterisks mark residues shared with bovine foregut lysozyme.

divergence of langurs and baboons (Figure 6.7), leading Stewart et al. (1987) to suggest that the ancestral Old World monkey may have been a leaf-eater. This would imply that the cercopithecine monkeys secondarily adapted to an omnivorous diet.

Using a maximum-likelihood method of reconstructing ancestral sequences, Yang et al. (1995) reexamined the lysozymes of the langur and bovine. The method these authors used yields posterior probabilities for each amino acid reconstruction; that is, the probability of that reconstruction being correct given the data set analyzed and the substitution model used. In this case, the substitution model was an empirically derived conservative model of amino acid replacement developed by Jones et al. (1992). By this method, Yang et al. (1995) concluded with high probability that four of the seven amino acid replacements reconstructed by Stewart et al. (1987) to have occurred on the branch leading to the langur (Figure 6.7C) occurred convergently on the branch leading to bovine. Their reconstruction also concluded that the two replacements occurring on the branch leading to both the langur and baboon (Figure 6.7C) also occurred convergently in the lineage leading to bovine.

The stomach lysozyme of the hoatzin is not closely related to that of the langur or bovine, but belongs to a separate group known as calcium-binding lysozymes (Kornegay et al. 1994). Because the calcium-binding lysozymes include mammalian representatives, the calcium-binding lysozymes must have diverged prior to the divergence of birds and mammals. Reconstructing ancestral sequences, Zhang and Kumar (1997) found that two amino acid replacements (aspartic acid at position 75 and asparagine at position 87) occurred convergently in bovine, langur, and hoatzin lineages, strongly suggesting an adaptive response to a digestive role for lysozyme.

MHC Nonclassical Class I Genes

In mammals, the class I MHC region includes not only the highly polymorphic, universally expressed loci discussed in Chapter 4, but also a number of related loci having much lower levels of polymorphism and more limited tissue expression. The polymorphic class I MHC loci are known as classical class I loci or class Ia loci, whereas the others are known as nonclassical class I loci or class Ib loci. In humans, the class Ia loci are *HLA-A*, *-B*, and *-C*; the class Ib loci include *HLA-E*, *-F*, and *-G*. From an evolutionary point of view, an interesting aspect of the class Ib loci is that they are not shared by mammals of different orders (Hughes and Nei 1989b). Phylogenetic analysis indicates that class Ib loci have evolved independently in different orders of mammals by duplication of class Ia loci. The fact that the class Ia loci of a New World primate are orthologous to the human class Ib gene *HLA-G* shows that the class Ia and class Ib genes are not mutually exclusive groups over evolutionary time (Watkins et al. 1990).

The function of the class Ib molecules is unknown, and has been controversial. One hypothesis is that the class Ib genes are a kind of expressed pseudogenes (Klein and Figueroa 1986; Howard 1987). On this hypothesis, the products of these genes, though expressed at low levels, play no essential biological function; and in evolutionary terms the genes are on the way out. Promoters of certain mouse class Ib genes include sequences homologous to elements known to be involved in regulating the universal expression of class Ia genes. These elements are highly conserved in class Ia genes, but have accumulated numerous mutations in the case of the class Ib genes (Hughes and Nei 1989b). Thus, class Ib genes seem to have evolved from class Ia genes whose promoters have mutated so as to reduce expression. However, there is evidence that some class Ib molecules bind peptides, suggesting that they may have some immune system role (Pamer et al. 1992).

Despite the distinct evolutionary origin of class Ib genes in different orders of mammals, there is evidence that a mouse class Ib molecular called H2-Qa-1^a has convergently evolved features of its peptide binding region (PBR) with the class Ib E locus of primates (Yeager et al. 1997). In a phylogenetic tree based on the α_1,α_2, and α_3 domains excluding the PBR codons, H2-Qa-1^a and the closely related mouse gene H2-T23^d cluster with mouse class Ia genes, while primate E genes cluster with primate Ia genes (Figure 6.8A). By contrast, in a phylogenetic tree based on the PBR only, H2-Qa-1^a and H2-\T$23^d$ cluster with the primate E genes (Figure 6.8B).

Reconstruction of ancestral sequences by the method of Yang et al. (1995) showed that six sites evolved convergently in the ancestor of H2-Qa-1^a and H2-T23^d and in the ancestor of primate E. Five of these residues are in the PBR (Table 6.3). One of these residues, an arginine at position 68, is unknown in any other class Ia or class Ib sequences except these primate and mouse class Ib genes, and the others are rarely seen in known class I sequences (Table 6.3). This convergent evolution in the PBR is particularly interesting because HLA-E and the mose H2-Qa-1 are known to bind similar peptides. Both of these molecules present peptides derived from the conserved leader sequences of MHC class Ia molecules, and the sequences are quite similar (Braud et al. 1997; DeCloux et al. (1997).

β *Tubulin and Benzimidazole Resistance*

At an accelerated pace in the present century, humans have altered the environment for numerous organisms by use of chemical substances designed to eliminate parasites or competitors considered undesirable from a human point of view. These substances include antibiotics and other drugs, insecticides, and herbicides. In many cases, the target species have evolved resistance to such toxicants, and in a number of these the mutations responsible for resistance have been studied at the molecular level (Taylor and Feyereisen 1969). Regarding the study of convergent evolution at the amino acid sequence level, toxicants provide examples of diverse organisms exposed to similar selection pressures. This is precisely the sort of case in which classical evolutionary theory would predict that convergent evolution would occur.

One interesting example involves the organic compounds known as benzimidazoles, including benomyl, which have been used as both veterinary and

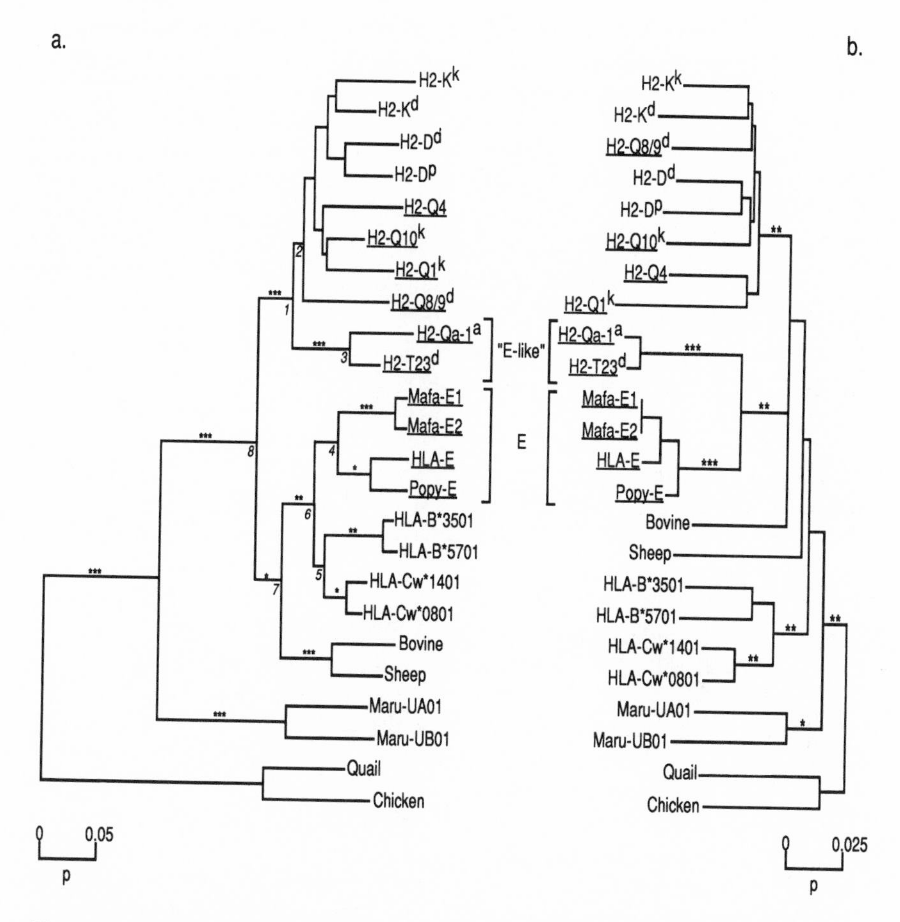

Figure 6.8. Phylogenetic trees of (a) remainder of α_1 and α_2 domains and (b) peptide-binding region (PBR) of MHC class Ib molecules of primates and rodents. The trees were constructed by the neighbor-joining method based on the proportion of amino acid difference (p) and rooted with wallaby (Maru-) and bird sequences. Species abbreviations are as follows: HLA—human; H-2—mouse; Mafa—crab-eating macaque; Popy—orangutan. The "E-like" sequences are mouse molecules resembling primate E in the PBR. Tests of significance of branch lengths: $^*p < .05$; $^{**}p < .01$: $^{***}p < .001$. From Yeager et al. (1997).

agricultural fungicides and as veterinary and medical antihelminthic agents. Benzimidazoles inhibit microtubule formation in fungi and worms without significantly affecting their hosts (Lacey 1990). Tubulin is a polymer of α and β tubulin subunits, and a number of benzimidazole-resistant mutations have been found to involve amino acid replacements in β tubulin.

Figure 6.9 shows phylogenetic tree of β tubulins of selected animals, plants, and fungi. The method of Yang et a. (1995) was used to reconstruct ancestral sequences for internal nodes of this tree. Shown on the tree are reconstructed

Table 6.3. Similarity between E-like (mouse) and E (primate) sequences among the 63 PBR (Bjorkman et al. 1987a,b; Saper et al. 1991; Garboczi et al. 1996) residues

		* * x *
		** xx 111111111111111111111111111
		x x2222345556666666666777777778888999112344444555555555566666667
		579245645789123456789012345670124579463335679012456789012356791
E-like	H2-Qa-1[a]	LYHFVVGVMPEYERETWKARDMGRNFRVNTLLYLWYCEYWSHKSMVDEHQQRAYLQGPVEWHY
	H2-T23[d]	. . T . I . . . . A . . . .
E	HLA-E	. . . . .S. . . . . S . . D . . . RS . . . TAQI. . . . . . .R . . . HEF . . . Q . . DAS . EH . . . . . .EDT . . . . . .
	Mafa-E1	. . . . .S . . . . DQ . . RS . . . TAQT . . . . . .R . . . HEF . . . Q . . DGS .EH . . . . . .EDT . . . . . .
	Mafa-E2	. . . . .S. . . . . .DQ . . RS . . . TAQT. . . . . . R . . . HEF . . . Q . . DGS .EH. . . . . . EDT. . . . . .
	Popy-E	. . . . .S. . . . . .D . . . RS . . . TAQT. . . . . . R . . . HEF . . . R . . DAC .EH. . . . . . EDT . . . R.
	position†	a a b b b bb ta a tabttb tcc fff f cadffef fettte dd tad a ta a
		b c b b c ebe f te d t t
		t d t

* = site found to be shared as the result of convergent/parallel evolution.

x = site found to be retained from ancestral sequences for both E and E-like.

† = location of residue within the PBR. a, b, c, d, e, f = pockets A, B, C, D, E, F (Saper et al. 1991); t = T-cell receptor directed (Garboczi et al. 1996). From Yeager et al. (1997).

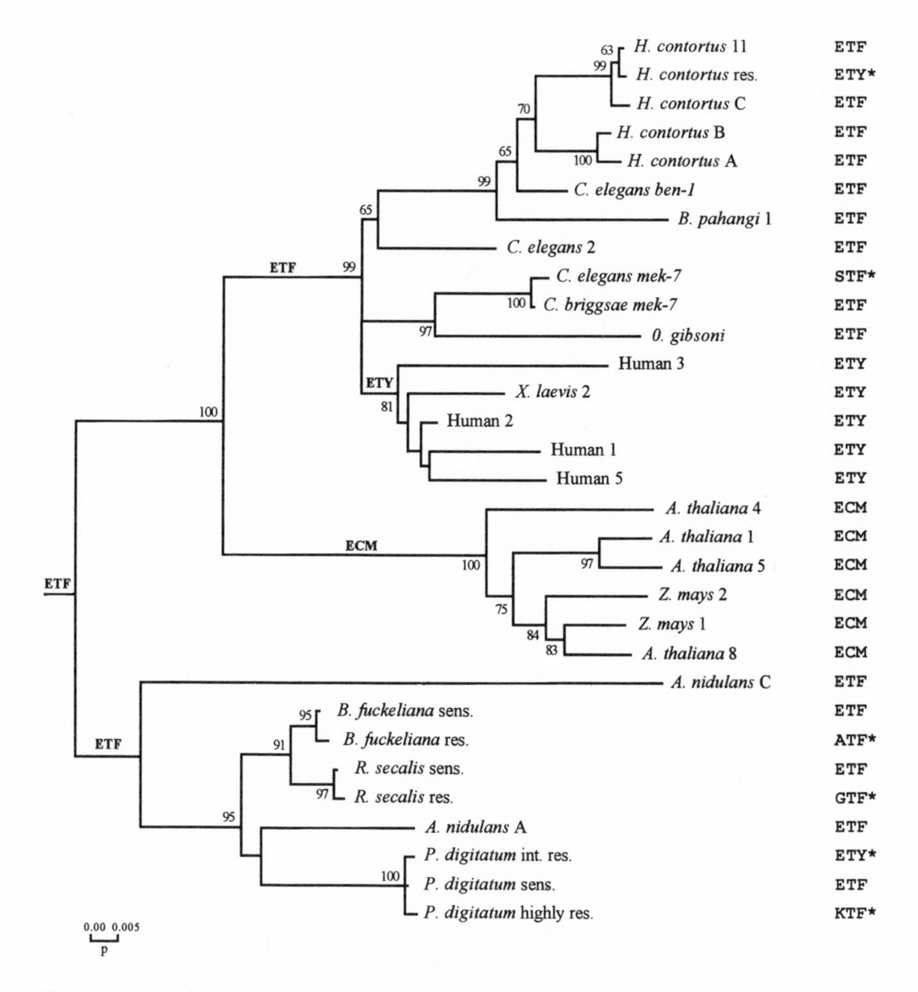

Figure 6.9. Phylogenetic tree of β tubulins, constructed by the neighbor-joining method based on the proportion of amino acid difference (p). For each sequence, the amino acid motif at residues 198–200 (numbered based on *Haemonchus contortus* sequence) is shown. For selected interior nodes, reconstructed ancestral amino acid sequences are shown. Asterisks indicate benzimidazole-resistant nematode and fungal sequences. Numbers on the branches are as in Figure 3.2. Species are as follows: nematodes, *Haemonchus contortus, Brugia pahangi, Caenorhabditis elegans, Caenorhabditis briggsae, Onchocerca gibsoni*; plants, *Arabidopsis thaliana, Zea mays*; fungi, *Aspergillus nidulans, Penicillium digitatum, Botryotinia fuckeliana, Rhychosporium secalis.*

sequences for residues corresponding to positions 198–200 of the β tubulin of the nematode worm *Haemonchus contortus* (Figure 6.9). The sequence of these residues is also shown for each extant sequence used in constructing the tree (Figure 6.9). Mutations at positions 198 and 200 are found in a number of independent benzimidazole-resistant mutants (Elard et al. 1996). Ancestral sequences at these positions were reconstructed with over 95% accuracy, given the data set used and the model of amino acid replacement assumed.

The ancestral multicellular eukaryote had ETF at these positions (Figure 6.9). In plants, this has become ECM, which is the motif found in modern plants; fungi, in contrast, retained the ancestral ETF motif. The motif ETY was introduced in the ancestor of vertebrates, while nematodes retained ETF. However, benzimidazole-resistant mutants in both nematodes and fungi have independently evolved the ETY motif (Figure 6.9). In addition, KTF, ATF, and GTF motifs have also evolved in independent resistant fungal mutants (Figure 6.9). In these latter three resistant mutants, the same site has been independently affected by mutations that replace the ancestral negatively charged glutamic acid with neutral (alanine, glycine) or positively charged (lysine) residues.

The nematode worm *Caenorhabditis elegans* is not a parasite of humans or domestic animals, but rather a free-living worm that has recently become an important model organism in genetics and development. Therefore, this species has not been purposefully exposed to benzimidazoles by humans, although one cannot rule out incidental exposure to these chemicals in the environment. In any event, experiments revealed benzimidazole-resistant and sensitive genotypes in *C. elegans* (Driscoll et al. 1989). Like many animals, including humans *C. elegans* has more than one β tubulin gene. Genetic experiments showed that benzimidazole resistance mapped to a single locus called *ben-1*, which turned out to encode a β-tubulin closely related to many other β tubulins known from nematodes (Figure 6.9). Like most other nematode β-tubulins, *ben-1* has ETY at positions 198–200. Benzimidazole-resistant mutations involved complete or partial deletion of *ben-1*.

Curiously, another *C. elegans* β-tubulin gene, *mek-7*, has the motif STF at residues 198–200. This gene is not closely related to *ben-1* (Figure 6.9). Yet the mutation from E to S at position 198 has evidently occurred recently in *mek-7* of *C. elegans*. Evidence for this is the fact that the orthologous gene from the closely related *C. briggsae* has E at 198, as do most β-tubulins (Figure 6.9). It seems most likely that *mek-7* has accumulated a benzimidazole-resistant sequence by chance, prior to any exposure of the worm to these toxicants. Coupled with deletion of the benzimidazole-sensitive *ben-1* gene, this neutral amino acid replacement enables some *C. elegans* to be resistant to benzimidazoles. If this is true, *mek-7* would represent a case in which a toxicant-resistant phenotype has arisen in the absence of selection. However, we do not have any evidence regarding population frequencies of these genotypes in *C. elegans*. Thus, at the present time we cannot rule out the possibility that recent natural selection by benzimidazoles in the environment has increased the frequency of either *ben-1*-null mutants or of *mek-7* alleles with S at position 198.

Do Immune System Genes Evolve Faster?

Murphy (1993) compared 615 putatively orthologous pairs of proteins from human and murine rodent (mouse or rat) and found a significantly higher mean amino acid difference in a group of proteins that he called "host defense receptors and ligands" than in other functionally defined categories of proteins (Figure 6.10). He attributed the more rapid evolution of host defense receptors and ligands to Darwinian selection. This selection, in turn, was hypothesized to arise from a phenomenon that Murphy called "molecular mimicry." The term molecular mimicry is generally used for a hypothetical mechanism that involves convergent evolution of parasite antigens to resemble those of the host, as an adaptation for evading host immune recognition (Damian 1964). No unambiguous cases of molecular mimicry are known (Doolittle 1994). In fact, Murphy (1993) actually has in mind the phenomenon of gene capture.

It is known that the genomes of numerous viruses include genes that are homologous to genes of vertebrates. These genes have evidently been incorporated or "captured" from the genome of the host. In DNA viruses such as poxviruses, these genes often correspond to immune system signaling molecules of vertebrates or their receptors (McFadden 1995). These stolen genes can affect the host's immune response by disrupting essential intercellular signaling mechanisms (e.g., Alcami and Smith 1992). Murphy (1993) hypothesized that avoidance of similarity to such captured foreign proteins is the basis for selection causing rapid evolution at the amino acid level of host defense receptors and ligands.

Murphy's (1993) hypothesis that gene capture is the source of selection on immune system receptors and ligands is simply an "adaptive story" in the sense

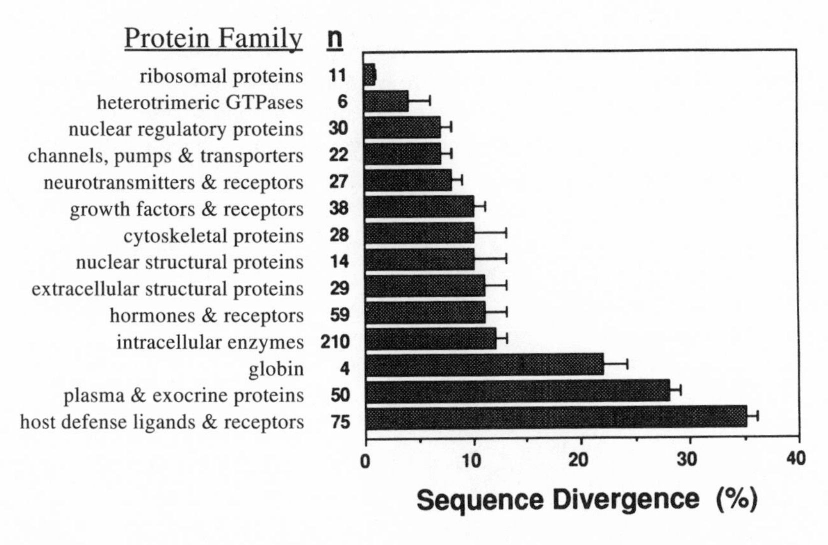

Figure 6.10. Percentage of amino acid difference ("sequence divergence") between humans and murine rodents for 14 functional categories of proteins. From Murphy (1993). Reproduced by permission.

of Gould and Lewontin (1979; see Chapter 1, The Adaptationist's Dilemma). No test of this hypothesis against alternatives was provided. What about Murphy's conclusion that immune system proteins have a faster rate of evolution? This conclusion can also be criticized. In the category host defense ligands and receptors, Murphy included a diverse array of molecules, such as interleukins, interferons, chemokines, and receptors belonging to a number of families including the immunoglobulin superfamily. Most of these molecules belong to families that have no representatives in any of the other functional categories surveyed by Murphy (1993) (Figure 6.10). Therefore, one cannot rule out the hypothesis that the higher rate of sequence divergence seen among host defense receptors and ligands simply reflects the fact that these molecules belong to families that are less constrained structurally than are those making up the other functional categories. Also, because Murphy (1993) examined only the proportion of difference at the amino acid level, he could not rule out the hypothesis that the observed differences were due to differences in mutation rate.

To overcome this problem, I analyzed immunoglobulin superfamily C2 domains from molecules expressed in immune system cells and those expressed in other tissues (Hughes 1997). These are homologous domains of approximately the same length, which form part of a wide variety of cell-surface receptors. Aside form the immune system, these receptors include several expressed in the nervous system and a few expressed in other systems besides the immune or nervous systems. I computed d_S and d_N for 107 C2 domains from 38 orthologous pairs of genes from humans and murine rodents (rat or mouse).

The results of these analyses were complex because of numerous intercorrelations among variables. For example, there was a strong negative correlation between the length of the extracellular domain of the protein (in number of amino acid residues) and d_N in the C2 domains of that protein (Figure 6.11). This correlation may occur because there is greater constraint on the evolution of C2 domains in proteins with large extracellular domains, often containing multiple C2 domains. There was no such correlation between extracellular length and d_S ($r = .063$; n.s.).

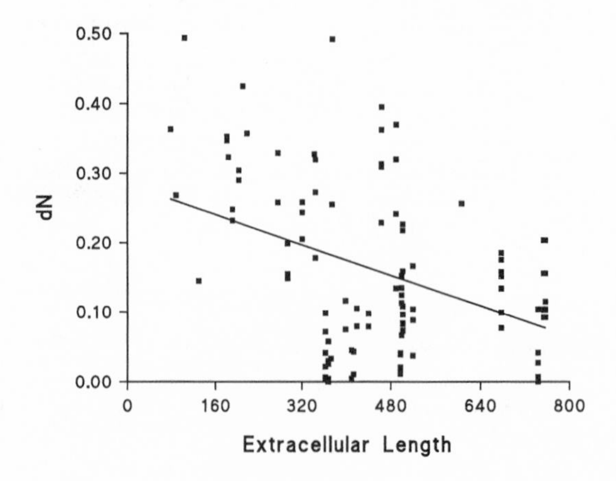

Figure 6.11. Relationship between the number of nonsynonymous substitutions per site (d_N) between humans and murine rodents in C2 domains and extracellular length (number of amino acids in consensus sequence) of the mature protein. The line is the linear regression line ($r = -.401$; $p < .001$). From Hughes (1997).

Because of this correlation, d_N was compared between different categories of expression by analysis of covariance, controlling statistically for the effect of the extracellular length of the protein. The results showed a highly significant effect of tissue expression (Figure 6.12). Mean d_N was highest for genes expressed in the immune system only (Figure 6.12). For genes expressed in the immune system and in other systems not including the nervous system and for genes expressed in the immune system, the nervous system, and other systems, mean d_N was nearly as high as that for genes expressed in the immune system only (Figure 6.12). However, mean d_N was much lower for genes expressed in the nervous system and/or other systems besides the immune system (Figure 6.12). Mean d_N for genes expressed in both the immune system and the nervous system was also significantly lower than that for genes expressed in only the immune system.

The results support the hypothesis that immune-system expressed C2 domains, particularly those expressed in the immune system only, evolve more rapidly at the amino acid level than those expressed in other systems. A similar analysis of d_S demonstrates that this is not an effect of a difference in mutation rate, because the d_S value for genes expressed in the immune system was not particularly high (Figure 6.12).

The relationship between expression pattern and evolutionary rate was further examined by computing correlation coefficients among the following variables: d_S, d_N, the length of the extracellular portion of the molecule (L) and the dummy variables corresponding to expression in immune system cells (Im),

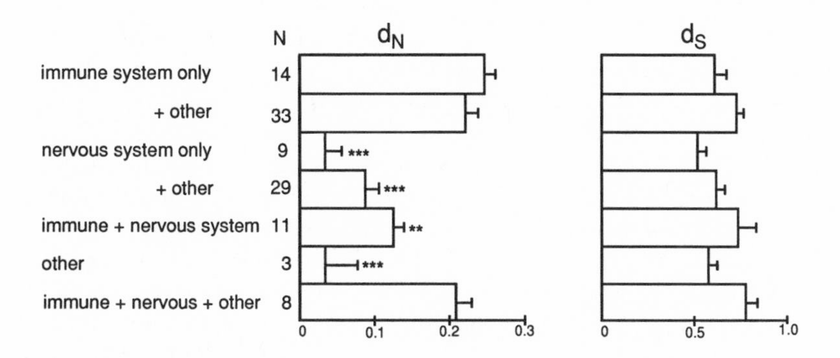

Figure 6.12. Evolutionary rates of C2 domains expressed in different cell types. (*Left*) Adjusted mean (±SE) number of nonsynonymous substitutions per site (d_N), adjusted for extracellular length, from analysis for covariance. In the analysis of covariance, there was a significant effect of the covariate ($p < .005$) and a significant effect of expression category ($p < .001$). (*Right*) Mean (±SE) number of synonymous substitutions per site (d_S). In a one-way analysis of variance in d_S, there was a significant effect of expression category ($p < .05$). Individual means were compared to those for genes expressed in the immune system only by the GT2 method (Sokal and Rohlf 1981), which uses a simultaneous level for multiple comparisons: $^{**}p < .01$; $^{***}p < .001$. From Hughes (1997).

Table 6.4. Partial correlations between d_N and d_S and five other variables, in each case simultaneously controling for all other variables

	With d_N	With d_S
d_S	0.366***	—
Im	0.578***	0.126
Neu	–0.018	0.040
Oth	0.327***	–0.006
Br	–0.334***	0.012
L	–0.332***	0.151

For definition of variables, see text. ***$p < .001$. From Hughes (1997).

expression in nervous system cells (Neu), expression in other cell types (Oth), and breadth of expression (Br). The pattern of correlation among these variables was quite complex. As is often found when d_S and d_N are compared for a number of genes (e.g., Wolfe and Sharp 1993), these quantities were positively correlated. There was a positive correlation between d_N and Im, as well as a negative correlation between d_N and Neu; but Im and Neu themselves were negatively correlated (Hughes 1997).

To control statistically for these numerous intercorrelations, I computed partial correlations between d_S and d_N and each other variable, simultaneously controlling for all other variables (Table 6.4). When this was done, there remained a positive correlation between d_S and d_N, but no other variable was significantly correlated with d_S (Table 6.4). However, there were several variables significantly correlated with d_N, including Im (Table 6.5). Thus, controlling statistically for other possibly relevant factors, we are left with the conclusion that for C2 domains expression in the immune system is associated with accelerated evolution at the amino acid level.

However, it is not clear what factors are responsible for this accelerated evolution. Most importantly, it is unclear whether it occurs because for some reason immune system expression leads to a relaxation of constraint or because of positive Darwinian selection. If it is the latter, a number of factors may be at work besides gene capture. For example, certain immune system receptors are used by viruses as means of entering cells. A well-known example is the use of the CD4 molecule, which includes two immunoglobulin C2 domains, as a receptor by the HIV-1 virus (Wang et al. 1990). If viruses have frequently made similar use of immune system receptors, this may have selected for amino acid replacements preventing viral entry. Finally, immune system receptors may be subject to positive selection due to their interaction with other molecules that are under positive selection driven by the need to recognize parasites, including MHC (Chapter 4), immunoglobulins (Chapter 7, Bifunctionality Preceding Gene Duplication), and T-cell receptors (Levinson et al. 1992). In this case, selection would act directly on foreign antigen receptors, and the evolutionary changes in these receptors would then select for changes in other immune system molecules.

7

Evolution of New Protein Function

Evolutionary biologists have argued that the duplication of genes, followed by changes in sequence to one or both copies, plays a fundamental role in adaptive evolution (Ohno et al. 1968; Ohno 1970, 1973; Kimura and Ohta 1974; Li 1983). For example, Li (1983, p. 14) writes: "Gene duplication is probably the most important mechanism for generating new genes and new biochemical processes that have facilitated the evolution of complex organisms from primitive ones." Gene duplication might occur by several different mechanisms, including: (1) unequal crossing over, which would cause tandem duplication of a single gene or a small number of closely linked genes (Figure 7.1); (2) chromosomal non-disjunction, leading to duplication of a chromosome; and (3) polyploidization, leading to duplication of the entire genome. In higher organisms in particular, most genes belong to multigene families or superfamilies, some of which may contain 1,000 members or more. The existence of multigene families is evidence of repeated events of gene duplication that has occurred over the history of life.

The purpose of the present chapter is to review evidence regarding gene duplication and its role in adaptive evolution. If, after duplication of a protein-coding gene, natural selection favors functional divergence of the two daughter genes, this will give rise to a form of directional selection favoring fixation of certain nonsynonymous mutations in one or both genes that adapt their protein products to distinct functions. However, the role of natural selection in the functional divergence of duplicate genes has been controversial. Indeed, the most popular hypothesis (Ohno 1973) for the evolution of new protein function after gene duplication explicitly denies any role of natural selection in the process.

In this chapter I discuss this widely cited hypothesis and recent evidence bearing on it. Among this is evidence that, in many cases, natural selection does indeed play a role in the diversification of members of multigene families. Such

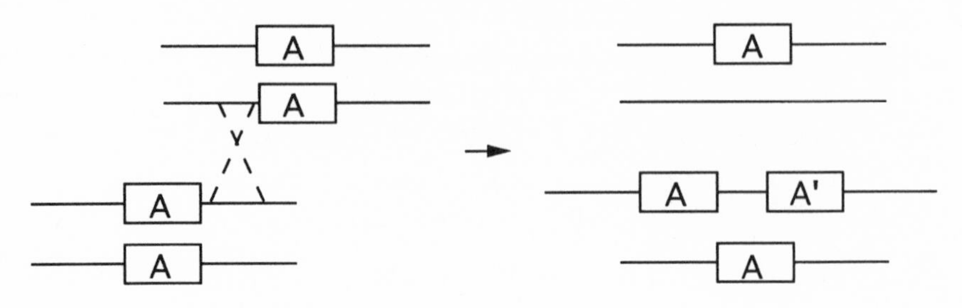

Figure 7.1. Gene duplication by unequal crossing over.

evidence, in turn, suggests an alternative model for the evolution of new protein function after gene duplication.

The MDR Model

The most widely cited model of the evolution of functionally novel proteins after gene duplication is based on the assumption that, once gene duplication has occurred, one of the two gene copies is redundant and thus is freed from all functional constraint. All mutations occurring in such a redundant gene will be selectively neutral. However, by chance, one or more mutations that preadapt the gene to a new function may occur and may then be fixed by random genetic drift. This model was described as follows by Ohno (1973):

> The mechanism of gene duplication provides a temporary escape from the relentless pressure of natural selection to a duplicated copy of a functional gene locus. While being ignored by natural selection, a duplicated and thus redundant copy is free to accumulate all manner of randomly sustained mutations. As a result, it may become a degenerate, nonsense DNA base sequence. Occasionally, however, it may acquire a new active site sequence, therefore a new function and emerge triumphant as a new gene locus.

I call this model for the evolution of new protein function the model of mutation during redundancy (the MDR model). Previously, I termed it the model of "mutation during nonfunctionality" (Hughes 1994a). This may be slightly misleading because, according to the model, the redundant gene may be "functional" in the sense of encoding a protein that is translated, although it performs no essential role in the cell. Becoming a true pseudogene—"a degenerate, nonsense DNA base sequence" in Ohno's words—is only one possible fate of the redundant gene copy according to this model.

There are several lines of evidence inconsistent with the MDR hypothesis that I will review in this chapter: (1) data from a tetraploid animal, the frog *Xenopus laevis*, suggests that, as long as both copies of a duplicate gene are expressed, both are subject to purifying selection (Hughes and Hughes 1993a). Thus, contrary to the MDR hypothesis, one gene copy is not "ignored by natural selection" and free to mutate at random; (2) patterns of nucleotide substitution indicate that functional diversification of members of multigene families has often occurred as a result of positive Darwinian selection, rather than random accumulation of mutations; and (3) a single gene may share two functions (Piatigorsky and Wistow 1991). Therefore, at least in these cases, the evolution of functional novelty precedes rather than follows gene duplication.

Functional Constraint on Duplicate Genes

The MDR model predicts that one of two duplicate gene copies will be redundant and thus free to accumulate nucleotide substitutions at random without purifying selection. In terms of equation 3.3, f_0 (the fraction of mutations that are neutral) is expected to be 100% or nearly 100% in such a redundant gene. Hughes and Hughes (1993a) tested this prediction by examining pairs of genes duplicated as a result of tetraploidization in the frog *Xenopus laevis*. An event whereby the genome of this organism was duplicated, giving rise to a tetraploid, is believed to have taken place 30–40 million years ago (Bisbee et al. 1977; Hughes and Hughes 1993a). The genome subsequently became rediploidized, meaning that at mitosis genes line up in pairs rather than in groups of four. In the case of about 50% of the pairs of duplicate genes, there is evidence that one of the two genes has been silenced, presumably as the result of a mutation preventing its expression (Graf and Kobel 1991). In the remaining cases, both duplicate genes are still expressed.

When d_S and d_N were estimated for 17 pairs of duplicate genes from X. *laevis*, it was found that d_S exceeded d_N in every case; and the difference was highly significant in most comparisons (Table 7.1). Because this is the pattern expected in the case of purifying selection, it is not consistent with the idea that one duplicate gene is totally unconstrained. Orthologous genes from humans and the mouse were also compared. Remarkably, the d_N values for the comparisons between the X. *laevis* duplicates showed a strong positive correlation with the d_N values for the human–mouse comparisons (Figure 7.2). Because the relative values of d_N for different protein-coding genes reflect the relative degree of functional constraint on the protein, these results indicate that the duplicate pairs of genes in X. *laevis* are subject to levels of constraint closely paralleling those of their mammalian homologues.

Because the MDR hypothesis predicts that after gene duplication one gene copy will be freed from constraint, this hypothesis predicts that one copy will evolve more rapidly at nonsynonymous sites than will the other. Relative rate tests were used to test this hypothesis in the case of the 17 X. *laevis* gene pairs. In a relative rate test, two sequences are compared to a reference sequence that is known to be less closely related to either of the two sequences than they are to each other (i.e., an outgroup to the two sequences). Figure 7.3 illustrates how the test works. Of the two sequences illustrated in Figure 7.3 (A and B), B has evolved more rapidly since their last common ancestor (C) than has A. Thus, the distance BC is greater than AC (Figure 7.3). If the ancestral sequence were available, we could test the hypothesis of equal rates of evolution (i.e., AC = BC) by comparing A and B directly with C. Ordinarily, this is not the case, however. But the distance between each sequence and the reference sequence

Table 7.1. Numbers of synonymous (d_S) and nonsynonymous (d_N) nucleotide substitutions per 100 sites (±SE) between duplicate genes of *Xenopus laevis*

Gene	d_S	d_N
Actin, skeletal	15.8 ± 2.6	0.0 ± 0.0***
Albumin	16.2 ± 2.3	5.9 ± 0.7***
Calmodulin	31.6 ± 7.1	0.0 ± 0.0***
Enkephalin A	14.2 ± 3.5	2.3 ± 0.7***
Furin	18.8 ± 2.3	1.5 ± 0.3***
Homeobox 2/2.3	14.5 ± 3.4	2.4 ± 0.7***
Insulin	12.0 ± 4.5	3.8 ± 1.3
Integrin β1	30.7 ± 2.9	1.0 ± 0.2***
N-CAM	14.7 ± 1.9	3.6 ± 0.5***
POMC	26.5 ± 4.9	3.6 ± 0.8***
Thyroid hormone receptor α	11.8 ± 0.2	0.8 ± 0.3***
Thyroid hormone receptor β	14.7 ± 1.9	3.6 ± 0.5***
Thyrotropin-releasing hormone	17.2 ± 4.1	7.9 ± 1.4*
Vimentin	14.2 ± 2.3	2.8 ± 0.5***
C-ets-1	21.5 ± 3.8	0.5 ± 0.3***
C-ets-2	32.8 ± 3.9	2.8 ± 0.5***
C-myc	26.5 ± 4.9	3.6 ± 0.8***
Mean	19.1 ± 1.7	2.5 ± 0.5***

Tests of the hypothesis that $d_S = d_N$: *$p < .05$; ***$p < .001$. Data from Hughes and Hughes (1993a).

(R) includes the branch CR (Figure 7.3). Therefore, the hypothesis AR = BR is equivalent to the hypothesis AC = BC because AR = AC + CR and BR = BC + CR.

In *X. laevis*, one of the two duplicated genes is designated A and the other B, but this designation is arbitrary. The MDR hypothesis predicts that one of each pair will evolve more rapidly at nonsynonymous sites, but it does not predict which one. In fact, the relative rate tests show no significant differences in rate of nonsynonymous substitution between the A and B copies (Table 7.2). Thus, there is no evidence that one of the two gene copies is less subject to functional constraint than is the other.

Why are both duplicate gene copies subject to purifying selection in this species? A similar question arose from earlier electrophoretic studies of tetraploid species. These studies suggested that the rate of silencing of duplicate loci has been surprisingly slow, in comparison with theoretical predictions for the rate of silencing of redundant gene loci (Bailey et al. 1978; Takahata and Maruyama 1979; Li 1980, 1982). Li (1982) pointed out that duplicate genes might undergo mutation in their regulatory regions so that they come to be expressed in different tissue types or at different stages of development. If this happens, the genes are no longer functionally redundant. Even if there is no differential expression, Takahata and Maruyama (1979) argued that there might be selection against all genotypes containing null alleles, but they did not explain how this might happen.

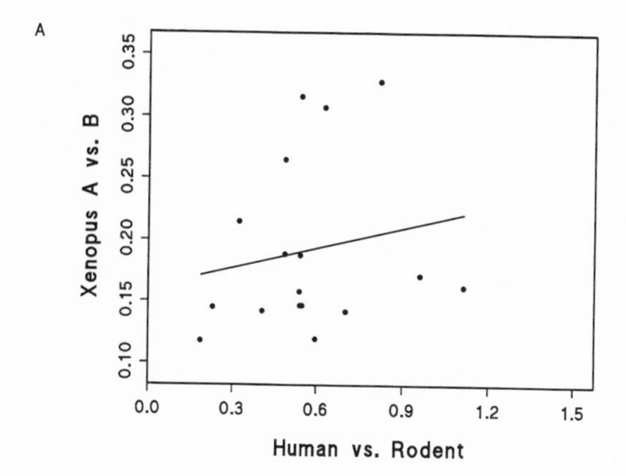

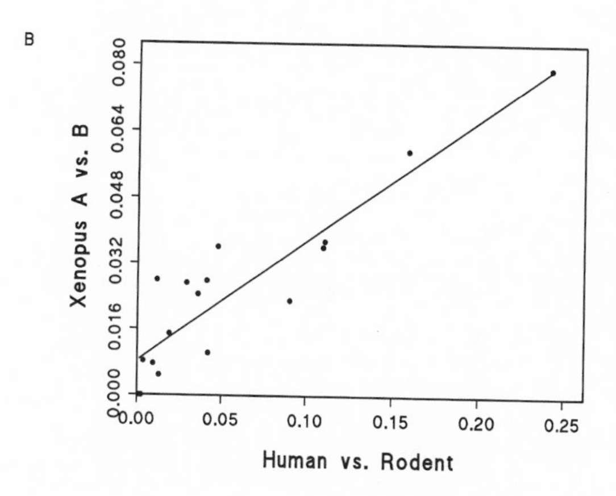

Figure 7.2. (**A**) Number of synonymous substitutions per site between *Xenopus laevis* duplicate genes plotted against that between human and mouse orthologues. The linear regression line is shown ($r = .185$; n.s.). (**B**) Number of nonsynonymous substitutions per site between *Xenopus laevis* duplicate gene pairs plotted against that between human and mouse orthologues. The linear regression line is shown ($r = .915$; $p < .001$). From Hughes and Hughes (1993a).

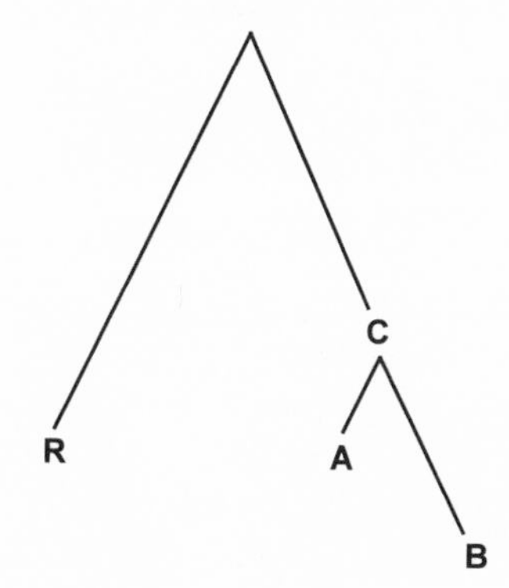

Figure 7.3. Relative rate test. The rate of evolution of A and B since their common ancestor (C) can be estimated by comparing each to a reference sequence (R).

Table 7.2. Relative rate tests comparing number of nonsynonymous substitutions per 100 sites between *Xenopus laevis* duplicate genes and reference sequence (human)

Gene	*Xenopus* A versus human	*Xenopus* B versus human	Difference
Actin, skeletal	2.2 ± 0.5	2.2 ± 0.5	0.0 ± 0.1
Albumin	58.8 ± 2.9	59.1 ± 2.9	–0.3 ± 1.2
Calmodulin	0.7 ± 0.3	0.7 ± 0.3	0.0 ± 0.0
Enkephalin A	24.7 ± 2.5	23.2 ± 2.4	1.5 ± 0.9
Furin	14.6 ± 1.1	14.7 ± 1.1	–0.2 ± 0.4
Homeobox 2/2.3	19.7 ± 2.2	20.0 ± 2.2	–0.3 ± 0.9
Insulin	35.6 ± 4.8	35.1 ± 4.8	0.5 ± 0.2
Integrin β1	12.2 ± 0.9	12.3 ± 0.9	–0.1 ± 0.2
N-CAM	19.5 ± 1.2	19.9 ± 1.2	–0.3 ± 0.6
POMC	32.1 ± 2.9	32.7 ± 2.9	–0.6 ± 1.2
Thyroid hormone receptor α	4.8 ± 0.8	5.4 ± 0.8	–0.6 ± 0.3
Thyroid hormone receptor β	4.4 ± 0.7	4.5 ± 0.8	–0.1 ± 0.3
Thyrotropin-releasing hormone	66.4 ± 5.8	70.1 ± 6.1	–3.7 ± 2.8
Vimentin	14.4 ± 1.3	15.4 ± 1.3	–1.0 ± 0.6
C-ets-1	5.0 ± 0.9	5.0 ± 0.9	0.1 ± 0.3
C-ets-2	21.7 ± 1.6	21.5 ± 1.6	0.2 ± 0.6
C-myc	24.8 ± 1.9	25.0 ± 1.9	–0.2 ± 0.7

Values are ±SE. No differences are significant; thus, no gene pair shows evidence of a rate difference. From Hughes and Hughes (1993a).

Null mutations—those that eliminate expression of the protein—can be placed in three broad categories: (1) mutations that prevent transcription (e.g., by damage to promoter regions required for transcription); (2) mutations that cause production of a defective transcript that either yields a greatly truncated protein product or is not translated at all; and (3) mutations in which a full-size protein is translated but that protein is in some way defective. Natural selection can be expected to act differently on these three types of null mutation. Clearly, if duplicated genes come to be expressed in different tissues or at different developmental stages, all three types of null mutation will be selectively disadvantageous. In the absence of differential regulation, the third type of mutation can still be disadvantageous, even though the first two types are not disadvantageous. If the protein in question interacts with other proteins, the presence of a defective protein even in small quantities may interfere with this interaction. If so, a null mutation of the third type will have a dominant deleterious effect; and, as hypothesized by Takahata and Maruyama (1979), there will be selection against any genotype bearing at least one allele.

These considerations lead to a general hypothesis regarding the role of natural selection on any duplicated gene locus, whether resulting from genome duplication or from duplication of a single gene. A mutation that totally eliminates expression will be selectively neutral; but, as long as the gene is expressed, it will be subject to purifying selection because of the deleterious effects of expressing a defective protein. This hypothesis would explain both of the following observations: (1) expressed X. *laevis* duplicate genes are subject to purifying selection; (2) there is in general no apparent selection against mutations leading to loss of expression of duplicate genes, either in X. *laevis* or in other tetraploid vertebrates (Hughes and Hughes 1993a). However, this hypothesis is not consistent with the MDR hypothesis.

Diversifying Selection in Multigene Families

On the MDR hypothesis, functional diversification of genes within multigene families occurs as a result of chance. On the other hand, there have been a substantial number of studies showing that natural selection has acted to favor amino acid differences between members of multigene families. The strongest evidence of this sort is based on comparisons of rates of synonymous and nonsynonymous nucleotide substitution.

Functional divergence between members of a multigene family is a form of directional selection, whereby selection favors amino acid replacements in one or both genes that adapt them to specific functions. However, once each has adapted to its new function, no new amino acid replacements are favored; but rather each gene will thereafter be subject to purifying selection. Therefore, comparison of d_S and d_N can generally only be used to test for adaptive divergence of duplicated genes in a relatively short time (say, 50 million years or less) after they have duplicated (Figure 7.4). Gradually, as purifying selection comes to predominate, d_S will catch up to d_N, and eventually exceed it, hiding any trace of past adaptive evolution (Figure 7.4). Thus, as we shall see, a pattern whereby the d_N/d_S ratio decreases over evolutionary time is a characteristic of multigene families that have diversified functionally.

In cases where adaptive diversification has occurred in the distant past and comparison of d_S and d_N does not reveal any evidence of it, it is possible that past diversifying selection has directionally changed chemical properties of the amino acid residues at certain sites. Evidence of such directional change in

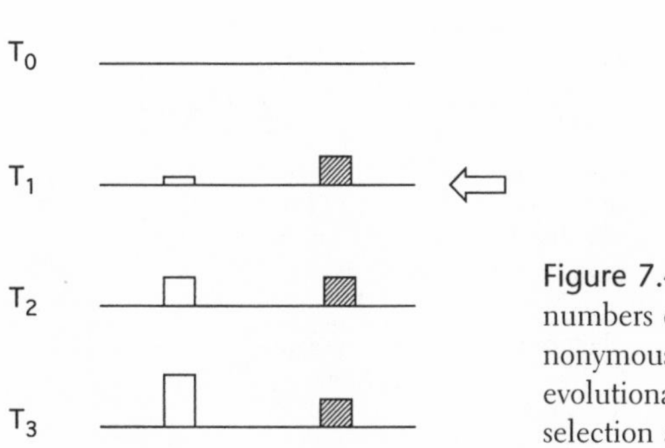

Figure 7.4. Hypothetical changes in numbers of synonymous (d_S) and nonsynonymous (d_N) substitutions per site over evolutionary time (T_0–T_4) under directional selection at the amino acid level.

amino acid residue properties can be obtained by comparing the proportion of conservative (p_{NC}) and radical (p_{NR}) nonsynonymous differences, using the method described in Chapter 2, Conservative and Radical Nonsynonymous Differences. For example, a pattern of $p_{NR} > p_{NC}$ with respect to amino acid residue charge in a given gene region indicates that amino acid replacements have occurred in that region in such a way as to change residue charge to a greater extent than expected under random substitution. Although not as strong evidence of positive selection $d_N > d_S$, such a pattern is suggestive of positive selection because such a nonrandom pattern of residue change is unlikely under neutral evolution.

Table 7.3 summarizes published cases in which DNA sequence analyses have provided evidence for adaptive diversification of members of multigene families. In this section I briefly discuss five of these; the immunoglobulin V_H variable region genes of mammals; olfactory receptors of vertebrates; mammalian defensins; the *jingwei* gene of *Drosophila*; and primate eosinophil ribonucleases.

Mammalian V_H Genes

The immunoglobulins or antibodies of vertebrates are specific recognition molecules involved in neutralizing foreign antigens. They are composed of heavy and light chains, each of which includes constant and variable regions. In mammals, each immunoglobulin chain is created by a somatic rearrangement of gene segments present in the germline (Tonegawa 1983). In production of heavy chains, the segments involved are V_H, D_H, and J_H, which encode the variable region; and C_H, which encodes the constant region. In mammals, there are two kinds of light chain (κ and λ), each of which is encoded by V, J, and C segments. In humans and the mouse, the primary antibody response results from immunogloblin chains generated solely by this process of rearrangement. In these species, the secondary antibody response, which produces antibodies

Table 7.3. Evidence of positive selection favoring divergence of genes in multigene families

Molecules	Organisms	Reference
1. $d_N > d_S$		
Serine protease inhibitors	Mammals	Hill and Hastie (1987)
Immunoglobulin V region	Mammals	Tanaka and Nei (1989)
Olfactory receptors	Catfish	Ngai et al. (1993)
Histatin	Human	Sabatini et al. (1993)
Interferon-ω	Ruminants	Hughes (1995b)
Defensins	Mammals	Hughes and Yeager (1997b)
Eosinophil cationic protein	Primates	Zhang et al. (1998)
2. $p_{NR} > p_{NC}$		
Integrin β-chains	Mammals	Hughes (1992b)
Olfactory receptors	Rat	Hughes and Hughes (1993b)
Cysteine proteases	Nematodes	Hughes (1994b)

highly specific for an abundant foreign antigen during the course of an infection, also involves a process of somatic mutation. In other mammals, the primary antibody repertoire may also be generated by means of somatic mutation (Diaz and Flajnik 1998).

The portions of the immunoglobulin variable region that are responsible for antigen binding are three complementarity-determining regions (CDR). Two of these (CDR1-2) are encoded in the V region gene segment, while the third (CDR3) is encoded by the D and J segments (Kabat et al. 1991). The remainder of the variable region is made up by the so-called framework (FR) regions. Because in humans and the mouse the primary antibody response occurs without somatic mutation, it seems a reasonable hypothesis that natural selection should favor diversification among V region segments, particularly in the CDRs. Tanaka and Nei (1989) tested this hypothesis with sequence comparisons among germline V_H gene segments of humans and the mouse.

Phylogenetic analyses have shown that V_H genes of mammals belong to three ancient families whose divergence predates the divergence of tetrapods from bony fishes (Ota and Nei 1994c). In comparisons among different families, synonymous sites are saturated; and estimations of d_S and d_N cannot be used to test for the adaptive diversification of these genes. Thus, Tanaka and Nei (1989) focussed on comparisons between closely related V_H genes.

Figure 7.5 shows plots of the uncorrected proportions of synonymous and nonsynonymous differences p_S and p_N) in CDR and FR regions for all pairwise comparisons among a set of closely related mouse V_H genes. The patterns for CDR and FR are strikingly different. In the FR, $p_S > p_N$ for all but a few comparisons among closely related sequences (Figure 7.5B). In the CDR, $p_N > p_S$ for a majority of comparisons (Figure 7.5A). The tendency for p_N to exceed p_S is particularly marked in comparisons among closely related sequences; i.e., in comparisons with low p_S values (Figure 7.5A). In comparisons between more distantly related sequences (i.e., when p_S is high), there seems to be a tendency for p_N values to level off. Thus, as p_S increases, p_S usually exceeds p_N (Figure 7.5A).

These results are consistent with the model of divergent evolution illustrated in Figure 7.4. It appears that natural selection favors divergence of duplicated V_H region genes in the CDR shortly after gene duplication. Then, after the V_H genes have diverged adaptively, this selection is relaxed, and purifying selection comes to predominate.

Rothenfluh et al. (1995) questioned the adaptive value of having a diversified repertoire of immunoglobulin V region genes, given that specific antibodies are produced by somatic mutation. However, this ignores the fact that in humans and the mouse—and probably in many other mammals as well—somatic mutation is not involved in the primary antibody response. Therefore, in these species, having a diversity of V regions for the primary immune response would seem to be advantageous. Interestingly, humans and the mouse have a much greater diversity of V_H, V_κ, and V_λ, than do pigs, sheep, and bovine; in the latter three species, somatic mutation is involved in generating the primary immune response (Sitnikova and Nei 1998). It would be interesting to examine the pattern of

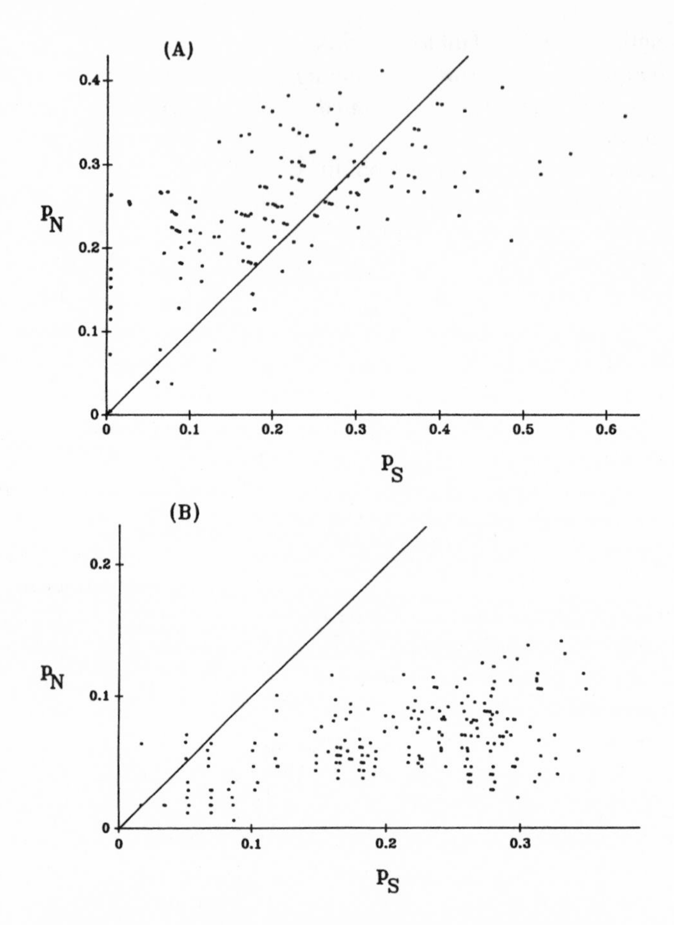

Figure 7.5. Proportion of nonsynonymous nucleotide difference (p_N) plotted against proportion of synonymous nucleotide difference (p_S) in CDR1-2 **(A)** and framework **(B)** regions of V_H region genes. In each case, a 45° line (Y = X) is shown. From Tanaka and Nei (1989). Reproduced by permission.

nucleotide substitution at V region genes in species that use somatic mutation to generate the primary antibody response in order to compare it with the pattern seen in Figure 7.5.

Olfactory Receptors

The mechanism by which animals recognize specific molecules in olfaction and taste remains poorly known. Important light was shed on this question by the discovery of an extensive family of olfactory receptor genes in vertebrates (Buck and Axel 1991). These receptors are members of a superfamily of transmembrane receptors called the G protein–coupled receptors, so called because their intracellular signal transduction mechanism involves proteins called G proteins (Iismaa and Shine 1992). Members of this family have an extracellular N-terminal domain and an intracellular C-terminal domain, and they pass through the cell membrane seven times (Figure 7.6). There are thus four extracellular domains (E1–E4), four intracellular domains (I1–I4), and seven helical transmembrane domains (TM1–TM4). These receptors are sometimes called heptahelicals because of their seven membrane-spanning domains (Mollon 1991).

In the rat, the olfactory receptor family is estimated to include about 1000 genes, only a small number of which have so far been sequenced (Buck and Axel 1991). Because fishes are thought to recognize a smaller range of odorant molecules than mammals (Hara 1975), Ngai and colleagues (1993) decided that the channel catfish, *Ictalurus punctatus*, would be a simpler system in which to study the evolution of this gene family. Indeed, this species was found to have fewer than 40 olfactory receptor genes in all, much fewer than a typical mammal (Ngai et al. 1993). Ngai et al. (1993) sequenced four closely related olfactory receptor genes from this species to examine adaptive evolution among them.

Because previous studies had suggested that ligands of G protein–coupled receptors often interact with the transmembrane domains, Ngai et al. (1993) examined the pattern of nucleotide substitution in the transmembrane domains (Table 7.4). In TM3 and TM4, mean d_N among sequences significantly exceeded mean d_S, whereas the reverse was true for the remainder of the gene (Table 7.4). Thus, these recently diverged genes appear to have been subject to positive selection favoring amino acid differences in TM3–4. Presumably, this selection has arisen because each gene has become adapted to binding a somewhat different type of odorant molecule.

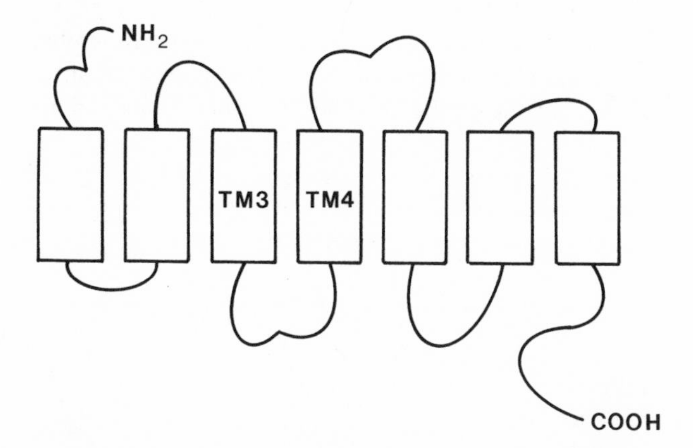

Figure 7.6. Schematic representation of G-protein-coupled receptor, showing the transmembrane domains (TM3–TM4) believed to be involved in odorant reception in the catfish. From Hughes (1993a).

Table 7.4. Mean numbers of synonymous (d_S) and nonsynonymous (d_N) nucleotide substitutions per 100 sites (±SE) in comparisons among three channel catfish (*Ictalurus punctatus*) olfactory receptor genes

Region	d_S	d_N
TM3–TM4	1.2 ± 1.7	9.6 ± 2.7**
Remainder	3.6 ± 1.1	1.1 ± 0.3*

Tests of the hypothesis that $d_S = d_N$: *$p < .05$; **$p < .01$. Data from Ngai et al. (1993).

The odorant molecules recognized by fishes are water soluble, and include amino acids, bile salts, steroids, and prostaglandins (Hara 1975). Mammals, by contrast, respond to volatile lipophilic molecules. Thus, it would not be surprising if selection acting on mammalian olfactory receptors were different. Buck and Axel (1991) sequenced 10 olfactory receptor genes from the rat. These genes are much more distantly related to each other than are the catfish genes, and synonymous sites are saturated in comparisons among them (Hughes and Hughes 1993b). Thus estimation of d_S and d_N could not be used to test for positive selection in this case. However, amino acid properties have changed in a nonrandom way, suggesting the action of natural selection.

The available rat olfactory receptor genes fall into three subfamilies (I–III; Figure 7.7). Subfamily III includes the majority of available sequences, and sequences in this subfamily are more closely related to each other than are those in the other two subfamilies. When the proportion of conservative (p_{NC}) and radical (p_{NR}) nonsynonymous nucleotide differences, with respect to residue polarity, were computed both within and between subfamilies, p_{NC} was generally greater than p_{NR} (Table 7.5). In all comparisons in the transmembrane regions and in most comparisons in the intracellular regions, p_{NC} was significantly greater than p_{NR} (Table 7.5). Thus, in these regions nonsynonymous differences occur

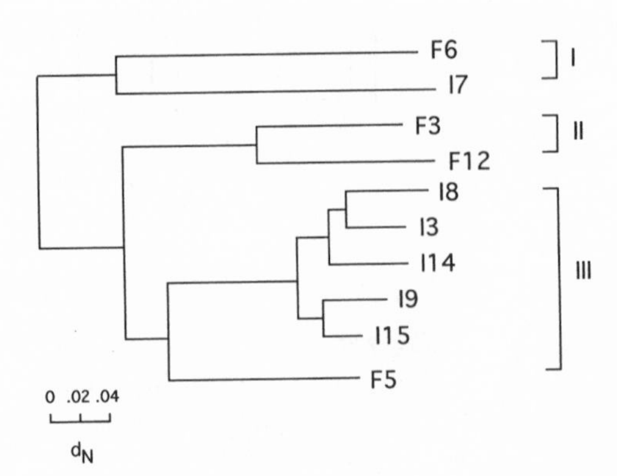

Figure 7.7. Phylogenetic tree of rat olfactory receptors, constructed by the neighbor-joining method based on the number of nonsynonymous substitutions per site (d_N), showing families (I, II, and III) of closely related sequences.

in such a way as to conserve residue polarity to a greater extent than would be expected under random substitution.

In the extracellular domains E1, E3, and E4, in the transmembrane domains, and in the intracellular domains, p_{NC} generally exceeded p_{NR}, and the difference was highly significant in several comparisons (Table 7.5). In E2, p_{NC} and p_{NR} were about equal for subfamilies I and II. By contrast, in E2 of subfamily III, p_{NR} was over four times as great as p_{NC}, a highly significant difference (Table 7.5). Thus, in family III, nonsynonymous differences causing a residue polarity change occur much more frequently than expected under random substitution.

The substitution pattern in these molecules is clarified by a plot of the proportion of amino acid differences, which involve a polarity change versus the proportion amino acid difference in the entire molecule (Figure 7.8). In E2, there is a strong negative relationship between these quantities (Figure 7.8A). Thus, in the comparison between two closely related sequences, all or nearly all amino acid differences in E2 involve a polarity difference. However, when two more distantly related sequences are compared, the proportion of amino acid differences that involve a polarity difference is much lower, avaraging around 40% when the two sequences are 60% different at the amino acid level (Figure 7.8A). In E1, E3, and E4, this relationship is reversed. In these domains, comparisons between closely related sequences generally involve a relatively low proportion of polarity differences; the proportion increases to 40–50% in the case of the most distant pairs of sequences (Figure 7.8B).

From these results, it appears that shortly after gene duplication a burst of nonsynonymous substitutions occurred, changing the polarity of residues in E2. Then, over time, changes in polarity occur less frequently so that as divergence

Table 7.5. Mean percent conservative (p_{NC}) and radical (p_{NR}) nonsynonymous nucleotide difference (±SE), with respect to amino acid residue polarity, in different domains of three subfamilies of rat olfactory receptors.

	E2		E1,E3,E4	
Subfamily	p_{NC}	p_{NR}	p_{NC}	p_{NR}
I	39.8 ± 9.0	45.1 ± 11.8	39.1 ± 6.4	23.8 ± 7.9
II	20.9 ± 6.8	13.7 ± 9.0	34.6 ± 5.9	7.2 ± 4.9***
III	14.2 ± 3.5	57.4 ± 12.4***	15.5 ± 2.9	16.1 ± 4.2

	TM1–7		I1–4	
Subfamily	p_{NC}	p_{NR}	p_{NC}	p_{NR}
I	39.7 ± 3.4	23.2 ± 3.6***	42.1 ± 5.6	26.4 ± 6.5
II	20.5 ± 2.8	8.6 ± 2.4**	29.6 ± 4.8	7.2 ± 4.0***
III	19.4 ± 1.7	11.5 ± 1.7**	20.1 ± 2.7	17.5 ± 3.4

Tests of the hypothesis that $p_{NC} = p_{NR}$: $^{**}p < .01$; $^{***}p < .001$. Data from Hughes and Hughes (1993b).

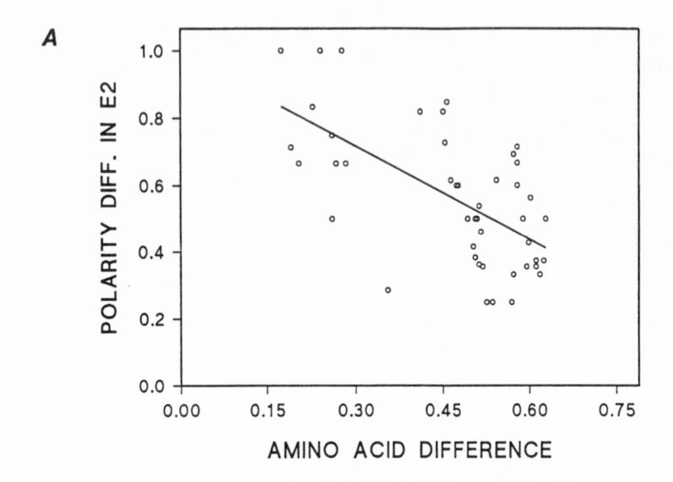

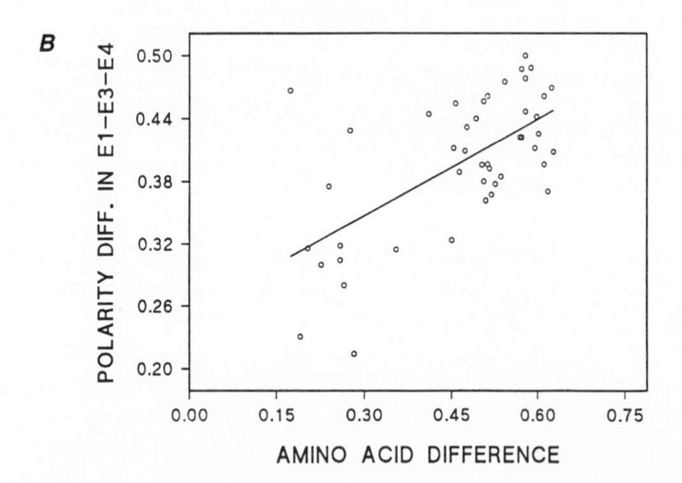

Figure 7.8. The proportion of amino acid differences that involve a polarity change in E2 (**A**) and in E1, E3, and E4 (**B**) plotted against proportion amino acid difference in the whole molecule for rat olfactory receptors. Linear regression lines are shown. From Hughes and Hughes (1993b).

between two sequences increases the proportion of polarity differences between them is no more than expected between two random sequences. (Assuming equal use of all amino acids, the proportion of polarity difference between two sequences chosen at random is expected to be about 48%). On the other hand, in E1, E3, and E4, the initial amino acid replacements occurring soon after gene duplication tend to be relatively conservative with regard to polarity. Subsequently, in these regions the proportion of amino acid differences involving a polarity difference increases over evolutionary time to the point where it is similar to that in E2.

These results suggest that E2 may be the focus for natural selection favoring functional differentiation of mammalian olfactory receptors. The fact that this focus is different from that of the catfish members of this family may reflect the difference between the types of odorant molecules recognized by the two species. In mammals, E2 may actually serve to bind odorants; alternatively, mammalian and fish receptors may turn out to bind odorants in the same domain, but E2 may serve some accessory function in mammals. In any event, the olfactory receptor data suggest a similar evolutionary pattern to that seen in the case of immunoglobulin V_H regions. Gene duplication is followed by a burst of amino acid changes that are selectively favored; thereafter, purifying selection comes to predominate.

Mammalian Defensins

The defensins of vertebrates are antimicrobial peptides that are stored in cytoplasmic granules of Paneth cells of the intestine, neutrophils, and macrophages (Ganz et al. 1989). The mature defensin is a highly cationic (positively charged) peptide of 29–34 amino acids; it is cleaved from a primary translation product consisting of a signal peptide (19 amino acids), a propiece (37–51 amino acids), and the mature peptide (Michaelson et al. 1992). Defensins are cytotoxic to mammalian and bacterial cells, presumably because of their ability to form pores in lipid bilayers (Kagan et al. 1990).

Because cationic peptides are often cytotoxic (Antohi and Brumfield 1984), the cationic character of defensins is thought to contribute to their cytotoxicity, although there are evidently other contributing factors (Michaelson et al. 1992). Observing that the propiece is anionic (negatively charged) Michaelson et al. (1992) proposed that the propiece neutralizes the cytotoxicity of the defensin until it is ready for antimicrobial attack. Cleavage of the propiece then activates the defensin. Consistent with this hypothesis, there is a negative correlation between the net charge of the propiece and that of the mature defensin (Figure 7.9A).

Phylogenetic analysis of mammalian defensins indicates that defensins have diversified as a result of recent gene duplication (Figure 7.10). For example, mouse and rat defensins form separate clusters, indicating that the gene duplications took place after these two species diverged (about 40 million years ago; Hedges et al. 1996). In the case of the mouse, the species for which the most sequence data were available, the defensins formed two well-differentiated families (labeled 1 and 2 in Figure 7.10). Numbers of synonymous and nonsynonymous substitutions were estimated for pairwise comparisons within species or, in the case of the mouse, within families (Figure 7.11). In the signal peptide d_S exceeded d_N for all comparisons (Figure 7.11A). In the propiece, d_S exceeded d_N for most comparisons, although d_N was slightly greater than d_S for a few comparisons between closely related sequences (Figure 7.11B). In the case of the mature defensin, however, d_N exceeded d_S for most comparisons between closely related sequences, often by ratios greater than 2:1 (Figure 7.11C). The

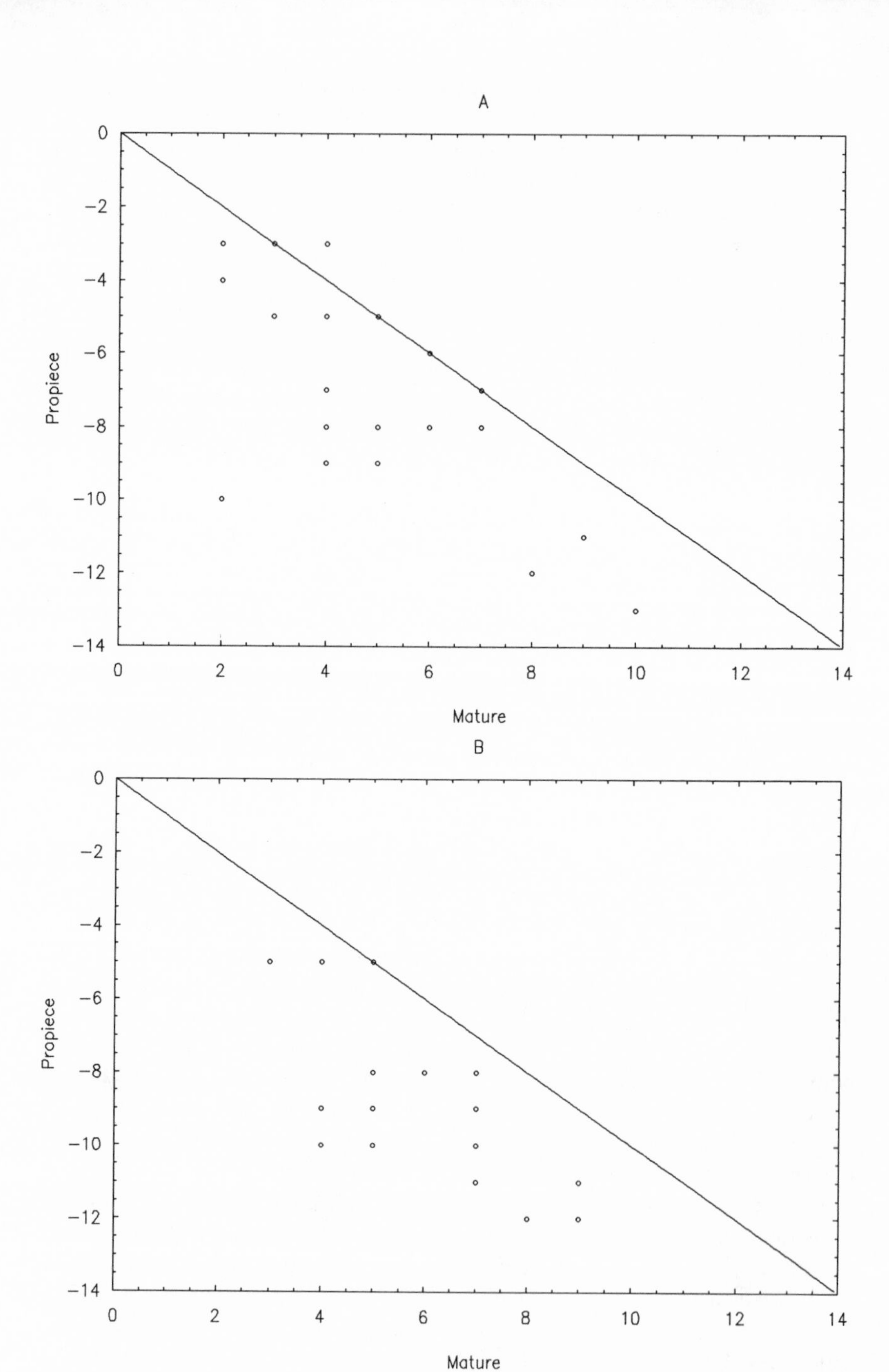

Figure 7.9. **(A)** Net charge in the propiece plotted against that in the mature defensin for 28 mammalian defensins ($r = -.742$; $p < .001$). The line is a 45° line ($Y = X$). **(B)** Net charge in the propiece plotted against that in the mature defensin for 22 reconstructed ancestral sequences ($r = -.755$; $p < .001$). The line is a 45° line.

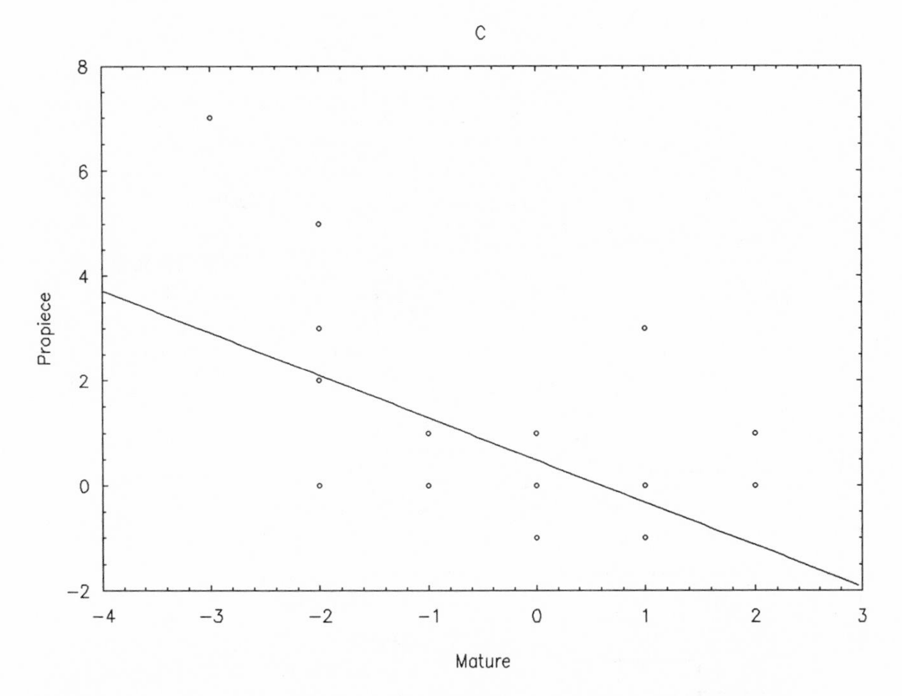

Figure 7.9. *Continued.* (C) Charge change in the propiece plotted against that in the mature defensin ($r = -.548$; $p < .001$). The line is the linear regression line Y = 0.486 – 0.808X. From Hughes and Yeager (1997b).

plot of d_N versus d_S in the mature defensins (Figure 7.11C) resembles that for immunoglobulin CDRs (Figure 7.5A).

When the proportions of conserved (p_{NC}) and radical (p_{NR}) nonsynonymous differences were computed in the mature defensins of rodents, with respect to residue charge, p_{NR} was found to exceed p_{NC}, often significantly (Table 7.6). Thus, amino acid differences occur in such a way as to change residue charge to a greater extent than expected under random substitution. Given the importance of positively charged residues for the defensin's function and the evident need to balance the positively charged residues in the mature defensin with negatively charged residues in the propiece, it is of interest to know how selection favoring charge diversity in the mature defensin might affect the amino acid composition of the propiece.

The coevolution of the mature defensin and the propiece were studied by reconstructing ancestral amino acid sequences (Yang et al. 1995). Figure 7.12 shows mean numbers of reconstructed amino acid changes per branch of the defensin phylogeny. There was a significantly greater number of changes that subtracted a positively charged residue in the mature defensin than in the propiece, while there were correspondingly more changes subtracting a negatively charged residue in the propiece than in the mature defensin (Figure 7.12). The relationship between net charge of the mature defensin and that of the

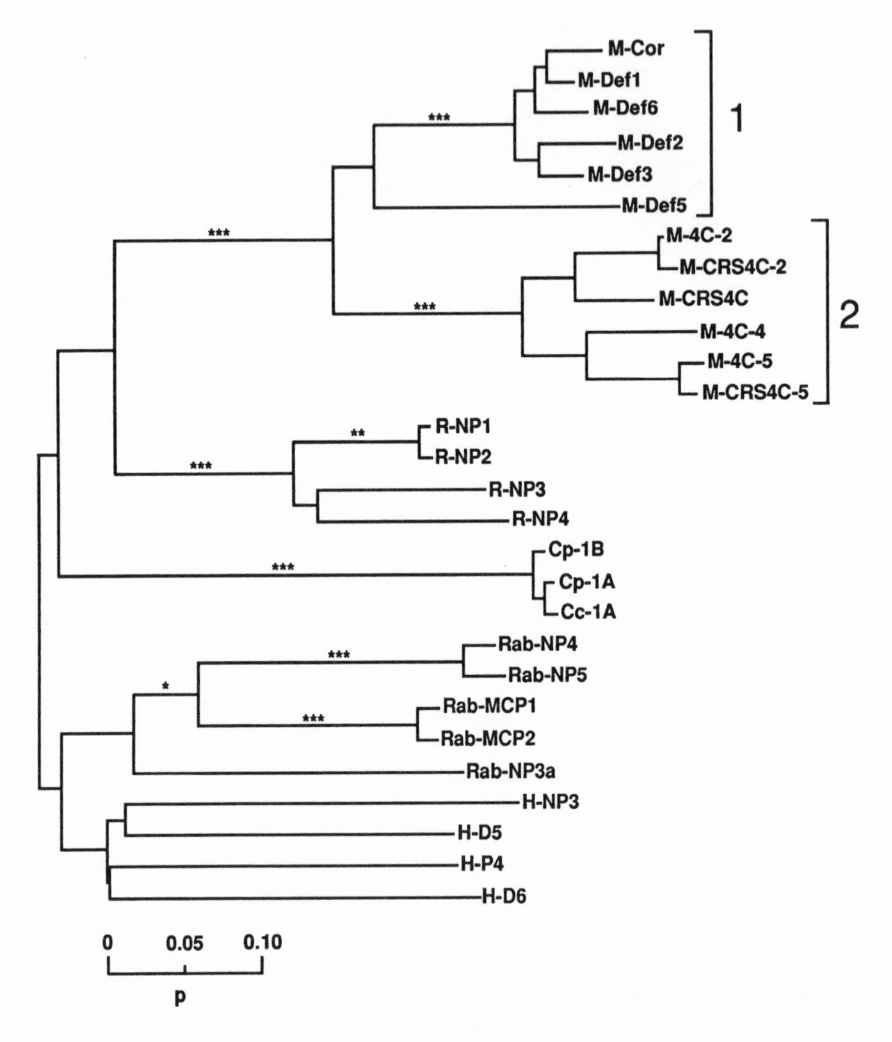

Figure 7.10. Phylogenetic tree of mammalian defensins, constructed by the neighbor-joining method based on the proportion of amino acid difference (p). Tests of significance of internal branches: $^{*}p < .05$; $^{**}p < .01$; $^{***}p < .001$. Prefixes of sequence symbols indicate the species as follows: Cc—*Cavia cutleri;* Cp—*Cavia porcellus;* H—human; M—mouse; R—rat; Rab—rabbit. From Hughes and Yeager (1997b).

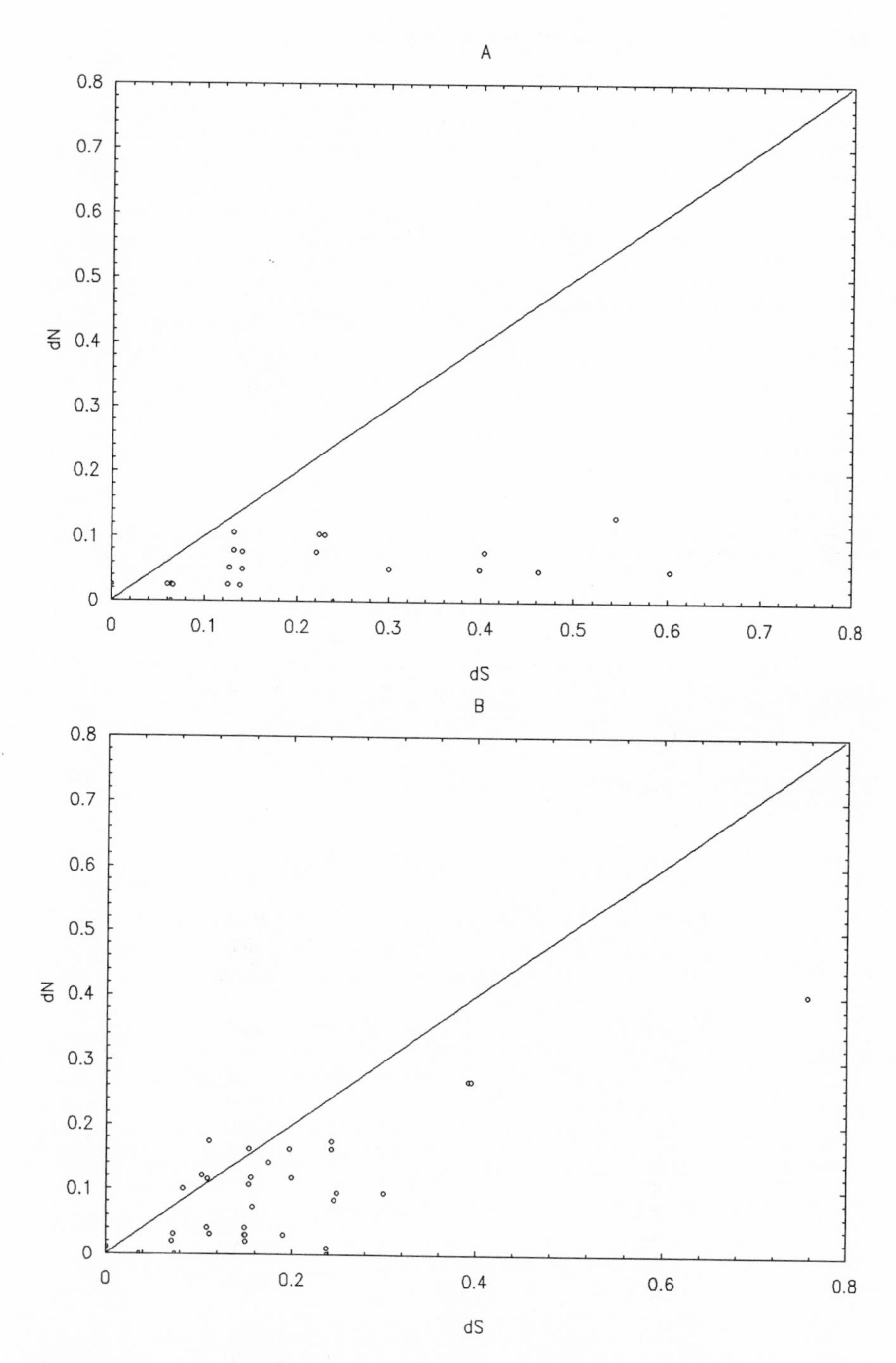

Figure 7.11. Plots of numbers of nonsynonymous nucleotide substitutions per site (d_N) against numbers of synonymous substitutions per site (d_S) in comparisons among closely related defensins (from the mouse, rat, guinea pig, and rabbit). **(A)** signal peptide; **(B)** propiece; **(C)** mature defensin. In each case, a 45° line is shown. From Hughes and Yeager (1997b).

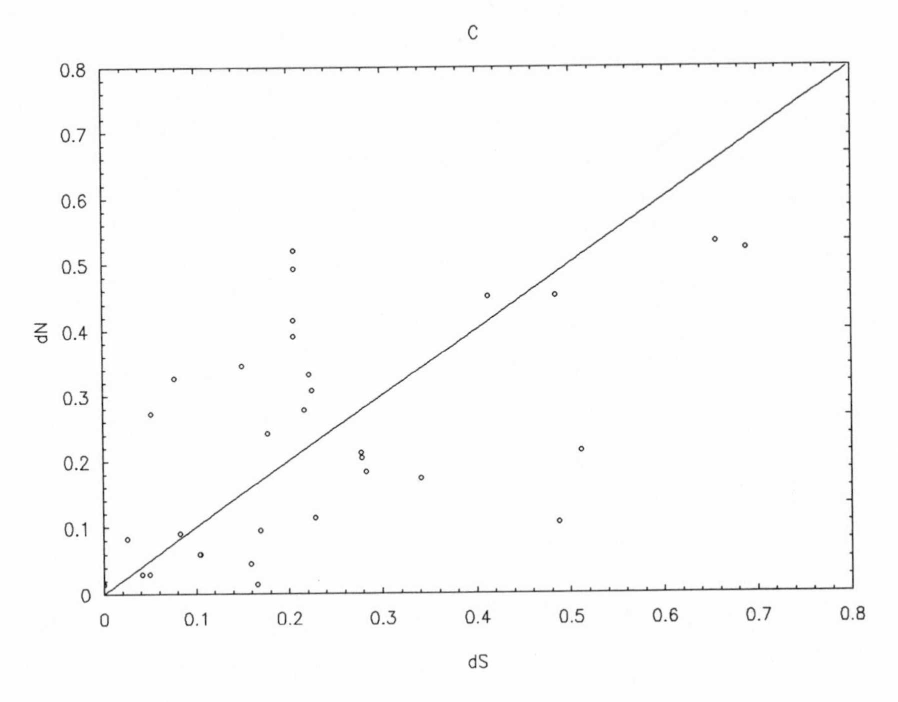

Figure 7.11. *Continued.*

Table 7.6. Mean percent conservative (p_{NC}) and radical (p_{NR}) nonsynonymous nucleotide difference (±SE) in comparison of mature defensins of rodents

Comparison	p_{NC}	p_{NR}
Mouse 1		
vs. mouse 1	30.9 ± 7.3	17.4 ± 6.7
vs. mouse 2	55.0 ± 8.5	68.8 ± 10.2
vs. rat	49.6 ± 7.7	44.6 ± 9.4
vs. guinea pig	41.6 ± 7.6	66.8 ± 9.7*
Mouse 2		
vs. mouse 2	7.6 ± 2.8	22.6 ± 6.1*
vs. rat	50.2 ± 8.2	82.3 ± 11.1*
vs. guinea pig	57.7 ± 8.2	69.3 ± 11.8
Rat		
vs. rat	15.7 ± 4.1	22.1 ± 6.6
vs. guinea pig	39.9 ± 7.1	55.3 ± 10.3
Guinea pig vs. guinea pig	1.4 ± 1.4	3.0 ± 3.0

Mouse families 1 and 2 are as illustrated in Figure 7.10. Tests of the hypothesis that $p_{NC} = p_{NR}$: $^{*}p < .05$. From Hughes and Yeager (1997b).

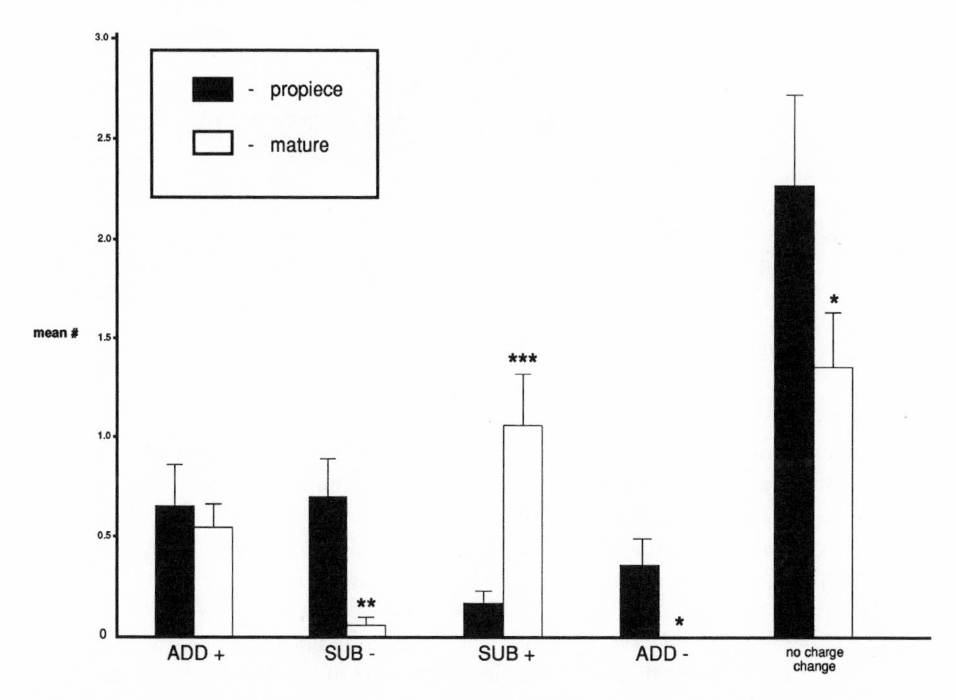

Figure 7.12. Mean numbers (with SE) of reconstructed amino acid changes per branch in the phylogeny of mammalian defensins: ADD+, a positive charge added; SUB–, a negative charge subtracted; SUB+, a positive charge subtracted; ADD–, a negative charge added. For each category, the hypothesis that the number of changes occurring in the propiece equaled that in the mature defensin was tested by paired sample *t*-test: $^{*}p < .05$; $^{**}p < .01$; $^{***}p < .001$. From Hughes and Yeager (1997b).

propiece (Figure 7.9B) was similar to that seen in extant sequences (Figure 7.9A). There was a correlation between net charge change in the mature defensin and that in the propiece (Figure 7.9C). As the mature defensin became less positively charged, the propiece became less negatively charged; and vice versa (Figure 7.9C). Because such coordinated change seems unlikely to have occurred by chance, these results support the hypothesis that defensins have diverged adaptively after gene duplication (Hughes and Yeager 1997b).

The jingwei *Gene in* Drosophila

Processed genes are genes that are reverse-transcribed from messenger RNA and incorporated in the genome. In eukaryotes, such genes are typically recognized by their lack of introns. Most are pseudogenes; but in a few cases processed genes have remained functional. In two species of *Drosophila*, Long and Langley (1993) found a processed gene related to the alcohol dehydrogenase (*Adh*) gene, which they called *jingwei*. The *jingwei* gene is remarkable in that it has captured the 5′ region of some unknown functional gene. The *jingwei* gene is thus a

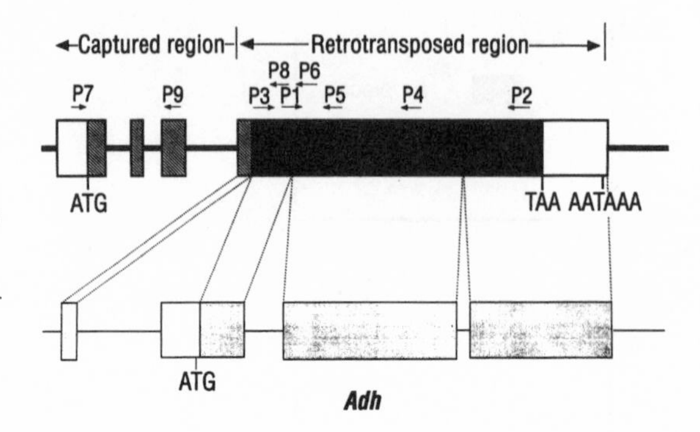

Figure 7.13. Structure of the *jingwei* gene (*top*) and its hypothetical origin through retrotransposition of *Adh* and its combination with a captured region of unknown origin. From Long and Langley (1993). Reproduced by permission.

chimera, consisting of three short exons derived from this unknown gene, followed by a large exon consisting of a small region from the unknown gene fused with the processed *Adh* gene (Figure 7.13). (The name *jingwei* comes from a figure in Chinese mythology who was tragically drowned but later reincarnated as a bird, reflecting the gene's apparent rescue from the usual fate of processed genes.) The events leading to the production of *jingwei* occurred in the ancestor of *D. teissieri* and *D. yakuba*. There is evidence that the *jingwei* gene is expressed, though at lower levels than *Adh*, but its function is unknown.

Figure 7.14 shows nucleotide changes in the evolution of *jingwei*, as reconstructed by Long and Langley (1993). According to this reconstruction, eight nonsynonymous changes and no synonymous changes occurred after the origin of *jingwei* and prior to the divergence of *D. teissieri* and *D. yakuba* (Figure 7.14). This suggests that positive selection acted on the *jingwei* gene soon after its origin (Long and Langley 1993). However, because the total number of

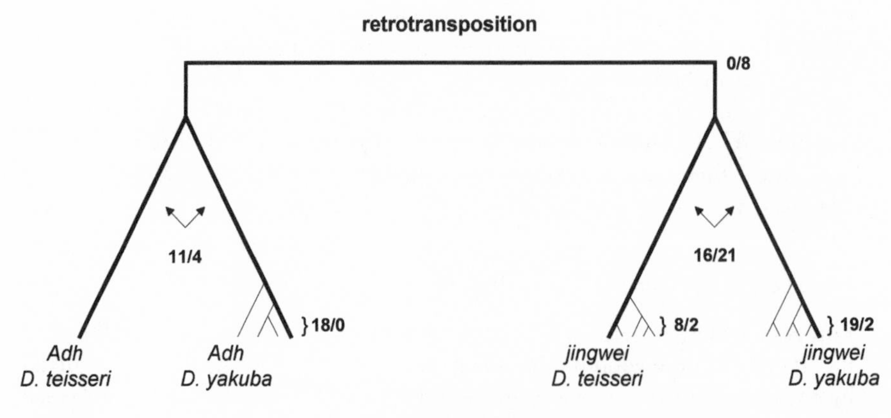

Figure 7.14. Reconstructed numbers of synonymous (*left*) and nonsynonymous (*right*) nucleotide substitutions between *Adh* and *jingwei* in *Drosophila teissieri* and *D. yakuba*. Numbers at the bottom are observed numbers of synonymous and nonsynonymous polymorphic sites within species. Redrawn with data from Long and Langley (1993).

changes is small, it is unclear whether the excess of nonsynonymous substitutions is too great to be attributed to chance alone.

In addition, the ratio of nonsynonymous to synonymous differences between the two species (21:16) is greater than the ratio of nonsynonymous to synonymous polymorphisms within the two species (4:27); by a χ^2 test, Long and Langley (1993) found difference to be significant at the 0.001 level. This approach to testing for positive selection is the same as that used by McDonald and Kreitman (1991) (see chapter 6, Adaptive Divergence between Species), and the results suggest that *jingwei* has diverged adaptively between these two species. However, this conclusion should be accepted with some caution because the function of *jingwei* is not known.

Eosinophil Ribonucleases

In mammals, the group of leukocytes known as eosinophils are characterized by granules containing a number of highly basic proteins; the presence of these proteins led to identification of this cell type over 100 years ago by staining with acid dyes (Hamann et al. 1991). In humans, two of these proteins are members of the ribonuclease gene family known as eosinophil-derived neurotoxin (EDN) and eosinophil cationic protein (ECP). EDN was named for its neurotoxic effect when injected into experimental animals, an effect having little to do with its natural role. Both EDN and ECP have ribonuclease activity and act as helminthotoxins; in other words, as defenses against parasitic worms. The ribonuclease activity of EDN is considerably more potent (by a factor of 50–100) than that of ECP, while the antihelminthic effect of ECP is about 10 times more potent than that of EDN (Ackerman et al. 1985; Slifman et al. 1986; Hamann et al. 1991). The biological function of the ribonuclease activity of EDN is apparently as a defense against RNA viruses (Domachowske and Rosenberg 1997). The role of eosinophils in defense against parasitic worms is well known; in addition to ECP and EDN, the eosinophil basic granule proteins include major basic protein and eosinophil peroxidase, both of which are antihelminthic agents, especially the former (Hamann et al. 1991).

Figure 7.15 shows a phylogenetic tree of EDN, ECP, and related ribonucleases. The tree was rooted with sequences of primate ribonuclease k6; a previous phylogenetic analysis of the entire ribonuclease family showed that k6 diverged from the ECP and EDN prior to the divergence of rodents and primates (data not shown). The phylogeny was thus consistent with the hypothesis that the primate ECP and EDN genes duplicated in the primate lineage after its divergence from rodents (Rosenberg et al. 1995; Zhang et al. 1998). The tree indicates that the related mouse genes have duplicated independently (Figure 7.15); these include the genes encoding ECP-1 and ECP-2, which are expressed in mouse eosinophils (Larson et al. 1996).

Zhang et al. (1998) applied a method to primate ECP and EDN genes whereby they estimated synonymous and nonsynonymous substitution on branches of a phylogeny of Old World primate ECP and EDN genes. On the branch leading to Old World primate ECP (corresponding to the branch labeled

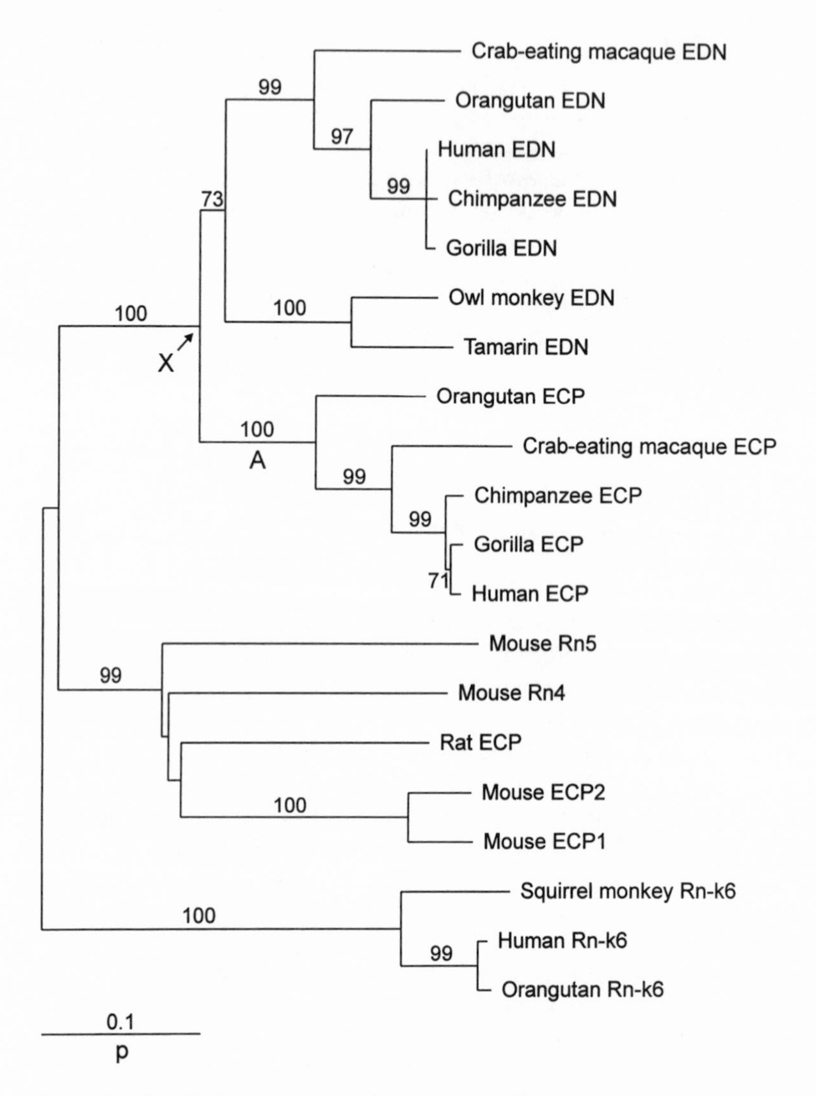

Figure 7.15. Phylogenetic tree of mammalian ribonuclease, constructed by the neighbor-joining method based on the proportion of amino acid difference (p). Numbers on the branches are as in Figure 3.2.

A in Figure 7.15), they found a significant excess of nonsynonymous over synonymous substitutions per site. This burst of nonsynonymous substitution evidently occurred as ECP adapted to its antihelminthic function.

Because in humans both EDN and ECP have both ribonuclease and antihelminthic activities, it seems likely that both of these functions were present in their ancestor (corresponding to node X in Figure 7.15). Rodent ECPs are known to be strongly antihelminthic (e.g., Lee 1991). Given the gene phylogeny, it is thus most parsimonious to conclude that an antihelminthic function was ancestral to primate EDNs and ECPs. What apparently happened in the Old World primates was that, after gene duplication, EDN specialized in antiviral ribonuclease activity, while ECP specialized as a defense against helminths and possibly other parasites.

Zhang et al. (1998) did not consider the possibility that the common ancestor of Old World primate ECP and EDN had an antihelminthic function. Instead, they assumed that this function arose de novo in Old World primate ECP. However, on their hypothesis, we must assume that an antihelminthic function arose independently at least three times in the phylogeny of Figure 7.15: in rodents ECPs, in Old World primate ECPs, and (to a limited extent) in EDN. Such a complicated scenario seems highly unlikely. It is much more reasonable to assume that the common ancestor of Old World primate ECP and EDN (node X) was a functional ribonuclease with antihelminthic activity, and that gene duplication enabled each of its two daughter genes to specialize.

Such a process of specialization might involve positive selection on one or both daughter genes. In the present case, if Zhang et al.'s (1998) reconstruction of the pattern of past nucleotide substitution is correct, this selection was focused mainly on amino acid replacements in ECP. Presumably, as positive selection increased the antihelminthic activity of ECP, mutations diminishing its ribonuclease activity would have been selectively neutral because EDN maintained the essential ribonuclease function. Thus, mutations diminishing ECP's ribonuclease function could have been fixed by genetic drift. At the same time, mutations leading to loss of antihelminthic activity by EDN would be permitted because that function was performed by ECP. On the other hand, it is possible that certain amino acid replacements enhancing antihelminthic activity might, by themselves, decrease ribonuclease activity or conversely, that certain replacements enhancing ribonuclease activity might diminish antihelminthic activity. If so, the reduction of ECP's ribonuclease activity might actually have occurred as a direct consequence of positive selection.

Bifunctionality Preceding Gene Duplication

The MDR hypothesis implies that a new gene function arises only after gene duplication and the process of random accumulation of mutations in one gene copy. Following this hypothesis, Kimura and Ohta (1974, p. 429) included the following statement in their list of "principles governing molecular evolution": "Gene duplication must always precede the emergence of a new gene having a new function." This statement is contradicted by the discovery of cases where a single gene can encode a protein having two different functions, functions that may differ strikingly from each other. This phenomenon was called "gene sharing" by Piatigorsky and Wistow (1991).

The most dramatic examples of gene sharing come from the eye lens crystallins of animals. In vertebrates, proteins from a wide variety of different families function as lens crystallins (Table 7.7). In some case, a single gene encodes a protein that serves as a crystallin as well as having some other function. For example, τ crystallin in the duck is encoded by the same gene that encodes α enolase. Thus, it appears that recruitment of different proteins as crystallins has occurred repeatedly in vertebrate evolution. Initially, as with the duck τ crystallin, a single gene presumably encodes a bifunctional molecule. After gene duplication, one gene copy can specialize as a crystallin, while the other retains the molecule's original function.

The process of gene duplication can be reconstructed in the case of δ crystallin of the duck, which is homologous to the enzyme argininosuccinate lyase (ASL) (Piatigorsky and Horwitz 1996). Both δ crystallin and argininosuccinate lyase are tetramers, with subunits encoded by two linked genes called δ1 and δ2. In the case of δ crystallin, products of the two genes are used equally in formation of tetramers; thus, all possible combinations are seen. A tetramer composed of four δ1 chains lacks ASL activity, but all other possible combinations have ASL activity at approximately equal levels (Piatigorsky and Horwitz 196). This suggests that, after gene duplication, the δ1 gene lost ASL activity while the δ2 gene retained it. In tissues other than the eye lens, the enzymatically active δ2 is expressed at much higher levels than δ1 (Li et al. 1995).

There are also duplicate δ1 and δ2 genes in the chicken (Nickerson et al. 1985, 1986). In the chicken embryo, the δ1 gene produces almost all (nearly 99%) of the lens crystallin (Parker et al. 1988). In other embryonic tissues, both genes are expressed, with δ2 predominating. In the adult chicken, there is a shift to δ2 expression, which is found at higher levels than δ1 even in the lens and at much higher levels in other tissues (Li et al., 1993). As in the duck, only δ2 is enzymatically active.

Table 7.7. Some animal lens crystallins and their evolutionary relationships

Crystallin	Taxonomic distribution	Related or identical*
α	All vertebrates	Small heat-shock proteins
β,γ	All vertebrates	*Myxococcus* protein S
δ	Some birds and reptiles	ASL*
ε	Some birds and reptiles	LDH-B*
ζ	Some mammals	Alcohol dehydrogenase
η	Some mammals	Cytoplasmic aldehyde dehydrogenase
λ	Some mammals	Hydroxyacyl CoA dehydrogenases
ρ	Frogs	NADPH-dependent reductases
τ	Many species	α-Enolase*
S	Cephalopods	Glutathione-S-transferases

Data from Piatigorsky and Wistow (1991).

Phylogenies of different portions of ASL sequence (Figure 7.16) indicates that the δ1 and δ2 genes duplicated prior to the divergence of the chicken and duck, but that the 3′ ends of the genes were subsequently homogenized within each species by interlocus recombination or "gene conversion" (Wistow and Piatigorsky 1990). The divergence time between the avian orders Anseriformes (including the duck and goose) and Galliformes (including the chicken) has been estimated at 112 million years (Kumar and Hedges 1998). This was used as a calibration to estimate divergence times of the 3′ ends of ASL sequences; separate estimates were made on the basis of d_S and d_N (Table 7.8). Because a greater number of sites are used, the estimates based on d_N may be somewhat more reliable. Using d_N, the homogenization of the duck δ1 and δ2 was estimated to have taken place at about 24 million years ago, while that of the chicken δ1 and δ2 was estimated at about 60 million years ago (Table 7.8). This is consistent with the above-mentioned evidence that the chicken seems to have progressed further in the direction of specialization of the two genes than has the duck.

The cephalopods (squid and octopuses) are molluscs that have independently evolved eyes structurally similar to those of vertebrates. Interestingly, squid have recruited glutathione-S-transferases to serve as lens crystallins by a process apparently similar to the recruitment of proteins as lens crystallins by vertebrates. Also interesting is evidence that squid crystallins may have evolved regulation convergently with vertebrate crystallins. AP-1 is a transcription factor involved in regulation of many vertebrate crystallin genes, and squid crystallin genes possess a consensus AP-1 binding sites in their promoter regions (Tomarev et al. 1994). This binding site appears to have evolved convergently in the squid genes (Tomarev et al. 1994).

Although the crystallins provide very dramatic examples of gene sharing, many other polypeptides are bifunctional or multifunctional. For example, the receptor for the cytokine IL-3 in humans is a heterodimer composed of two distinct polypeptides called α and β (Callard and Gearing 1994). The β chain also combines with other chains to form receptors for the cytokines IL-3, IL-5, and GM-CSF. Another recently discovered example involves a molecule known as neuropilin(Soker et al. 1998). Neuropilin is expressed in the axons of neurons and recognizes a molecule called semaphorin III, which directs the growth of axons. The same receptor is expressed in endothelial cells lining blood vessels. In the latter case, neuropilin acts as a receptor for VEGF, a protein that stimulates growth of blood vessels. Furthermore, numerous enzymes are known to have the property called "substrate ambiguity" by Jensen (1976), meaning that they can act on a number of different substrates.

An Alternative Model of the Evolution of Novel Proteins

In the last three sections I have discussed evidence that runs counter to the predictions of the MDR model. Indeed, no case has yet been reported that involves evolution of a new gene function by the mechanism postulated by the MDR model. We must thus conclude that evolution of a new gene function in accordance with the scenario postulated by this model is, at best, very rare,

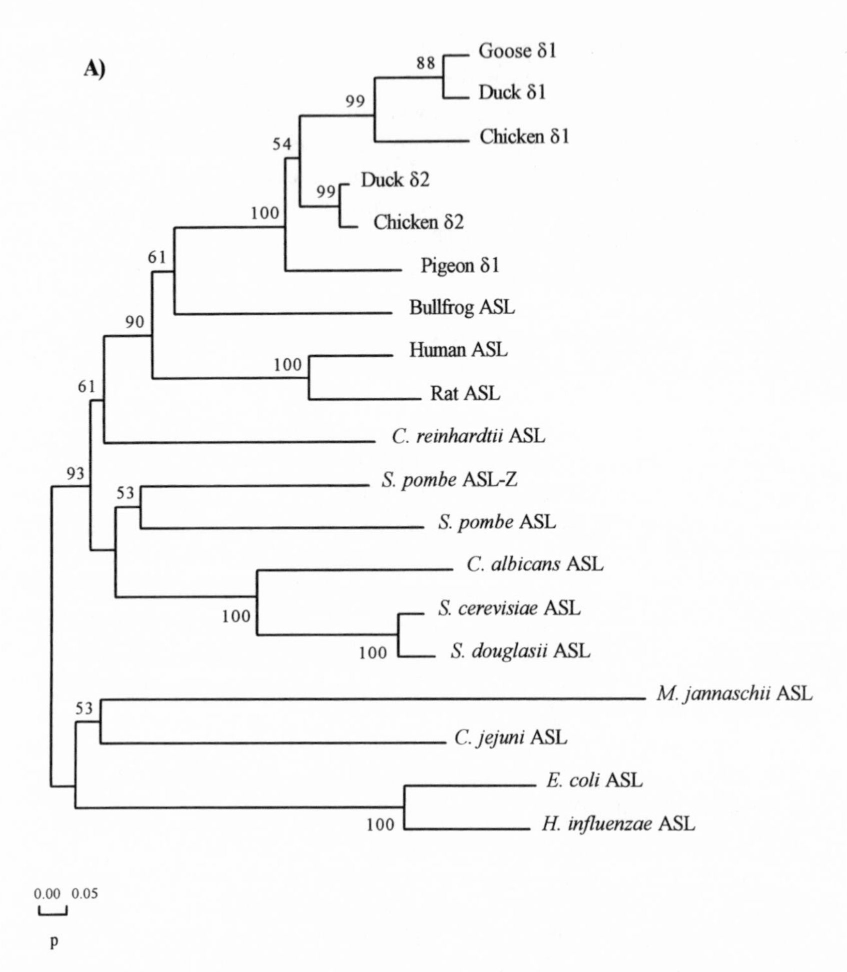

Figure 7.16. Phylogenetic trees of (**A**) the 5′ end (exons 2–5) and (**B**) the 3′ end (exons 6–17) of avian δ crystallins and other eukaryotic members of the ASL family, constructed by the neighbor-joining method based on the proportion of amino acid difference (p). Numbers on the branches are as in Figure 3.2. The topology in the 5′ end indicates that duck and chicken δ1 and δ2 genes duplicated before these species diverged. In the 3′ end, the chicken and duck δ1 and δ2 cluster together, as do duck δ1 and δ2, suggesting subsequent homogenization of this portion of the gene within each species.

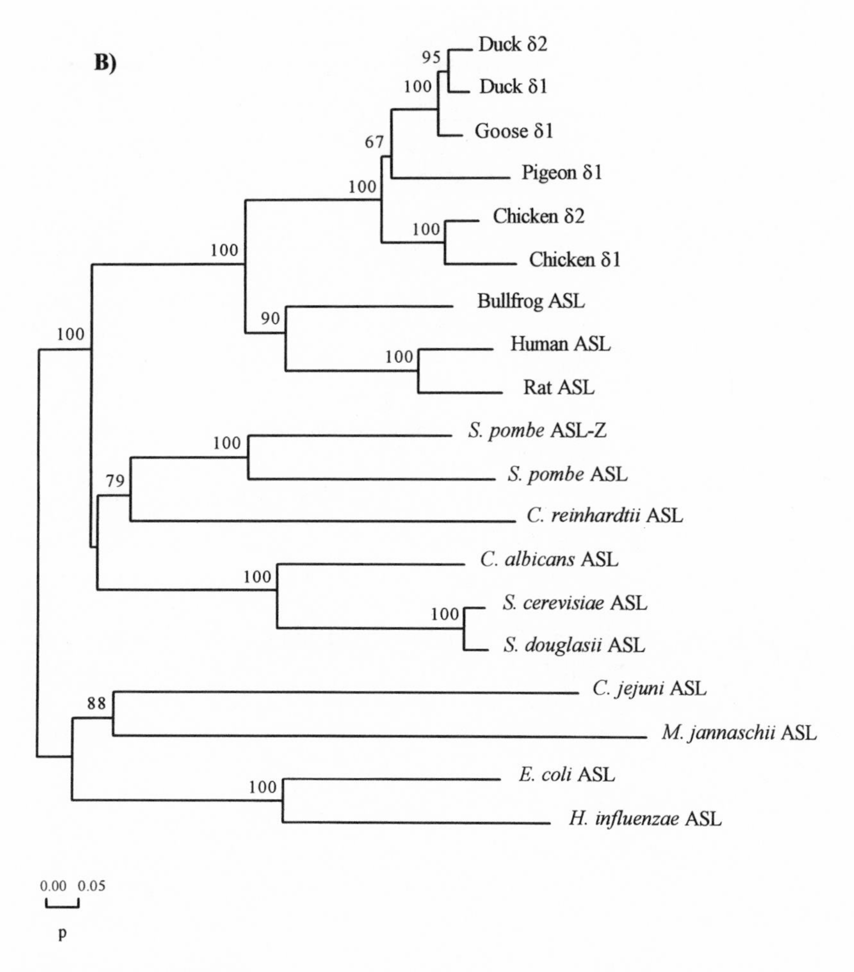

Figure 7.16. *Continued.*

and may never happen. It is, of course, dangerous to say "never" when dealing with evolutionary biology; because of the opportunism of the evolutionary process, evolution may proceed by a variety of unexpected pathways. At any rate, the available data suggest that the MDR model represents an unlikely scenario for the evolution of new gene function. Therefore, there is need for an alternative model to understand the evolution of new gene function. In this section I briefly outline such a model, one that is consistent with available data on the evolution of multigene families (Hughes 1994a).

In this model, gene duplication that leads to the production of two genes encoding functionally distinct proteins is ordinarily preceded by a period of gene sharing; that is, a period in which a single generalist gene performs two distinct functions. Once gene duplication occurs, it becomes possible for the products of the two duplicate genes to specialize so that each of them performs

Table 7.8. Numbers of synonymous (d_S) and nonsynonymous (d_N) nucleotide substitutions per 100 sites (±SE) in exons 6–17 and divergence time estimates for avian δ crystallin genes.

Comparison	d_S	Divergence time estimate (millions of years ago)	d_N	Divergence time estimate (millions of years ago)
Anseriformes vs. Galliformes[a]	25.0 ± 3.2	112 (calibration)	5.8 ± 0.8	112 (calibration)
Duck δ1 vs. δ2	3.7 ± 1.2	16 ± 6	1.3 ± 0.4	24 ± 8
Chicken δ1 vs. δ2	11.2 ± 2.2	50 ± 10	3.1 ± 0.6	60 ± 12

[a]The divergence time between the avian orders Anseriformes (including duck and goose) and Galliformes (including chicken) was estimated at 112 million years ago by Kumar and Hedges (1998); this value was used as a calibration for other time estimates.

only one of the functions performed by the ancestral gene (Figure 7.17). Such specialization might be achieved by a change in the regulation of expression of one or both of the daughter genes. Thus, in a multicellular organism, each daughter gene might come to be expressed in a more restricted set of tissues than the ancestral gene. Alternatively, the ancestral gene might encode a protein with a broad functional specificity; for example, an enzyme that catalyzes a number of related reactions or a receptor that binds a number of related signaling molecules. After gene duplication, this broad function could be divided among more specific molecules.

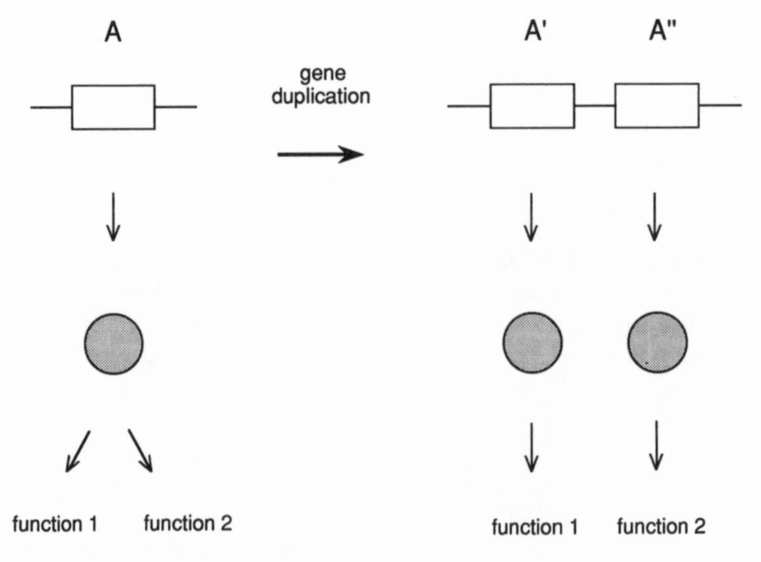

Figure 7.17. Model of evolution of functionally distinct proteins after duplication of a bifunctional ancestral gene. From Hughes (1994a).

The functional specialization of daughter genes often requires the activity of positive Darwinian selection, although positive selection is not invariably present. When the daughter genes come to differ in terms of expression, this natural selection presumably acts on regulatory elements. A possible example of this is the apparent convergent evolution of an AP-1-binding region in the promoter of squid crystallin genes mentioned in the previous section. When the protein products of the daughter genes come to differ functionally, this will result from natural selection favoring nonsynonymous substitution in one or both genes. The gene families considered in the section "Diversifying Selection in Multigene Families" provide evidence that such diversifying selection at the amino acid level has occurred.

A similar model of evolution was implied by Goodman et al. (1975) in the case of vertebrate hemoglobin. They hypothesized that originally the hemoglobin molecule was a homotetramer, consisting of four identical polypeptide chains. Subsequently, gene duplication gave rise to the separate α and β chains found in hemoglobins today. Jensen (1976) and Jensen and Byng (1981) also implied a period of gene sharing before gene duplication when they proposed that two enzymes that are specialized to catalyze two separate reactions may evolve, after gene duplication, from an enzyme capable of catalyzing both reactions. Orgel (1977) also drew attention to the possibility that functional differentiation might precede gene duplication.

When a protein is bifunctional prior to gene duplication, it might perform one of its two functions much less efficiently than the other. If both functions are beneficial to the organism from the point of view of fitness, selection will immediately favor amino acid replacements in one gene that improve its performance of the less efficiently performed function. A dramatic change in efficiency may not require a large number of amino acid replacements. For example NADP-dependent isocitrate dehydrogenase (IDH) of the bacterium *Escherichia coli* has a 7,000-fold preference for NADP rather than NAD as a coenzyme (Chen et al. 1995). When five amino acid replacements were experimentally introduced into the coenzyme-binding region of the enzyme, this preference was reversed to a 200-fold preference for NAD (Chen et al. 1995). The evolution of NADP-dependent IDHs in eubacteria seems to have been an important adaptive advance, enabling them to grow on acetate (Dean and Golding 1997). A plausible ancestral molecule was an NAD-dependent IDH with low affinity for NADP.

Walsh (1995) used methods of theoretical population genetics to study the evolution of new functions by duplicate genes. He showed that the probability that a duplicate locus becomes a pseudogene depends on s, the selection coefficient for advantageous mutants; ρ, the ratio of advantageous to null mutations; and effective population size (N_e). Specifically, a duplicate gene is likely to become a pseudogene unless ρS is much greater than 1, where $S = 4\ N_e s$. If, as proposed here, selection favoring efficient performance of two functions occurs prior to duplication of a bifunctional gene, selection favoring functional specialization will be present immediately after gene duplication. Clearly, in this case there is a better chance of the duplicate gene's evolving a new function rather than becoming a pseudogene than there would be if such selection were not present until some time after gene duplication.

According to this hypothesis there are three possible fates for a duplicate gene:

1. The gene may become a pseudogene due to a "null" mutation that eliminates expression of a protein product. This is predicted to be the fate of the vast majority of duplicate genes. There are three main sources of evidence that silencing is frequently the fate of duplicate genes. First, there is the widespread presence of pseudogenes in multigene families of eukaryotes (Li 1997). Second, phylogenetic reconstructions have indicated that several vertebrate gene families evolve by a "birth-and-death" process, whereby genes duplicate and are silenced, leading to a turnover of loci that eliminates orthologous relationships among species over time (Nei and Hughes 1992). The best known examples of this process are the class I MHC loci (Hughes and Nei 1989b) and the immunoglobulin V region genes (Ota and Nei 1994c; Sitnikova and Nei 1998). Finally, loci duplicated by polyploidization are gradually silenced (Li 1982).
2. In a small number of cases, duplicate genes will evolve distinct functions. In the present hypothesis, those functions will generally be ones performed at least partially by the ancestral gene. Positive selection is frequently involved in favoring amino acid replacements that better adapt the products of daughter genes to their new roles.
3. If a duplicate gene neither adapts to a more specialized function nor is silenced by a "null" mutation, it will continue to be subject to purifying selection because a high proportion of possible nonsynonymous substitutions will have a dominant deleterious effect. As discussed previously in this chapter (Functional Constraint on Duplicate Genes), the fact that duplicate genes of *Xenopus laevis* are subject to purifying selection supports the occurrence of such selection. Note that even though functionally redundant duplicates are subject to purifying selection as long as they are expressed, there is likely to be no selection against a null mutation that removes expression of a redundant duplicate gene.

As an alternative to positive selection in the evolution of new gene function, Zhang et al. (1998) mention what they call the "Dykhuizen-Hartl effect." This term was used by Kimura (1983), citing a paper by Dykhuizen and Hartl (1980), for cases of "preadaptation" at the molecular level. In such a case, a neutral substitution occurs, which then fortuitously becomes beneficial following a change of environment. Such effects, no doubt, sometimes occur. The benzimidazole-resistant *mek-7* gene of *Caenorhabditis elegans* discussed in chapter 6, Convergent Sequence Evolution, may be an example of a neutral substitution that incidentally confers a benefit in a changed environment. In the case of duplicate genes, this effect would occur if one or more neutral substitutions are fixed in one of the daughter genes that preadapt it for a new function it can assume upon an environmental change.

However, the general importance of the Dykhuizen-Hartl effect in the case of duplicated genes is questionable. First, there will be a relatively short time window for the preadapting neutral mutations to occur and to become fixed. If one gene is silenced by a null mutation, this opportunity will presumably be at an end. Thus, if it occurs, the effect is most likely to involve only one or a few amino acid replacements, which can be fixed relatively rapidly. But one or a few selectively neutral amino acid replacements are unlikely to have profound effects on protein structure and function and, thus, are unlikely to lead to significant changes in function. Indeed, for this reason, the Dykhuizen-Hartl effect seems unlikely to play more than a very limited role in evolution in general.

Specialization after Duplication of a Generalized Ancestral Gene

We now know of a number of cases in which phylogenetic analyses strongly suggest that two present-day more specialized genes have evolved, after gene duplication, from a bifunctional ancestor. As discussed earlier in this chapter (Bifunctionality Preceding Gene Duplication), the ECP and EDN genes of Old World primates probably evolved specialized functions after duplication of a more generalized ancestral gene. I will discuss two other examples in this section.

As mentioned, the β chain of the human IL-3 receptor is a multifunctional protein encoded by a single gene; the same β chain is common to the IL-3, IL-5, and GM-CSF receptors (Callard and Gearing 1994). This β chain is often called the cytokine receptor common β chain or CYRβ. In the mouse, there has been a duplication of the homologous gene. Phylogenetic analysis indicates that this duplication occurred after the divergence of rat and mouse (Figure 7.18). One of the two mouse genes encodes the β chain of the IL-3 receptor,

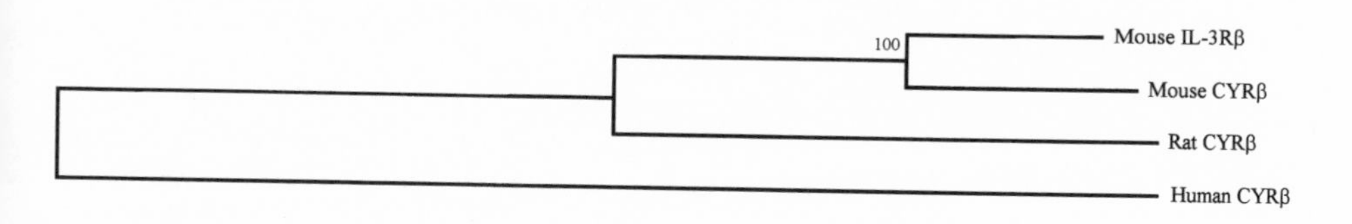

Figure 7.18. Phylogenetic tree of mammalian common cytokine receptor β chains (CYRβ) and mouse IL3Rβ, constructed by the neighbor-joining method based on the proportion of amino acid difference (*p*). Numbers on the branches are as in Figure 3.2.

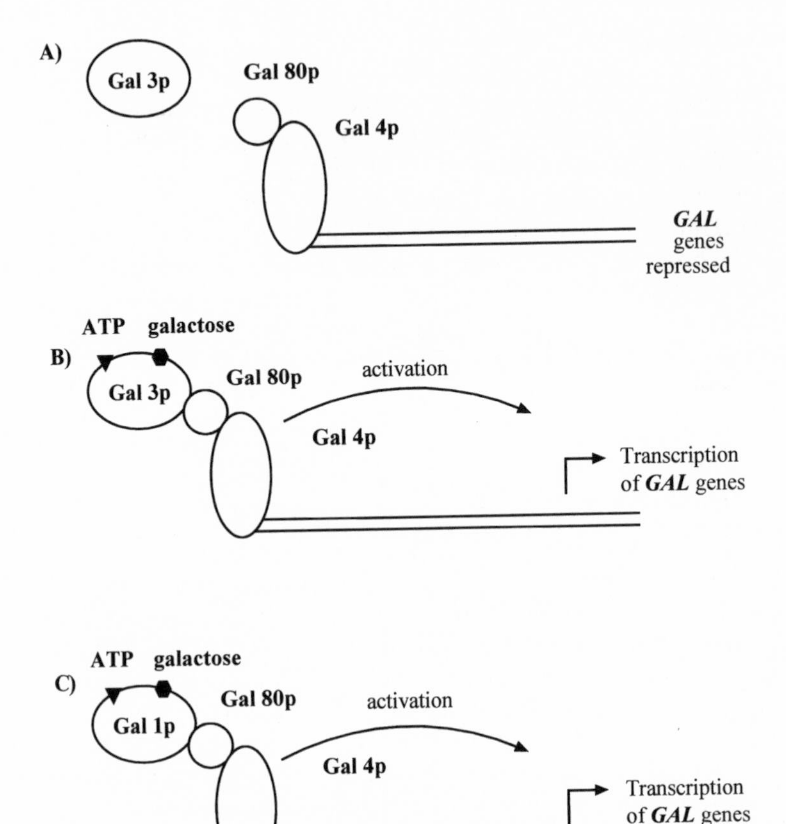

Figure 7.19. Function of Gal3p. (**A**) In the absence of ATP and galactose, Gal3p does not associate with regulatory proteins, and the GAL genes are repressed. (**B**) ATP and galactose bind Gal3p, creating a conformational change that causes it to bind Gal80p, initiating transcription of the GAL genes. (**C**) In *Kluyveromyces lactis*, Gal1p plays the role of Gal3p.

Figure 7.20. (facing page) Phylogenetic tree of galactokinases, constructed by the neighbor-joining method based on the proportion of amino acid difference (p). Numbers on the branches are as in Figure 3.2. Species included are the following: bacteria, *Actinobacillus pleuropneumoniae, Haemophilus influenzae, Escherichia coli, Salmonella typhimurium, Mycobacterium tuberculosis, Streptomyces lividans, Bacillus subtilis, Lactobacillus helveticus*; fungi, *Saccharomyces cerevisiae, Saccharomyces carlbergensis, Kluyveromyces lactis*.

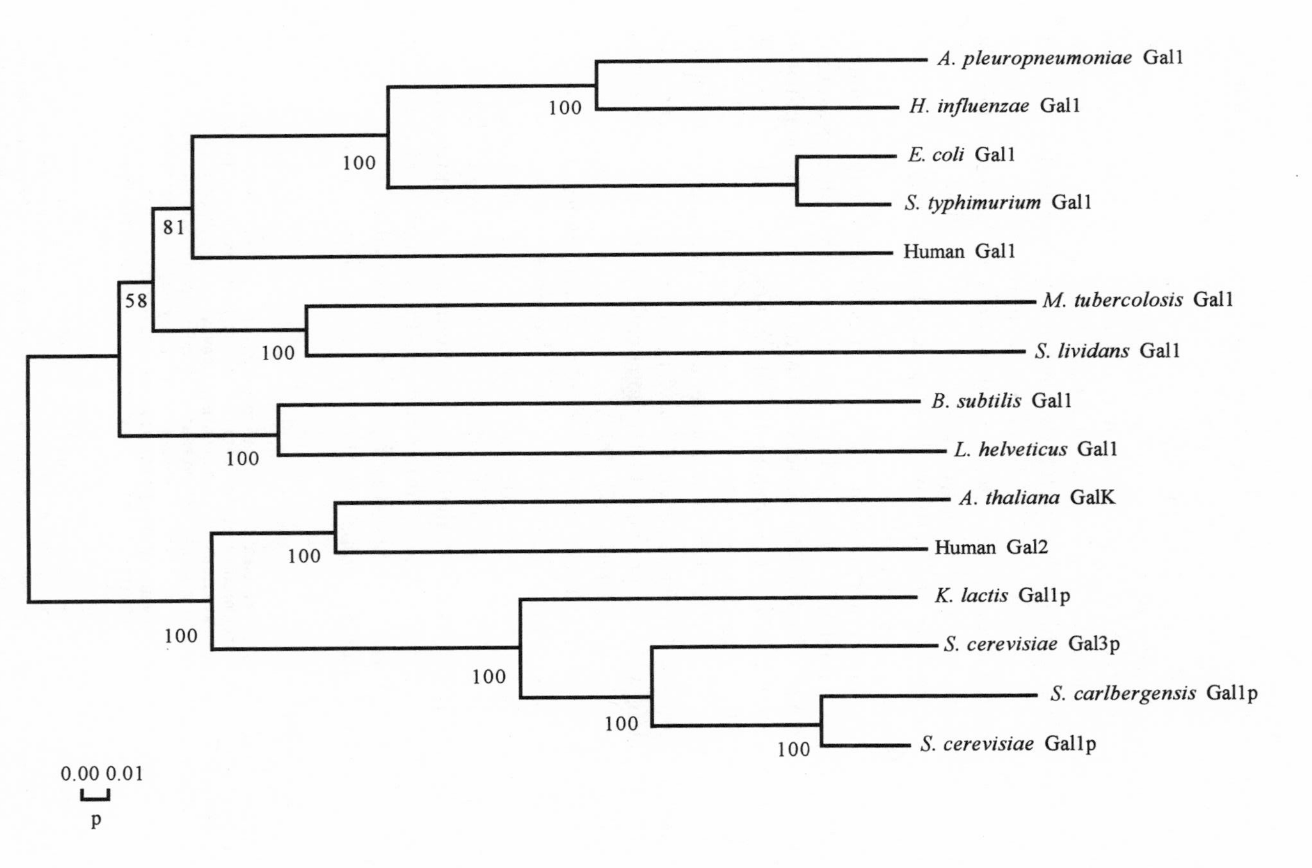

A. pleuropneumoniae Gal1
H. influenzae Gal1
E. coli Gal1
S. typhimurium Gal1
Human Gal1
M. tubercolosis Gal1
S. lividans Gal1
B. subtilis Gal1
L. helveticus Gal1
A. thaliana GalK
Human Gal2
K. lactis Gal1p
S. cerevisiae Gal3p
S. carlbergensis Gal1p
S. cerevisiae Gal1p
100
100
81
58
100
100
100
100
100
100
100
100
0.00 0.01
p

while the other is part of the IL-5 and GM-CSF receptors. The ancestral mouse gene, like the human gene and evidently the rat homologue, most likely had all three functions; but, after gene duplication in the mouse, these functions have been divided between two more specialized daughter genes.

The *GAL1* and *GAL3* genes of yeast (*Saccharomyces cerevisiae*) provide another example. The *GAL* genes, including *GAL1* and *GAL3*, encode proteins involved in galactose metabolism. They are inducible by galactose and repressible by glucose. Their regulation has been the subject of numerous studies (Johnston 1987). The protein product of *GAL1*, called Gal1p, is a galactokinase, which catalyzes the production of galactose-1-phosphate from ATP and galactose, the initial step in the metabolism of galactose. Gal3p, the product of the *GAL3* gene, showed homology to Gal1p, but lacks galactokinase activity (Bajwa et al. 1988; Bhat et al. 1990).

The major transcription factor for the *GAL* genes is Gal4p. In the uninduced state, Gal4p is inhibited by Gal80p (Figure 7.19B). Gal3p, in the presence of galactose and ATP, binds Gal80p, thereby removing the inhibition and initiating transcription of the *GAL* genes (Figure 7.19B) (Yano and Fukasawa 1997). In a related fungus species, *Kluyveromyces lactis*, there is no *GAL3* gene, and Gal1p functions both as a galactokinase and to remove the inhibition of Gal4p by Gal80p (Figure 7.19C) (Meyer et al. 1991; Zenke et al. 1996). In *S. cerevisiae* mutants lacking *GAL3* expression, Gal1p apparently is able to take over this regulatory role (Yano and Fukasawa 1997).

A phylogenetic tree of galactokinases of eukaryotes and prokaryotes revealed two major clusters (Figure 7.20). One includes bacterial galactokinases, along with human Gal1; the other includes human Gal2, a plant galactokinase, and the fungal Gal1p and Gal3p (Figure 7.20). Because both clusters include human genes, the phylogeny suggests that these two human genes represent a very ancient gene duplication that occurred prior to the divergence of eukaryotes and eubacteria. Among the fungal genes, the *K. lactis* Gal1p clusters outside Gal1p and Gal3p of *Saccharomyces* (Figure 7.20). Gal1p from *S. cerevisiae*

```
S. cerevisiae     Gal3 PGRVNLIGEHIDYCDFSVLPLAIDVDMLCAVKILDEK-NPSITLTNADPKFAQRKFDLPLDG
S. cerevisiae     Gal1 PGRVNLIGEHIDYCDFSVLPLAIDFDMLCAVKVLNEK-NPSITLINADPKFAQRKFDLPLDG
S. carlbergensis  Gal1 PGRVNLIGEHIDYCDFSVLPLAIDFDMLCAVKVLNEK-NPSITLINADPKFAQRKFDLPLDG
K. lactis         Gal1 PGRVNLIGEHIDYCQFSVLPMAIENDLMLACRLTSESENPSITLTNHDSNFAQRKFDLPLDG
A. thaliana       GalK PGRVNLIGEHIDYEGYSVLPMAIRQDTIIAIRKCEDQKQ--LRIANVNDKYTMCTY--PADP
Human             Gal1 PGRVNLIGEHTDYNQGLVLPMALELMTVLVGSPRKDGLVSLLTTSEGADEPQRLQFPL--PT
Human             Gal2 PGRVNIIGEHIDYCGYSVLPMAVEQDVLIAVEPVKTY-A--LQLANTNPLYP--DF--STSA
                       ***** **** **    *** *

S. cerevisiae     Gal3 SYMAIDPSVSEWSNYFKCGLHVAHSYLKKIAPERFN-NTPLVGAQIFCQSDIPTGGGLSS--
S. cerevisiae     Gal1 SYVTIDPSVSDWSNYFKCGLHVAHSFLKKLAPERFA-SAPLAGLQVFCEGDVPTGSGLSSSA
S. carlbergensis  Gal1 SYVTIDPSVSDWSNYFKCGLHVAHSFLKKLAPERFA-SAPLAGLQVFCEGDIPTGSGLSSSA
K. lactis         Gal1 SLIEIDPSVSDWSNYFKCGLLVAQQFLQ----EKYNFKGPVHGMEIYVKGDIPSGGGLSSSA
A. thaliana       GalK DQ-EIDLKNHKWGHYFICAYKGFHEYAKSKG---VNLGSPV-GLDVLVDGIVPTGSGLSSSA
Human             Gal1 AQRSLEPGTPRWANYVKGVIQYYP---AAPLP----------GFSAVVVSSVPLGGGLSSSA
Human             Gal2 NNIQIDKTKPLWHNYFLCGLKGIQEH--------FGLSNLT-GMNCLVDGNIPPSSGLSSSA
                                  *  *                           *             ****++
```

Figure 7.21. Conserved region of eukaryotic galactokinases: * residues conserved in all sequences; + conserved in all sequences except yeast Gal3p.

clusters with that from the closely related *S. carlbergensis*. This indicates that the duplication of *GAL1* and *GAL3* genes took place after the divergences of *Kluyveromyces* from *Saccharomyces* but before the divergence of *S. cerevisiae* and *S. carlbergensis*. The fact that the bifunctional *K. lactis* Gal1p forms an outgroup to *Saccharomyces* Gal1p and Gal3p supports the hypothesis that the ancestor of *Saccharomyces GAL1* and *GAL3* also encoded a bifunctional protein and that functional specialization occurred after gene duplication.

The loss of galactokinase activity by Gal3p must have resulted from certain changes in its amino acid sequence. The three-dimensional structure of galactokinases is not known; but a comparison of the 14 available sequences of functional eubacterial and eukaryotic galactokinases reveals 29 aligned residues that are 100% conserved in all 14 sequences. Surprisingly, 28 of these are also conserved in Gal3p. The residue that is not conserved forms part of a two-residue deletion unique to Gal3p (Figure 7.21). This deletion occurs in a region of high sequence conservation in the functional galactokinases (Figure 7.20). The strong conservation of this region suggests that it may be functionally important and that the deletion in Gal3p may be responsible for its loss of enzyme activity.

It is uncertain whether this deletion was selectively favored or whether it was simply a neutral mutation fixed by chance. Even if the latter is true, this would not be an example of the Dykhuizen-Hartl effect mentioned in the previous section. That effect requires an environmental change, after which a previously neutral mutation becomes advantageous. In the present case, no environmental change is likely to have been involved. A mutation removing the galactokinase activity of Gal3p would not ever need to become advantageous, because Gal1p continues to perform this function. However, it is possible that having a regulatory protein distinct from the functional enzyme did confer some selective advantage in *S. cerevisiae*.

8

Adaptive Characteristics of Genomes

In previous chapters, I have considered studies of adaptive evolution at the DNA level that have focused on a single locus or a number of related loci within a multigene family. Other studies of adaptive evolution have considered characteristics of an organism's genome as a whole or of some portion of the genome, such as a chromosome or chromosomal segment. Bennett (1971) introduced the term *nucleotype* to refer to aspects of an organism's genome that affect its phenotype independently of the information content of the nucleotide sequence itself. It has been proposed that traits of genomes such as genome size or nucleotide content can influence such phenotypic characters as developmental rate or metabolic rate, as well as on the overall stability of the genome itself. Similarly, the linkage relationships among genes whose products interact functionally may have some effect on cellular efficiency. Thus, natural selection can presumably act on nucleotypic characters through their overall phenotypic effects.

The purpose of the present chapter is to review evidence regarding the adaptiveness of nucleotypic or genomic characteristics. I consider the following topics: (1) nucleotide content bias and the nonrandom use of synonymous codons; (2) genome size; and (3) the clustering of genes on chromosomes. Here, the null hypothesis is typically that observed characteristics of genomes are the result of chance processes that have taken place over evolutionary history. For example, the nucleotide content of a genome, rather than being adaptive, might represent the cumulative effect of a bias in DNA repair enzymes (Filipski 1987). Acting over a long period of time, such a bias might be expected to introduce a compositional bias into the genome, especially at sites where no functional constraint is present. Deciding whether observed patterns are adaptive or reflect some other process can sometimes be quite difficult in the case of genome-level phenomena.

Nonrandom Codon Usage in *E. Coli*

If synonymous mutations are selectively neutral, one would expect f_0 (in terms of equation 3.3) to equal 1 at synonymous sites. If so, and if the mutation itself does not show a bias toward particular nucleotides, all of the codons encoding a given amino acid should be used about equally. However, in the vast majority of organisms synonymous codon usage is not equal. Rather, there are pronounced biases toward one or a few of the possible codons encoding each amino acid. At a time when sequence data were available for only a small number of species, Grantham et al. (1980) proposed that each species has a characteristic pattern of codon usage. Subsequent data have supported this view. Yet patterns of codon usage can differ dramatically between species. One major determinant of codon usage patterns is the proportion of G or C at third codon position, where most of the synonymous mutations involve a change from G or C to A or T (Table 2.1), and this quantity can provide a simple index of the type of codon bias seen in a species. For example, in available genes of the malaria parasite *Plasmodium falciparum*, G + C content at third positions is 16.4%, whereas in *Entamoeba histolytica*, another protist parasite, it is 85.3%.

In addition to differences between species, in well-studied species there are known to be differences among genes with respect to the pattern of synonymous codon usage. In the well-studied bacterium *Escherichia coli* and some other unicellular organisms, examination of such differences between genes has given support to a hypothesis regarding the adaptive significance of codon usage patterns.

Soon after sequences of *E. coli* genes became available, it was observed that there is a relationship between codon usage and tRNA abundance (e.g., Ikemura 1981). In other words, preferred codons seemed to be those corresponding to the anticodons of the most abundant tRNAs. Furthermore, it was observed that this pattern is most clear cut in the case of genes encoding abundant proteins; that is, genes expressed at high levels. This relationship is illustrated for the case of *E. coli* in Table 8.1. In this table, codon bias is expressed in terms of a quantity called relative synonymous codon usage (RSCU) (Sharp et al. 1986).

The RSCU value for a given codon is the observed frequency of that codon divided by the frequency expected under the assumption of equal usage of all the synonymous codons for that amino acid. For the jth codon encoding the ith amino acid

$$\mathrm{RSCU}_{ij} = \frac{Xij}{1/ni} \sum_{j=1}^{ni} Xij \tag{8.1}$$

where Xij is the number of occurrences of the jth codon of the ith amino acid, and ni is the number (from 2 to 6) of alternative codons for the ith amino acid. (Because, in the universal genetic code, methionine and tryptophan are each encoded by only one codon, it is not meaningful to compute RSCU values for these amino acids.)

In Table 8.1, RSCU values are computed for the 18 amino acids encoded

Table 8.1. Relative synonymous codon (RSCU) values for 165 *Escherichia coli* genes

		Gene group				
Amino acid	Codon	VH	H	M	L	R
Phe	TTT	0.46	0.60	0.72	1.11	1.30
	TTC	1.54	1.40	1.28	0.89	0.70
Leu	TTA	0.11	0.17	0.39	0.74	0.88
	TTG	0.11	0.36	0.55	0.79	0.81
	CTT	0.22	0.33	0.49	0.54	0.77
	CTC	0.20	0.45	0.57	0.64	0.49
	CTA	0.04	0.07	0.11	0.18	0.11
	CTG	5.33	4.62	3.89	3.12	2.94
Ile	ATT	0.47	0.96	1.14	1.64	1.56
	ATC	2.53	2.03	1.78	1.24	1.13
	ATA	0.01	0.01	0.08	0.12	0.31
Val	GTT	2.24	1.51	1.23	0.98	1.08
	GTC	0.15	0.53	0.69	0.89	1.18
	GTA	1.11	0.88	0.65	0.60	0.39
	GTG	0.50	1.09	1.43	1.53	1.35
Tyr	TAT	0.39	0.67	0.91	1.18	1.13
	TAC	1.61	1.33	1.09	0.82	0.87
His	CAT	0.45	0.57	0.76	1.14	1.12
	CAC	1.55	1.43	1.24	0.86	0.88
Gln	CAA	0.22	0.35	0.54	0.66	0.80
	CAG	1.78	1.65	1.46	1.34	1.20
Asn	AAT	0.10	0.35	0.54	0.91	1.13
	AAC	1.90	1.65	1.46	1.09	0.87
Lys	AAA	1.60	1.45	1.53	1.51	1.45
	AAG	0.40	0.55	0.47	0.49	0.55
Asp	GAT	0.61	0.94	1.09	1.28	1.26
	GAC	1.39	1.06	0.91	0.72	0.74
Glu	GAA	1.59	1.45	1.44	1.37	1.30
	GAG	0.41	0.55	0.56	0.63	0.70
Ser	TCT	2.57	1.75	1.32	0.83	0.80
	TCC	1.91	1.75	1.35	0.83	0.85
	TCA	0.20	0.26	0.55	0.59	0.89
	TCG	0.04	0.48	0.84	0.95	0.93
	AGT	0.22	0.24	0.43	0.87	0.76
	AGC	1.05	1.52	1.51	1.93	1.78
Pro	CCT	0.23	0.47	0.45	0.55	0.54
	CCC	0.04	0.07	0.15	0.52	0.81
	CCA	0.44	0.50	0.72	0.75	0.71
	CCG	3.29	2.97	2.68	2.19	1.95
Thr	ACT	1.80	0.97	0.77	0.62	0.48
	ACC	1.87	2.37	2.06	1.78	1.93
	ACA	0.14	0.13	0.33	0.48	0.41
	ACG	0.18	0.53	0.84	1.13	1.17
Ala	GCT	1.88	0.93	0.79	0.53	0.50
	GCC	0.23	0.68	0.88	1.24	1.15
	GCA	1.10	0.92	0.86	0.74	0.70
	GCG	0.80	1.47	1.48	1.49	1.65

—continued

Table 8.1. *Continued*

		Gene group				
Amino acid	Codon	VH	H	M	L	R
Cys	TGT	0.67	0.76	0.87	0.79	1.14
	TGC	1.33	1.24	1.13	1.21	0.86
Arg	CGT	4.39	3.86	3.33	2.17	1.90
	CGC	1.56	2.00	2.16	2.76	2.70
	CGA	0.02	0.06	0.18	0.29	0.54
	CGG	0.02	0.03	0.21	0.57	0.65
	AGA	0.02	0.02	0.09	0.13	0.11
	AGG	0.00	0.03	0.04	0.09	0.11
Gly	GGT	2.28	2.23	1.80	1.34	1.31
	GGC	1.65	1.50	1.67	1.74	1.55
	GGA	0.02	0.08	0.20	0.33	0.39
	GGG	0.04	0.19	0.33	0.59	0.75

Gene groups (number of genes) are as follows: VH = very highly expressed (27); H = highly expressed (15); M = moderate codon bias (57); L = low codon bias (58); R = regulatory or repressor (8). Data from Sharp and Li (1986).

by two or more synonymous codons in 165 *E. coli* genes categorized by expression. Codon usage bias is highest, as indicated by the highest RSCU values, for the highly expressed genes and lowest for the regulatory or repressor genes, which are expressed at very low levels. These results are consistent with the hypothesis that codon bias corresponds to tRNA abundance as an adaptation for efficiency in translation. Presumably, the selection of efficiency of translation is much stronger in the case of highly expressed genes, of which each cell must translate multiple copies. More efficient translation will mean more rapid reproduction, and thus greater fitness (Bulmer 1991). In the case of a gene encoding a repressor protein—of which only a single molecule may be needed per cell—there is presumably no strong pressure for efficient translation.

The most obvious alternative to an adaptive explanation for nonrandom usage of synonymous codons is that the pattern of codon usage results from the mutational bias. A mutational bias—for example, a bias toward G and C rather than A and T—would lead to a nucleotide composition bias. Because mutations in the first and second positions of codons generally change the amino acid whereas the majority of mutations at the third position are synonymous, a composition bias might be expected to have more of an effect at the third position than at other codon positions.

Data from 42 species of eukaryotes and 33 species of prokaryotes were used to compare nucleotide content at different codon positions (Figure 8.1). The proportion of G and C (pGC) at both the first and third positions was strongly correlated with pGC at the third position (Figure 8.1). Thus, each species seems to have its own overall nucleotide composition bias that is reflected at all codon positions. Further analyses suggested that, as expected, this nucleotide content bias had a different effect at the different codon positions. A general linear

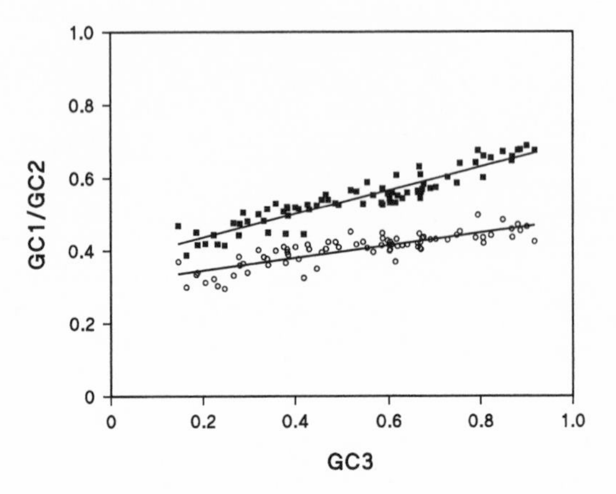

Figure 8.1. GC content (pGC) at first (*solid symbols*) and second (*open circles*) codon positions plotted against that at the third codon position for 36 species of prokaryotes and 42 species of eukaryotes. The lines are linear regression lines: for the first position, $Y = 0.371 + 0.325X$ ($r = .932$; $p < .0001$); for the second position, $Y = 0.310 + 0.174X$ ($r = .831$; $p < .0001$). Test of parallelism of the lines: $p < .001$.

model analysis of covariance was used to test for parallelism between the two linear relationships (1) between pGC at the first position and pGC at the third position, and (2) between pGC at the second position and pGC at the third position. The hypothesis of parallelism was rejected (Figure 8.1). Indeed, the linear regression line for the former relationship has a slope (0.325) that is nearly twice the slope (0.174) of the latter relationship. Thus, with increasing pGC at the third position, pGC at the first position increases faster than that at the second position. This phenomenon may at least in part reflect the greater constraint at second positions than at first positions, arising from the fact that all mutations at the second position are nonsynonymous. Furthermore, it may reflect the fact that certain amino acids with A or T in the second position are necessary for protein function, and thus, are retained even as the overall pGC in the genome becomes very high.

For both eukaryotes and prokaryotes, pGC at the first position was not significantly different from that at the third position, while at the second position it was significantly lower than at the third position (Figure 8.2A). This, again, apparently reflects selective constraints at the second position that prevent pGC there from reaching the level seen at the other codon positions.

Genomes may deviate from equal usage of the four nucleotides either in the direction of high pGC or in the direction of low pGC (Figure 8.1). As a measure of how biased GC content is away from 50% GC, I used the following quantity:

$$bGC = |pGC - 0.50| \tag{8.2}$$

For both prokaryotes and eukaryotes, mean bGC was significantly lower in positions 1 and 2 than in position 3 (Figure 8.2B). These results undoubtedly reflect the greater functional constraint at positions 1 and 2, which are less free than position 3 to respond to a mutational bias affecting pGC. Thus, it seems likely that genome-wide GC content biases will affect codon usage.

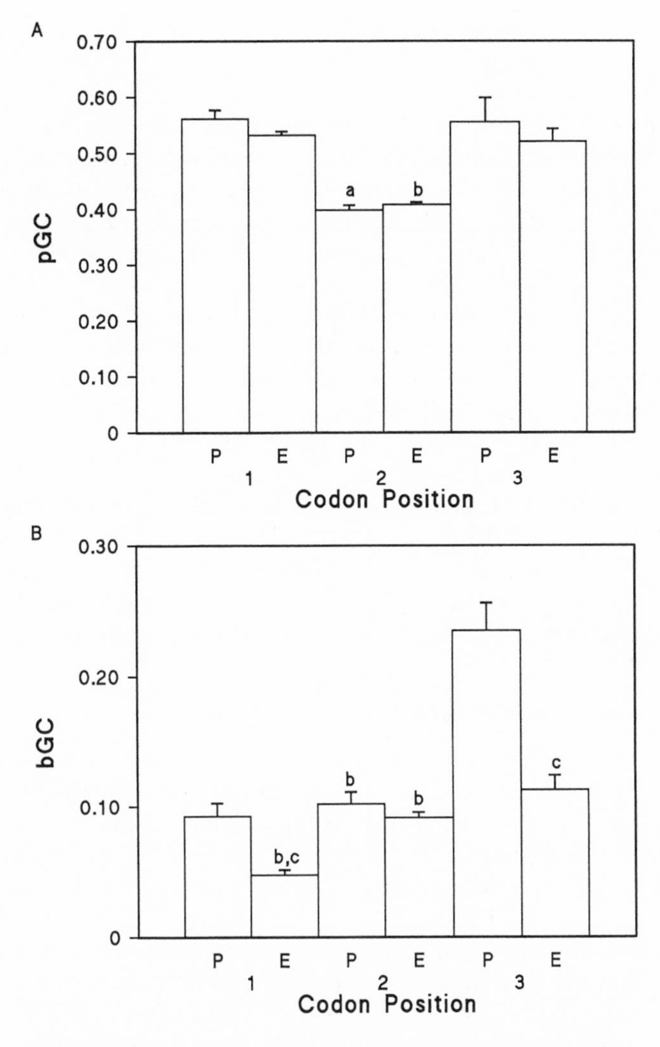

Figure 8.2. (**A**) Mean pGC (with SE) at first, second, and third codon positions in prokaryotes (P) and eukaryotes (E). (**B**) Mean bGC (with SE) at first, second, and third codon positions in prokaryotes (P) and eukaryotes (E). Based on 36 species of prokaryotes and 42 species of eukaryotes. Results of statistical tests: (a) significantly different from position 3 ($p < .0005$); (b) significantly different from position 3 ($p < .0001$); (c) significantly different from prokaryotes ($p < .001$).

If a mutational bias can produce codon bias, it might be argued that tRNA abundance has evolved to match the mutationally driven codon bias, rather than the other way around. However, this alternative explanation is probably not satisfactory in the case of *E. coli*. A mutational bias would probably take one of two forms: either a bias toward A and T or a bias toward G and C. Thus, if a mutational bias was the driving force, we would expect to see that the "preferred" codons were all codons ending in A or T (in the case of a mutational bias toward A and T) or codons ending with C or G (in the case of a mutational bias toward C and G). We see neither trend in *E. coli* (Table 8.1). Furthermore, it is hard to imagine how a mutational bias would correlate with level of expression.

Interestingly, at both the first and third codon positions mean bGC was significantly greater in the case of prokaryotes than in the case of eukaryotes (Figure 8.2B). I discuss this and some other differences between prokaryotes and eukaryotes with respect to codon bias below (this chapter, Prokaryotic versus Eukaryotic Codon Usage).

Patterns of codon preference correlated with tRNA abundance have been reported in other unicellular organisms, such as the yeast *Saccharomyces cerevisiae* (Ikemura 1985). Now that the complete yeast genome as well as complete genomes of several bacterial species have been sequenced, it is to be expected that codon bias in these species can be examined more thoroughly.

Codon Usage in Multicellular Organisms

The role of natural selection on codon usage in multicellular organisms is expected to be different from that in unicellular organisms. First of all, the distinction between genes of high and low expression is complicated in multicellular organisms by the fact that different genes are expressed in different tissues. For example, in vertebrates, insulin is expressed at a high level by islet cells of the pancreas but not at all by other cells. Furthermore, the relationship between translational efficiency and fitness is much less straightforward in multicellular organisms than it is in a single-celled, asexually reproducing bacterium. In the latter case, efficient translation will presumably directly impact reproductive success; a bacterial clone that translates its abundant proteins more efficiently than another clone will also presumably have a faster growth rate.

Despite these obvious differences, multicellular organisms also use synonymous codons in a nonrandom fashion. Numerous explanations, both adaptive and nonadaptive, have been proposed to account for the observed patterns, but as yet no consensus has emerged.

Isochores in Vertebrates

Regarding vertebrates, one important discovery is that of so-called isochores (Cuny et al. 1981; Bernardi et al. 1985). Bernardi (1993, pp. 186–187) defines this term as follows: "Vertebrate genomes are mosaics of *isochores*—namely, of long, compositionally homogeneous DNA segments that can be subdivided into a small number of families characterized by different GC levels." The genomes of birds and mammals, in contrast to those of other vertebrates, are characterized by a high frequency of GC-rich isochores, in which most genes are located. The pGC at the third position of genes reflects the overall pGC of the isochore (D'Onofrio and Bernardi 1992).

Bernardi and his co-workers have proposed that the existence of GC-rich isochores in birds and mammals is an adaptation to endothermy (which they call "warm-bloodedness"); for a review, see Bernardi (1993). G-C base pairs (involving three hydrogen bonds) are more heat stable than A-T base pairs (involving two double bonds). Therefore, these authors view the presence of GC-rich isochores in birds and mammals as an adaptation to the high body temperatures of "warm-blooded" vertebrates.

There are many problems with this hypothesis, including the following:

1. The idea that the cells of "warm-blooded" vertebrates are inevitably exposed to higher temperatures than the "cold-blooded" vertebrates is physiologically naive. Birds and mammals are correctly described as endothermic, meaning that they maintain a relatively constant body temperature as a result of metabolically generated heat. Other vertebrates are generally ectothermic, meaning that they are dependent on heat from external sources to maintain their body temperatures. But it need not be true that average body temperatures of ectotherms are any lower than those of endotherms. An ectotherm living in a stable warm environment will have as high and constant a body temperature as that of an endotherm. Furthermore, among the "cold-blooded" vertebrates are species that are effectively endothermic in certain tissues or at certain stages of their life cycles (Schmidt-Nielsen 1979). For example, tuna and other large fast fish have effectively endothermic swimming muscles; and an incubating snake of the boa family uses muscular contraction to maintain a body temperature well above that of its environment.
2. As mentioned (Chapter 1), the hypothesis that the GC content of bird and mammal genomes is an adaptation to endothermy is impossible to test directly. We do not have the technology to genetically engineer an AT-rich bird or mammal. Thus, for testing this hypothesis, we are forced to rely on comparative data. For example, we might test whether a similar GC-richness characterized organism other than birds or mammals that live at high temperatures. Examples include endothermic insects, of which there are numerous examples; endoparasites of birds and mammals; or microorganisms living in extreme environments such as hot springs. If some of these species have, convergently with birds and mammals, evolved a genome characterized by GC-rich isochores, this would lend some support to Bernardi's hypothesis. However, it would certainly not prove the hypothesis, because different factors may be at work in widely divergent species of organisms, including different phylogenetic histories. By contrast, if other organisms living at high tempera-

tures do not have GC-rich genomes, that does not really disprove Bernardi's hypothesis. These species, because of some aspect of their evolutionary history, may not have ever evolved the same adaptation as is seen in birds and mammals.

Because of these difficulties, the most prudent course is to test Bernardi's hypothesis against a suitable null model. In this case, the null hypothesis is the hypothesis that the isochore structure of bird and mammal genomes is not an adaptive trait. If so, the most likely explanation for this pattern is some type of mutational bias. Regional differences in GC content might then result from differences in the biases of DNA-repair enzymes used in different regions (Filipski 1987). Alternatively, regional differences in GC content might be explained by differences in nucleotide availability during replication of germline DNA (Wolfe et al. 1989). GC-rich isochores are located in regions of the genome that replicate early in the germline cell cycle, at which point the GC content of the pool of available nucleotides is high. AT-rich isochores, on the other hand, replicate later, when the GC content is lower. This difference in nucleotide availability during replication could, over time, yield regional GC content differences.

So far it has not been possible to decide between the adaptive and mutational explanations for the GC-content pattern seen in birds and mammals. One difficulty with an adaptive explanation is the fact that bird and mammal genomes include numerous genes with low GC content (Li 1997). Regarding Bernardi's hypothesis that high GC content is an adaptation to endothermy, comparative data do not provide consistent evidence (Table 8.2). Thermophilic bacteria include species with extraordinarily high GC content, like *Thermus aquaticus*, but also species with intermediate and low GC content (Table 8.2). Other bacteria cover a similarly broad range. Likewise, protists parasitic on endothermic vertebrates include species with both low and high GC content (Table 8.2). Finally, on the basis of genes available in the database, third position GC content of endothermic vertebrates is lower than that of some bony fish, which are mainly ectothermic (Table 8.2).

Thus, Bernardi's hypothesis that the isochore structure of bird and mammal genomes is an adaptation to endothermy does not have any strong support. It is possible that this pattern of genome arrangement is adaptive, but for some other reason than that advocated by Bernardi. If it is not adaptive, we still have no clear evidence regarding what mutational mechanism could be responsible for maintaining it.

Drosophila

Very few comparative studies of codon usage have been made in any animal phylum outside the vertebrates. Not surprisingly, the one invertebrate species that has been well studied is the fruitfly, *Drosophila melanogaster*. Shields et al. (1988) argued that codon usage in *Drosophila* is not selectively neutral because of a difference they observed between *Drosophila* and human genes. In human genes, GC content at third positions in coding regions is correlated with that

Table 8.2. GC content (%GC) in coding regions of selected species

	Codon position		
	1	2	3
Prokaryotes			
Archaebacteria			
Halobacterium halobium	66.1	42.1	80.7
Methanobacterium thermoautotrophicum[a]	55.3	39.6	56.7
Methanococcus jannaschii[a]	41.5	29.5	24.8
Eubacteria			
Bordatella pertussis	67.2	48.5	84.8
Escherichia coli	58.8	40.7	55.4
Mycobacterium tuberculosis	67.6	49.7	79.4
Mycoplasma genitalium	41.8	30.3	23.2
Thermus aquaticus[a]	67.5	42.5	91.6
Eukaryotes			
Protists			
Plasmodium falciparum	38.7	29.9	16.4
Toxoplasma gondi	57.2	44.8	58.6
Fungi			
Candida albicans	44.3	38.3	27.9
Saccharomyces cerevisiae	44.6	36.6	38.0
Plants			
Monocots			
Rice *Oryza sativa*	57.0	43.4	67.3
Maize *Zea mays*	58.3	43.7	67.6
Dicots			
Arabidopsis thaliana	51.3	40.4	43.0
Pea *Pisum sativum*	52.0	40.3	38.3
Potato *Solanum tuberosum*	51.6	39.6	38.4
Animals			
Caenothabditis elegans	49.7	38.8	38.5
Drosophila melanogaster	56.5	42.0	66.3
Silkmoth *Bombyx mori*	52.9	42.4	54.5
Zebrafish *Danio rerio*	52.6	43.3	58.7
Salmon *Oncorhyncus mykiss*	54.5	40.6	66.8
Clawed frog *Xenopus laevis*	52.6	40.9	49.2
Chicken *Gallus gallus*	55.5	41.5	60.1
Mouse *Mus musculus*	55.7	42.8	59.8
Human *Homo sapiens*	56.0	42.4	60.3

Data complied from the CUTG database (Nakamura et al. 1998).

[a]Indicates thermophilic bacteria.

in introns of the same genes; presumably, this relationship reflects isochore structure. Shields et al. (1988) did not find such a correlation in *Drosophila*, leading them to conclude that some other factor besides regional GC content—presumably natural selection. However, a subsequent study using a larger number of sequences revealed a significant correlation between GC content at third positions and in introns (Kliman and Hey 1994).

On the other hand, "preferred" codons for each amino acid were found to be used more frequently at sites encoding amino acids conserved among three *Drosophila* species (Akashi 1994). Because these conserved amino acids are likely to be functionally important, use of "preferred" codons may be an adaptation that enhances translational accuracy at functionally critical residue positions (Akashi 1994). Moriyama and Powell (1997) studied the relationship between codon bias and tRNA abundance in *Drosophila*. This is a complicated issue, because *Drosophila* tRNA pools change over the course of development and differ from one tissue to another. Nonetheless, for 10 amino acids for which the most abundant tRNAs in the adult are known, codon bias in *D. melanogaster* matched tRNA abundance (Moriyama ad Powell 1997). Still, because all of the "preferred" codons ended in G and C, it remains a possibility that mutational pressure determines codon bias and that tRNA abundance has adapted to codon bias rather than the other way around.

Akashi's (1994) hypothesis of selection for translational accuracy at functionally important codons may conceivably be applicable to many multicellular eukaryotes. Although, for the reasons mentioned previously, it seems difficult to imagine that multicellular organisms are consistently subject to selection for enhanced translational efficiency as may occur in microorganisms, selection for translational accuracy may be a general phenomenon. Certainly this hypothesis deserves to be tested in other groups of eukaryotes.

Prokaryotic Versus Eukaryotic Codon Usage: Some General Trends

To compare the overall patterns of codon usage in prokaryotes and eukaryotes, I examined data on codon usage in 36 species of prokaryotes and 42 species of eukaryotes from the CUTG database (Nakamura et al. 1998). As mentioned previously, one difference between these two major groups is that at both first and third positions the degree of content bias (measured as absolute deviation from a pGC of 0.5) was greater in prokaryotes at both first and third codon positions (Figure 8.2B).

One simple index of the degree of codon bias for a particular amino acid is simply the RSCU of the codon for the amino acid that has the highest RSCU (RSCUmax). RSCUmax values were compared between prokaryotes and eukaryotes for amino acids with twofold degenerate codon sets (Figure 8.3), fourfold degenerate codon sets (Figure 8.4), sixfold degenerate codon sets (Figure 8.5). In each case, a general linear model approach was used to test for three effects: an amino acid effect (i.e., a difference among amino acids); a "superkingdom" effect (i.e., a difference between the prokaryote and eukaryote "superking-

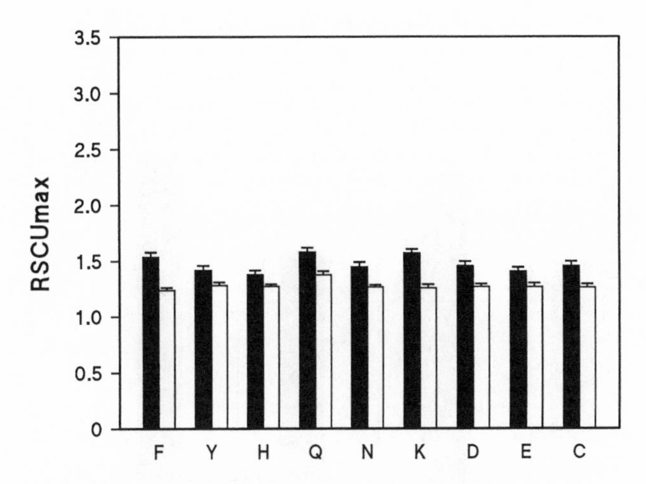

Figure 8.3. Mean RSCUmax (with SE) for amino acids with twofold degenerate codon sets for 36 species of prokaryotes (*solid bars*) and 42 species of eukaryotes (*open bars*). Overall means were as follows: prokaryotes 1.475 ± 0.037; eukaryotes 1.275 ± 0.029. A two-way analysis of variance was used to test for effects of superkingdom ($p < .001$), of amino acid ($p < .001$), and their interaction ($p < .02$).

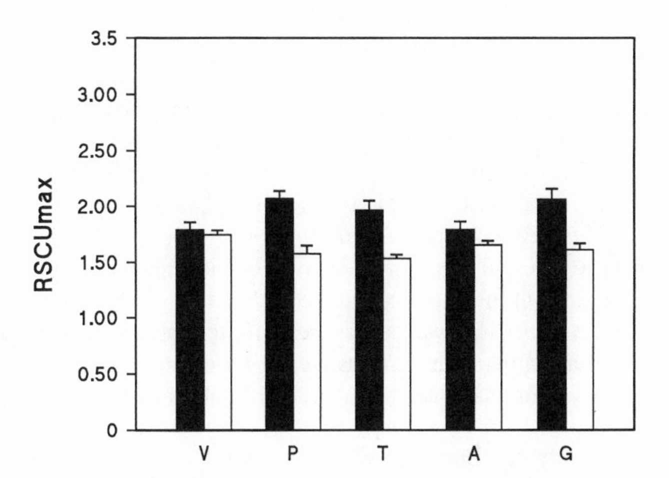

Figure 8.4. Mean RSCUmax (with SE) for amino acids with fourfold degenerate codon sets for 36 species of prokaryotes (*solid bars*) and 42 species of eukaryotes (*open bars*). Overall means were as follows: prokaryotes 1.941 ± 0.076; eukaryotes 1.621 ± 0.052. A two-way analysis of variance was used to test for effects of superkingdom ($p < .001$), of amino acid (n.s.), and their interaction ($p < .001$).

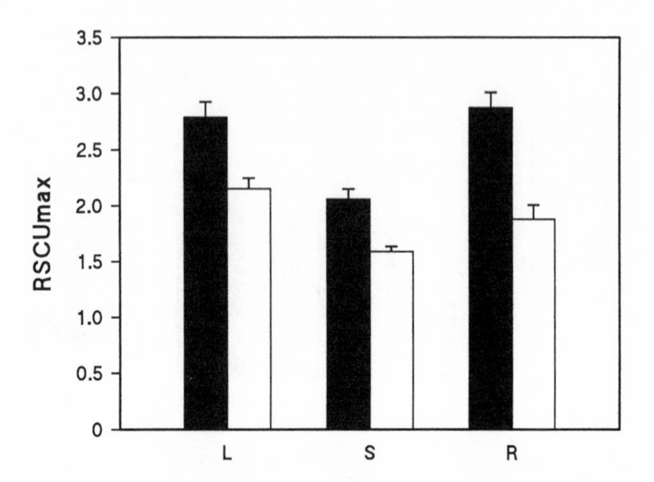

Figure 8.5. Mean RSCUmax (with SE) for amino acids with sixfold degenerate codon sets for 36 species of prokaryotes (*solid bars*) and 42 species of eukaryotes (*open bars*). Overall means were as follows: prokaryotes 2.578 ± 0.133; eukaryotes 1.874 ± 0.101. A two-way analysis of variance was used to test for effects of superkingdom ($p < .001$), of amino acid ($p < .001$), and their interaction ($p < .05$).

doms"); and an interaction between the amino acid effect and the superkingdom effect. A significant interaction would mean that differences among amino acids with respect to RSCUmax occur in a different way in the two superkingdoms.

For all three groups of amino acids there was a highly significant superkingdom effect. Mean RSCUmax values were consistently higher for prokaryotes than for eukaryotes (Figures 8.3–8.5). The twofold (Figure 8.3) and sixfold (Figure 8.5) degenerate amino acids also showed significant amino acid effects, but the fourfold degenerate amino acids did not (Figure 8.4). In all cases, there was a significant superkingdom effect by amino acid interaction; the effect was strongest in the case of fourfold degenerate amino acids (Figure 8.4). Thus, the data show that codon usage was, on average, more biased in prokaryotes than in eukaryotes; that, in general, bias was greater for certain amino acids than for others; and that amino acids can show different patterns of bias in prokaryotes and in eukaryotes.

To examine why the biases are different for different amino acid, I divided both prokaryotic and eukaryotic species into two groups on the basis of GC content at the third position: (1) GC-rich species (19 prokaryotes and 25 eukaryotes) were species with third-position pGC > .5; and (2) GC-poor species (17 prokaryotes and 17 eukaryotes) were species with third-position pGC < .5. Mean third-position pGC in GC-rich prokaryotes (.775 ± .026 S.E.) was significantly higher than that in GC-rich eukaryotes (.625 ± .011) ($p < .001$). However, there was no significant difference between mean third-position pGC of GC-poor prokaryotes (.310 ± .022) and that of GC-poor eukaryotes (.367 ± .019).

The differences among amino acids are most easily seen in the case of sixfold degenerate amino acids. In prokaryotes, mean RSCUmax for each of the three sixfold degenerate amino acids was greater than 2.0; but in eukaryotes, mean RCSUmax for Ser and Arg were less than 2.0. In both prokaryotes and eukaryotes, the extent of bias for Ser was less than that for Leu or Arg (Figure 8.5). The amino acid showing the most striking difference between prokaryotes and eukaryotes was Arg. In prokaryotes, mean RSCUmax for Arg was higher than that for any other amino acid, but this was not true in eukaryotes. This difference in arginine codon usage can mainly be explained by differences with regard to usage of specific codons. A striking contrast was observed between prokaryotes and eukaryotes in their use of the codons CGC and AGG (Table 8.3). In both GC-rich and GC-poor species, RSCU for CGC in prokaryotes was about twice that in eukaryotes, while RSCU for AGG in eukaryotes was about twice that in prokaryotes (Table 8.3).

Somewhat analogous patterns were found in the case of fourfold degenerate codon sets (Figure 8.4). Overall mean RSCUmax was greater for prokaryotes than for eukaryotes, but there were no significant differences among amino acids (Figure 8.4). RSCUmax in prokaryotes was particularly high in the case of proline, threonine, and glycine (Figure 8.4). In the case of proline, mean RSCU for CCG was significantly higher for GC-rich prokaryotes than for GC-rich eukaryotes; but no such difference was seen in mean RSCU for CCC (Table 8.3). In the case of threonine, mean RSCU for ACG was significantly higher in prokaryotes than in eukaryotes in both GC-rich and GC-poor categories (Table 8.3). In the case of glycine, mean RSCU for GGC was significantly higher in prokaryotes than in eukaryotes in both GC-rich and GC-poor categories (Table 8.3).

In the case of twofold degenerate sites, there were strong differences among amino acids and between superkingdoms, as well as a significant superkingdom by amino acid interaction effect (Figure 8.3). This latter effect seems to be

Table 8.3. Mean RSCU (±SE) for selected codons in GC-rich and GC-poor species of prokaryotes and eukaryotes

Amino acid	Codon	GC-rich prokaryotes (19)[a]	Eukaryotes (25)	GC-poor prokaryotes (17)	Eukaryotes (17)
Arginine	CGC	2.98 ± 0.19	1.42 ± 0.07***	0.92 ± 0.15	0.43 ± 0.05**
	AGG	0.47 ± 0.18	1.18 ± 0.06***	0.57 ± 0.10	1.12 ± 0.12**
Proline	CCC	1.28 ± 0.13	1.22 ± 0.05	0.41 ± 0.06	0.53 ± 0.10
	CCG	2.00 ± 0.13	0.79 ± 0.08***	0.57 ± 0.12	0.38 ± 0.05
Glycine	GGC	2.37 ± 0.13	1.42 ± 0.05***	0.82 ± 0.10	0.52 ± 0.06*
	GGG	0.66 ± 0.09	0.73 ± 0.52	0.58 ± 0.07	0.47 ± 0.05
Threonine	ACC	2.23 ± 0.11	1.49 ± 0.05***	0.74 ± 0.10	0.84 ± 0.06
	ACG	1.18 ± 0.10	0.72 ± 0.52***	0.52 ± 0.07	0.36 ± 0.04

[a]Numbers in parentheses are numbers of species. *t*-Tests of the equality of means between prokaryotes and eukaryotes: *$p < .05$; **$p < .01$; ***$p < .001$.

mainly caused by the relatively high (>1.5) RSCUmax values in prokaryotes for phenylalanine, lysine, and glutamine. These amino acids were examined further by comparing the RSCU for codons ending in T and A in GC-poor species of prokaryotes and eukaryotes (Table 8.4). For all three of these amino acids, RSCU for the codons ending with T and A was significantly higher in GC-poor prokaryotes than in GC-poor eukaryotes (Table 8.4).

Overall, the differences between prokaryotes and eukaryotes with respect to RSCUmax can thus be explained by three main factors: (1) the stronger GC-content bias at the third position in prokaryotes (Figure 8.2); (2) a tendency on the part even of GC-rich eukaryotes to avoid codons involving the dinucleotides CG (in CGC, CCG, and ACG) and GC (in CGC and GGC) (Table 8.3); (3) a tendency on the part even of GC-poor eukaryotes to avoid strings of successive Ts and As (as in TTT, AAA, and CAA) (Table 8.4). All of these factors can easily be explained by mutational patterns without the need to invoke natural selection. The fact that the GC-content biases are greater in prokaryotes than in eukaryotes can be explained by the fact that prokaryotic taxa have diverged from each other more anciently than have eukaryotic taxa and, thus, have had time to accumulate differences as a result of mutational biases. Furthermore, because the genomes of prokaryotes are much simpler than those of eukaryotes, the entire prokaryotic genome may be exposed to a similar mutational bias, whereas in eukaryotes, different regions of the genome may be subject to different biases, as has been proposed in the case of birds and mammals (Filipski 1987; Wolfe et al. 1989).

The fact that the dinucleotide CG (often written CpG to emphasize the polarity of the phosophodiester linkage) is underrepresented in vertebrate genomes is well known. This is usually attributed to the fact that CpG dinucleotides are often methylated, which in turn increases their mutability (Bird 1986). In the species examined here, avoidance of CpG in coding regions was seen in all eukaryotes in comparison to prokaryotes, but this tendency was strongest in vertebrates. Table 8.5 shows mean RSCU values from GC-rich species for synonymous codons including CpG: TCG (serine), CCG (proline), ACG (threonine), and GCG (alanine). Mean RSCU values for each of these codons were highest for prokaryotes, next for eukaryotes other than vertebrates, and lowest for vertebrates (Table 8.5).

Table 8.4. Mean RSCU (±SE) for selected AT-rich codons in GC-poor prokaryotes and eukaryotes

Amino acid	Codon	Prokaryotes	Eukaryotes
Phenylalanine	TTT	1.48 ± 0.04	1.13 ± 0.05***
Lysine	AAA	1.46 ± 0.03	1.08 ± 0.06***
Glutamine	CAA	1.40 ± 0.04	1.26 ± 0.05*

Species analyzed are as in Table 8.3. *t*-Tests of the equality of values for prokaryotes and eukaryotes: *$p < .05$; ***$p < .001$.

Table 8.5. Mean RSCU (±SE) for CpG-containing codons in GC-rich species of prokaryotes, vertebrates, and other eukaryotes

			Eukaryotes	
Amino acid	Codon	Prokaryotes	Other	Vertebrates
Serine	TCG	1.66 ± 0.17***	0.98 ± 0.10*	0.39 ± 0.02
Proline	CCG	2.00 ± 0.13***	1.03 ± 0.11*	0.54 ± 0.03
Threonine	ACG	1.18 ± 0.10***	0.90 ± 0.07*	0.52 ± 0.03
Alanine	GCG	1.29 ± 0.09***	0.85 ± 0.07**	0.46 ± 0.02

Species analyzed are as in Table 8.3. *t*-Tests of the equality of a given value with that for vertebrates: $^{*}p < .05$; $^{**}p < .01$; $^{***}p < .001$.

Natural Selection on Codon Usage

Given that there is good evidence of adaptive codon usage at least in some microorganisms, the question has arisen how natural selection might affect this trait. The problem is that, when estimates have been made of the strength of the selection coefficient favoring the optimal codon at a single codon position, the selection coefficients turn out to be very small. For example, Li (1987) suggested values such as 10^{-5} or 10^{-6} might be reasonable. Because a very slightly advantageous allele essentially behaves like a neutral allele (chapter 3, Adaptive Evolution), it seems unclear how such very slightly advantageous mutations might be fixed by natural selection. Of course, the effective population sizes of microorganisms may be very large; thus, the condition that the selection coefficient exceed $1/2Ne$ may be met even for selection coefficients of 10^{-6} or lower. For example, the effective population size of *E. coli* has been estimated at about 2×10^{8}. The requirement for a very large effective population size in order to have effective selection on synonymous codon usage might explain why the evidence for adaptive codon usage is much stronger in the case of microorganisms than in the case of multicellular organisms. Perhaps the latter simply lack the requisite population sizes. On the other hand, as mentioned previously, there is another, perhaps stronger, reason why we should not expect to see adaptive codon usage in multicellular organisms; namely, there may not be a direct relationship between efficiency of translation and fitness in these organisms.

Kimura (1981) noted that selection on codon usage can be modeled by a type of model that I will refer to (following Nei 1987) as minute gene effect models. The models consider some abstract character related to the fitness of an individual, which results from the small additive phenotypic effects of a large number of loci. Under such models, the selection coefficient at any one of these loci contributing to the phenotypic character can be so small that the locus is effectively neutral even though there is either stabilizing selection or directional selection acting on the phenotypic character itself (Kimura and Crow 1978; Milkman 1978, 1982).

The applicability of these models to most phenotypic characters is questionable. They make the assumption, usual in quantitative genetics, that a large number of loci have small additive effects on the character in question; but in reality, this assumption is probably often violated in the case of quantitative characters such as body mass, milk yield, and the like. For example, in the case of certain quantitative characters, there may be nonadditive interactions between loci. Furthermore, a small number of major loci may explain most of the variation in quantitative traits. However, the assumptions of these models, though perhaps unrealistic in the case of the characters for which the models were originally designed, turn out to be reasonable in the case of a genomic character such as genome-wide GC content or codon usage. Because these genomic characters result directly from summation of the properties of individual nucleotides or codons, individual sites contribute to the character in a perfectly additive way. Furthermore, in the case of the traits traditionally considered by quantitative geneticists, some proportion of the variance is contributed by environmental effects. But there is no environmental component to these genomic characters.

Minute effect selection should be capable of leading to an optimal value of some genomic character such as GC content or codon usage without appreciable "cost of selection" (or "substitutional load"). However, in practice, it may be difficult to differentiate between such selection and mutational biases because both forces may often act in the same direction. Only when they act in opposite directions—as seems to be true with codon usage in highly expressed genes of *E. coli*—is there good reason to prefer an adaptive explanation.

Kimura (1983) has also made the point that, even in the case of microorganisms such as *E. coli*, adaptive evolution of codon usage patterns is primarily something that has occurred in the past. At the present time, selection on codon usage in such species primarily takes the form of purifying selection, eliminating mutant allelic forms of highly expressed genes that have low frequencies of preferred codons. Thus, in these species, a purifying selection is expected to be present even at synonymous sites in highly expressed genes. Comparisons of *E. coli* with the related *Salmonella typhimurium* provide evidence for purifying selection at synonymous sites. Highly expressed genes with strong codon bias show much lower levels of synonymous substitution between these two species than do other genes (Sharp 1991).

Genome Size

Different species of organisms differ greatly with respect to both genome size and number of genes (Table 8.6). Genome sizes may be estimated by several different methods. In eukaryotic organisms, the genome size is usually expressed in terms of total base pairs (bp) in the haploid genome or in terms of the mass (in pg) of the nuclear DNA content (usually in diploid somatic cells). In general, genome sizes and gene numbers are related. For example, bacteria have both much smaller genomes than mammals and many fewer genes (Table 8.6). When the logarithm of estimated gene number is plotted against the logarithm of genome size for the species in Table 8.6, there is a significant linear correlation (Figure 8.6). But the correlation is far from perfect, and inclusion of other species would probably make it even less so. For example, certain amoebas have

Table 8.6. Estimated gene numbers and genome sizes of selected organisms

		No. genes	Genome size (bp)
Prokaryotes	*Mycoplasma genitalium*	473	5.8×10^5
	Haemophilus influenzae	1,760	1.83×10^6
	Bacillus subtilis	3,700	4.2×10^6
	Escherichia coli	4,100	4.7×10^6
	Myxococcus xanthus	8,000	9.45×10^6
Fungi	*Saccharomyces cerevisiae*	5,800	1.35×10^7
Protoctista	*Cyandioschyzon merolae*	5,000	1.17×10^7
	Oxytrichia similis	12,000	6.0×10^8
Animalia	*Caenorhabditis elegans*	14,000	1.8×10^8
	Drosophila melanogaster	12,000	1.65×10^8
	Fugu rubripes	70,000	4.0×10^8
	Mus musculus	70,000	3.3×10^9
	Homo sapiens	70,000	3.3×10^9
Plantae	*Nicotiana tabacum*	43,000	4.5×10^9

Data from Miklos and Rubin (1996).

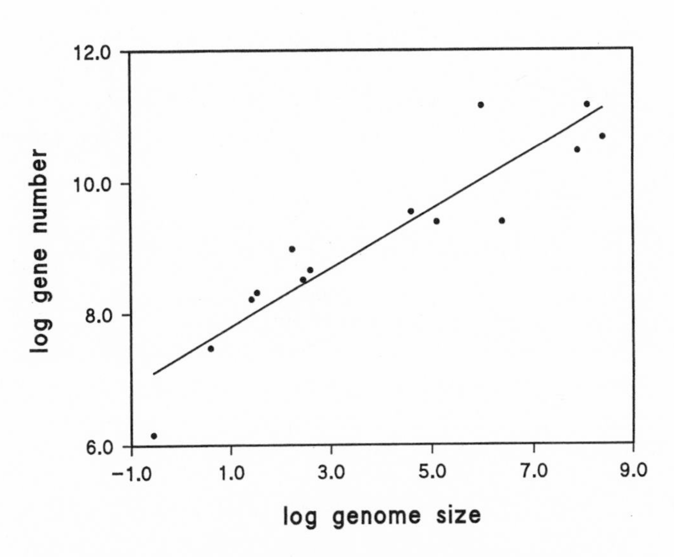

Figure 8.6. Natural logarithm of gene number plotted against that of genome size for 14 species of prokaryotes and eukaryotes. The line is the linear regression line ($r = .932$; $p < .001$). Data from Miklos and Rubin (1996).

genome sizes over 6×10^{14} bp, more than 100 times as large as the largest mammalian genomes (Table 8.8). Presumably, these protists have gene numbers comparable to those of other unicellular eukaryotes, which are much lower than those of mammals (Table 8.6). Even among vertebrates, the pufferfish *Fugu rubripes* is estimated to have approximately as many genes as a mammal, but its genome size is only about one-eighth that of a mammal (Table 8.6).

There also seems to be a rough correlation between genome size and the complexity of an organism; for example, multicellular organisms often have genome sizes greater than those of unicellular organisms (Table 8.6). Again, however, the correlation is far from perfect. The term *C-value paradox* was coined to refer to cases where seemingly phenotypically simple organisms have genome sizes larger than those of more complex organisms. The term *C-value* referred to the "constant" or "characteristic" value of genome size within a species. It was generally believed to show little variation in contrast to the great variation among species. Although there has been relatively little study of within-species variation in genome size, the available data suggest that C-values can be far from constant in some species (Table 8.7). Therefore, C-value is something of a misnomer, at least for certain species.

The C-value paradox is best illustrated by comparisons among multicellular eukaryotes (Table 8.8). For example, within vertebrates, considerably larger genome sizes are found in some fish and amphibians than in the apparently more phenotypically complex mammals (Table 8.8). Likewise, many plants, though morphologically simpler than animals, have larger genomes (Table 8.8).

The C-value paradox is no longer much of a paradox. We now know that much of the eukaryotic genome is made up of noncoding DNA, and presumably most of the DNA in paradoxically large genomes is noncoding. This is nicely illustrated by the difference between humans and pufferfish. If these two species have about the same number of genes, as is believed, but the genome size of

Table 8.7. Examples of within-species variation in genome size (pg per diploid cell)

Species	*n*	Range	Mean	Reference
Broad whitefish (*Coregonus nasus*)	111	4.39–10.51	7.07	Lockwood and Derr (1992)
Atlantic salmon (*Salmo salar*)	31	5.55–5.80	5.70	Lockwood and Derr (1992)
Channel catfish (*Ictalurus punctatus*)	115	1.951–2.002	1.977	Tiersch et al. (1990)
Pocket gopher (*Thomomys bottae*)	24	8.42–11.43	9.79	Sherwood and Patton (1982)
Neotropical bat (*Corollia brevicauda*)	9	4.97–6.93	5.86	Burton et al. (1989)
Florida scrub jay (*Aphelocoma coerulescens*)	39	3.045–3.195	3.129	Tiersch and Mumme (1993)
Japanese quail (*Coturnix coturnix*)	49	2.78–2.85	2.81	Tiersch and Wachtel (1991)

Table 8.8. Estimated genome size ranges (in bp) for selected groups of multicellular eukaryotes

Animals	
Sponges	4.9×10^7–5.4×10^7
Molluscs	8.8×10^8–5.2×10^9
Insects	9.8×10^7–7.4×10^9
Echinoderms	5.3×10^8–3.2×10^9
Vertebrates	
Jawless vertebrates	6.4×10^8–2.8×10^9
Sharks and rays	1.5×10^9–1.6×10^{10}
Bony fishes	3.8×10^8–1.4×10^{11}
Amphibians	9.3×10^8–8.4×10^{10}
Reptiles	1.2×10^9–5.3×10^9
Birds	1.7×10^9–2.3×10^9
Mammals	1.4×10^9–5.7×10^9
Plants	
Gymnosperms	4.1×10^9–7.7×10^{10}
Angiosperms	5.0×10^7–1.3×10^{11}

Data from Li and Graur (1991).

the former is much larger, the difference must be made up by noncoding DNA in the former species.

On the other hand, when two related species are compared, it may sometimes be true that the species with the larger genomes does indeed have more genes. For example, the frog genus *Xenopus* includes 16 species, with genome sizes ranging from 3.5×10^9 bp to 1.6×10^{10} bp (Kobel and Du Pasquier 1986). These differences have arisen by numerous events of polyploidization within the past 40 million years or so. The ancestral chromosome number for the genus seems to have been 18, but there are species with 36, 72, and 108 chromosomes. As discussed (chapter 7, Functional Constraint on Duplicated Genes), in one member of this genus, the tetraploid *Xenopus laevis*, it is known that many of the genes duplicated by polyploidization are still functional. Thus, within this genus, species with larger genomes probably do have more genes. In evolutionary terms, however, this situation is expected to be transitory, because eventually duplicate genes that have not evolved separate functions will be silenced by mutation.

If differences in genome size are mainly explainable by differences in the amount of noncoding DNA, which is presumably largely nonfunctional "junk," it would seem unlikely that most such differences are adaptive. However, adaptive hypotheses regarding genome size have been proposed. The genomes of birds are generally smaller than those of other vertebrates, and the range of genome sizes seen in birds is considerably narrower than that in other vertebrate classes (Table 8.8). Tiersch and Wachtel (1991), comparing data on genome sizes of 135 species of birds from seven orders, found a range of genome sizes from 2.10 to 3.85 pg per nucleus. This range is considerable smaller than that seen within at least one fish species, the broad whitefish (Table 8.6). These observations have led to the hypothesis that avian genome size is adaptive to the high rate of oxidative metabolism in birds, which results primarily from the demands of flight (Wachtel and Tiersch 1993).

This hypothesis is based on the observation that cell size and nuclear genome size are correlated in vertebrates (Szarski 1976, 1983). In keeping with this trend, bird cells are generally smaller than mammalian cells. Smaller cells are advantageous in an animal with a high rate of oxidative metabolism because a smaller cell has a greater surface area per volume of cytoplasm, thus facilitating gas exchange.

However, there are other possible explanations for the small and relatively uniform genome sizes of birds. One of these is that some chance event occurred that led to loss of some DNA in the genome of an avian ancestor, and that this reduced genome size was fixed by chance, perhaps during a population bottleneck in the origin of birds. Even if this reduced genome size were selectively somewhat deleterious, it might be fixed by chance if the ancestors of modern birds went through a severe population bottleneck (Ota and Nei 1995).

Many biologists formerly believed that birds went through an extreme population bottleneck about 65 million years ago at the end of the Cretaceous period, the time of the extinction of the dinosaurs (Alvarez 1983). All of the living orders of birds were believed to have diverged since that time. Indeed, there are some rather imaginative speculation that birds would be particularly vulnerable to the effects of the meteor impact that allegedly caused mass extinction (Alvarez 1983; Wyles et al. 1983). In any event, this view is no longer tenable. Molecular analyses clearly demonstrate that at least some of the living orders of birds diverged over 100 million years ago, in the mid-Cretaceous period (Kumar and Hedges 1998).

To test whether genome reduction in birds occurred as the result of natural selection or as the result of a chance event, Hughes and Hughes (1995c) compared introns and exons from orthologous genes of humans and chickens. The results showed that chicken introns are significantly smaller than corresponding human introns (Figure 8.7). Exons, by contrast, were nearly identical in length between the two species (Figure 8.7). When the logarithm of the chicken intron length was plotted against the logarithm of the length of the corresponding

---

Figure 8.7. (facing page) (**A**) Mean length of 111 homologous introns and 141 homologous exons from 41 genes of human (open columns) and chicken (hatched columns). Paired-sample *t*-tests of the significance of the difference between humans and chickens: for introns, $p < .005$; for exons, n.s. (**B**) Natural logarithm of chicken intron length plotted against that of human intron length. The slope of the regression line (0.37) was significantly different from zero ($p < .00001$) and from 1.0 ($p < .00001$). (**C**) Natural logarithm of chicken exon length plotted against that of human exon length. The slope of the regression line (0.90) was significantly different from zero ($p < .0001$) and from 1.0 ($p < .01$). The slopes of the regression lines for introns and exons are significantly different from each other ($p < .0001$). From Hughes and Hughes (1995c).

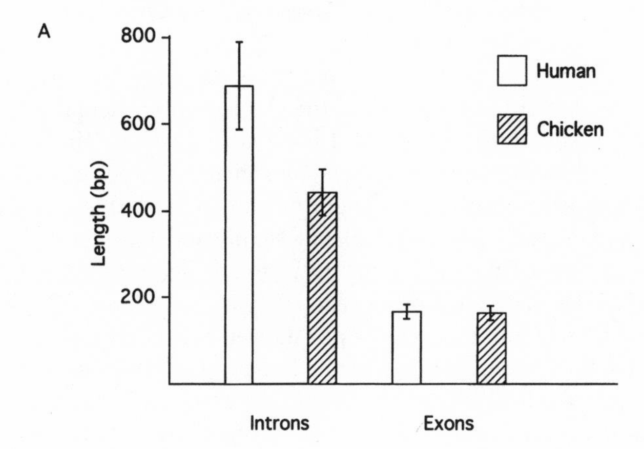
A
800
600
400
200
Length (bp)
Human
Chicken
Introns
Exons

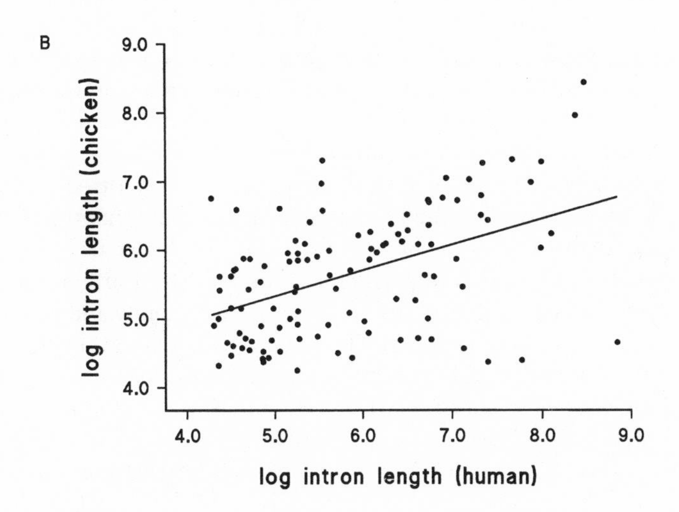
B
9.0
8.0
7.0
6.0
5.0
4.0
log intron length (chicken)
4.0
5.0
6.0
7.0
8.0
9.0
log intron length (human)

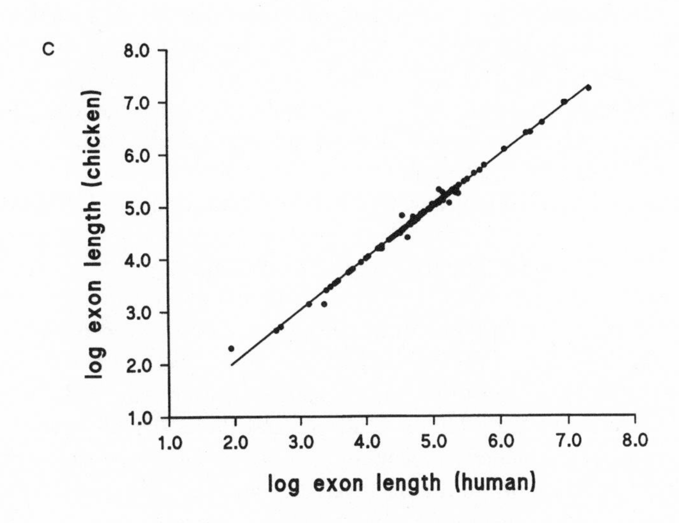
C
8.0
7.0
6.0
5.0
4.0
3.0
2.0
1.0
log exon length (chicken)
1.0
2.0
3.0
4.0
5.0
6.0
7.0
8.0
log exon length (human)

human intron, the slope of the line was 0.37 (Figure 8.7). This slope was significantly less than 1.0, meaning that the chicken intron length increases as a function of human intron length at a less than linear rate. In other words, intron lengths in the two species have a "negatively allometric" relationship. The relationship between log exon length in humans and chickens also had a slope (0.90) significantly less than 1.0 but significantly greater than that for introns. The strong negative allometry in intron lengths indicates that DNA loss in chickens has occurred disproportionately in long introns. Alignment of corresponding human and chicken introns showed that DNA loss has occurred by multiple deletion events scattered through the intron rather than by deletion of a single large stretch of DNA.

The fact that numerous deletion events, particularly in large introns, have contributed to the reduced genome size of birds is not consistent with the hypothesis of a single large deletion event fixed by chance during a population bottleneck. Rather, it suggests a pattern of long-term directional selection favoring reduced genome size. The selection operating on genome size was presumably consistent with the models of minute effect selection discussed earlier in this chapter (Natural Selection on Codon Usage).

Other lines of evidence support the hypothesis that reduced genome size in birds is an adaptation to flight. For one thing, the genomes of bats are considerably smaller than those of other mammals (Burton et al. 1989), suggesting that this trait has evolved in parallel in birds and bats. Furthermore, within birds themselves, there is a relationship between flying ability and genome size. I classified 39 avian families for which genome sizes were available into four categories with regard to flight ability: strong fliers, moderate fliers, weak fliers, and flightless families (Figure 8.8). Mean genome sizes were greatest in flightless families, intermediate in weak fliers, and smaller in strong fliers (Figure 8.8). The analysis illustrated in Figure 8.8 assumes statistical independence of the families analyzed. Because most avian families probably diverged from each other in the distant past (over 50 million years ago), these families share relatively little of their evolutionary history. Thus, the assumption that they are statistically independent may be approximately valid.

A type of analysis that avoids the problem of nonindependence due to shared evolutionary history is one in which traits are mapped on a known phylogeny (Harvey and Pagel 1991). I conducted such an analysis using the phylogeny derived from DNA-DNA hybridization by Sibley and Ahlquist (1990). Although there are some groups of birds whose positions in this phylogeny are poorly resolved, the families for which genome size data are available are not among them. Assuming this phylogeny, it is possible to identify six branches that correspond to periods of evolutionary time during which reduction or loss of flying ability occurred (Figure 8.9). When flightless taxa or taxa with reduced flying ability were compared to sister taxa, the former in each case had larger mean genome sizes (Table 8.9). These results suggest that, after an initial genome size reduction in the ancestors of modern birds, genome sizes have increased in taxa lacking strong flying ability. Such an increase is possibly due to a relaxation of selective pressures arising from the metabolic demands of active flight.

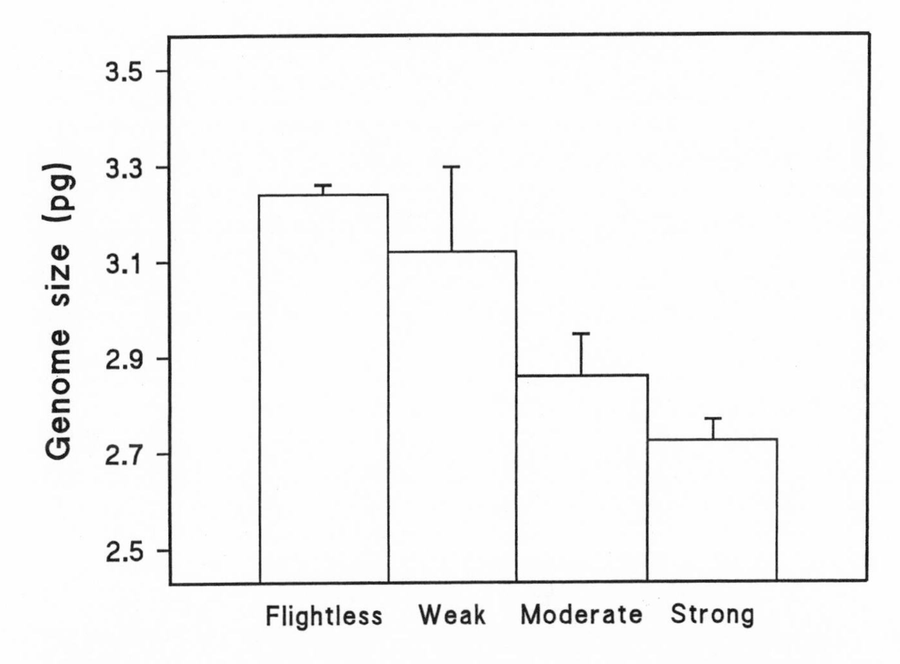

Figure 8.8. Mean genome size for families of birds categorized according to flight ability: flightless (two families); weak fliers (five families); moderate fliers (15 families); strong fliers (17 families). Genome size data are from Tiersch and Wachtel (1991) and data on flight ability from Van Tyne and Berger (1959). One-way analysis of variance, $p < .02$.

The insects include many flightless species that have descended from flying ancestors. Loss of flight has occurred independently many times in the insects, having happened more than once in most major orders. In addition, flying insect species show great differences in flight activity. For these reasons, a broad comparative study of genome size in insects might be of considerable interest. It is worth noting that the best known insect genetically, *Drosophila melanogaster*, is an active flier and has a genome size that falls toward the low end of the range of genome sizes known in insects (Tables 8.6 and 8.8).

For 30 families of birds for which data on genome size and incubation and fledgling times were available, there was no significant correlation between genome size and developmental time ($r = .154$ for incubation period and .156 for fledgling period, both not significant). In salamanders, where the range of genome sizes is much greater than in birds, several studies have observed a significant correlation between developmental time and genome size (Sessions and Larson 1987; Pagel and Johnstone 1992). The results in salamanders suggest that noncoding DNA can accumulate in salamander species unless there is selection against it, as appears to be the case when selection favors rapid development.

Cavalier-Smith (1985) suggested that natural selection can actually favor

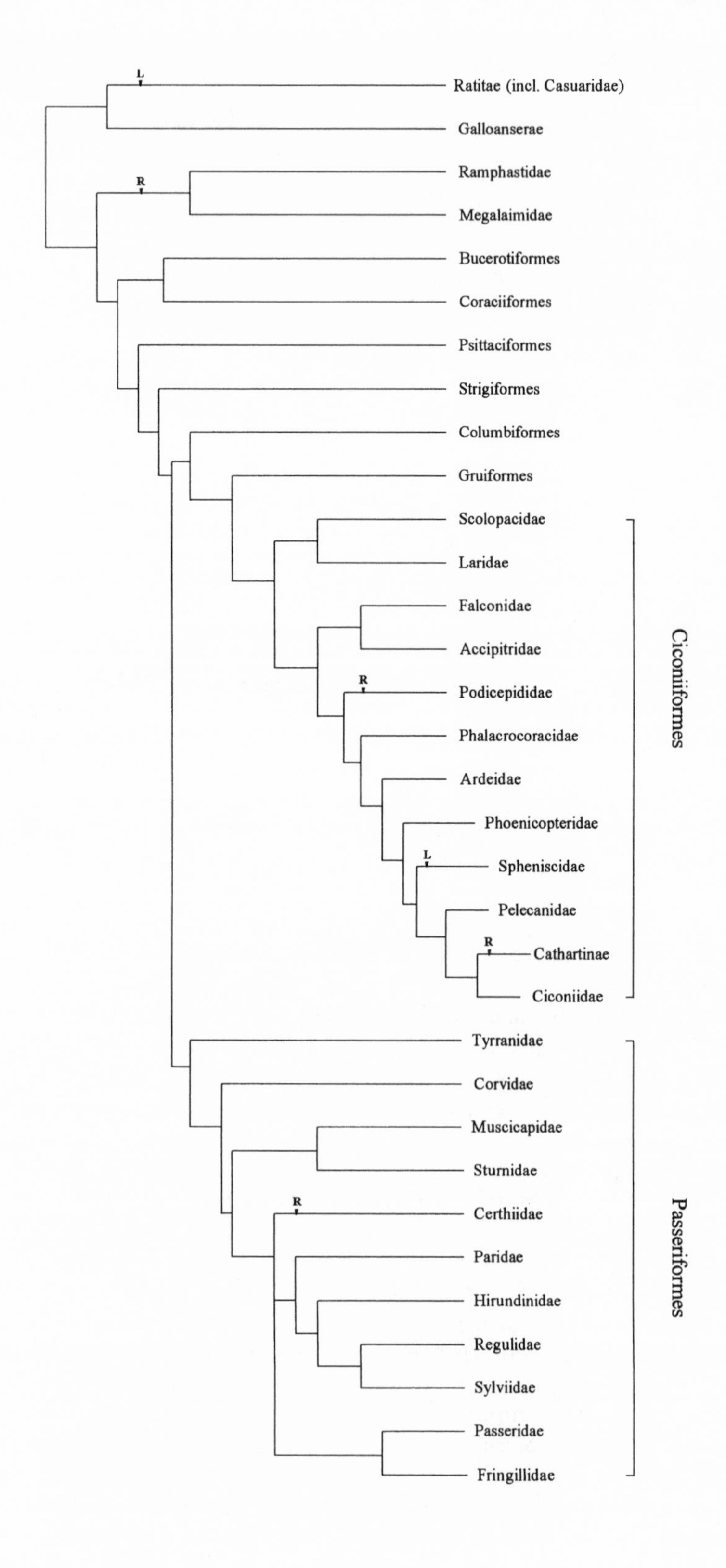
L
Ratitae (incl. Casuaridae)
Galloanserae
R
Ramphastidae
Megalaimidae
Bucerotiformes
Coraciiformes
Psittaciformes
Strigiformes
Columbiformes
Gruiformes
Scolopacidae
Laridae
Falconidae
Accipitridae
R
Podicepididae
Phalacrocoracidae
Ardeidae
Phoenicopteridae
L
Spheniscidae
Pelecanidae
R
Cathartinae
Ciconiidae
Ciconiiformes
Tyrranidae
Corvidae
Muscicapidae
Sturnidae
R
Certhiidae
Paridae
Hirundinidae
Regulidae
Sylviidae
Passeridae
Fringillidae
Passeriformes

Table 8.9. Comparison of mean genome sizes of flightless and weakly flying avian taxa with sister groups

	Taxon	Genome size (pg)	Sister group	Genome size (pg)
Flightless	Casuaridae (cassowaries)	3.22	Galloanserae	2.67
	Spheniscidae (penguins)	3.26	Pelecanidae, Ciconiinae	2.80
Weak fliers	Ramphastidae/Megalaimidae (toucans and barbets)	3.49	Bucerotiformes, Coraciiformes, Psittaciformes, Strigiformes, Columbiformes, Gruiformes, Ciconiiformes, Passeriformes	2.77
	Podicepididae (grebes)	3.03	Phalocrocoracidae, Ardeidae, Phoenicopteridae, Threskiornithidae, Ciconiinae, Pelecanidae	2.70
	Cathartinae (New World vultures)	3.02	Ciconiinae	2.85
	Certhiidae (creepers)	2.75	Paridae, Hirundinidae, Regulidae, Sylviidae	2.69

Sister groups were chosen based on the DNA-DNA hybridization phylogeny of Sibley and Ahlquist (1990) (see Figure 8.9). Genome sizes were significantly lower for flightless and weakly flying taxa compared to sister taxa (paired sample *t*-test, $p < .05$).

large genome size when large cell size is favored. The genome, he proposed, serves as a "nucleoskeleton" determining cell volume. In salamanders, however, Pagel and Johnstone (1992) found that the correlation between genome size and developmental rate was statistically independent of the relationship between genome size and cell size. They argued that this is evidence against Cavalier-Smith's hypothesis that large genome size is an adaptive trait. Rather, they argued that large genome size in salamanders is due to the accumulation of junk DNA. However, in certain salamander species, rapid development is favored; in these species, natural selection favors reduction of genome size through loss of noncoding DNA.

Genome Duplication in the Vertebrates

Observing that genome duplication by polyploidization has occurred recently in several different taxa of fish and amphibians, Ohno (1970) proposed that

Figure 8.9. (facing page) Phylogeny of bird taxa for which genome size data are available, based on DNA-DNA hybridization data of Sibley and Ahlquist (1990). Hypothetical events of flight loss (L) and reduction (R) are indicated.

genome duplication has played an important role in vertebrate evolution. In particular, he hypothesized that two rounds of genome duplication occurred early in vertebrate history. This hypothesis has become very popular recently (Holland et al. 1994; Kasahara et al. 1996; Sidow 1996). As evidence in favor of this hypothesis, Sidow (1996, p. 715) mentions the following: "When comparing *Drosophila* with vertebrates, one finds an uncanny consistency in the multiple by which vertebrate developmental regulator genes outnumber their *Drosophila* homologues; it is often the number four (e.g., *Hox* clusters, *Cdx, MyoD, 60A, Notch, elav, btd/SP* . . .) and sometimes two (e.g., *Wnt-5, decapentaplegic, Eve*) or three (e.g., *Msx, hedgehog* . . .)." As further evidence for this hypothesis, Sidow states that vertebrates are estimated to have approximately four times as many genes as does *Drosophila*, an estimate that he attributes to Miklos and Rubin (1996). In fact, the values presented by Miklos and Rubin (1995) and reproduced here in Table 8.6 include estimated gene number for vertebrates that is about 5.8 times that for *Drosophila*. Sidow (1996) further hypothesizes the gene families "with only two vertebrate paralogs lost one copy after the first genome duplication; those vertebrate genes with three lost one paralog after the second genome duplication" (Sidow 1996).

Despite widespread citation of the hypothesis that vertebrates underwent two rounds of genome duplication early in their history (the 2R hypothesis), it has not been subjected to rigorous testing by phylogenetic analysis. As pointed out by Sidow in the passage cited above, the occurrence of families having two or three paralogues in vertebrates is explainable on the 2R hypothesis; but these families cannot really be used to test the 2R hypothesis because their occurrence is also consistent with several alternative explanations. Even the occurrence of four paralogues in vertebrates cannot in itself be taken as supporting the 2R hypothesis. For example, if the four vertebrate paralogues are shown by phylogenetic analysis to have duplicated prior to the origin of vertebrates, then, clearly, their duplication cannot have been part of the hypothetical genome duplications early in vertebrate history. An example of a phylogeny of this sort is shown in Figure 8.10C. In this example, the duplication occurred prior to the divergence of protostomes (including insects) from deuterostomes (including vertebrates).

Even when four vertebrate paralogues can be shown to have diverged after the origin of vertebrates, their phylogenetic relationship must exhibit a specific topology in order to be counted as supporting the 2R hypothesis. This topology is illustrated in Figure 8.10A. I will call this a topology of the form (AB) (CD), because the four genes form two clusters, with A being a sister group to B, and C a sister group to D. An alternative topology is one in which one of the four paralogues diverged prior to the other three (Figure 8.10B). This topology can be symbolized as (A) (BCD). Note that the topology of the relationships among B, C, and D is not relevant to the question of support for the 2R hypothesis. Clearly, a topology of the (A) (BCD) type does not support the 2R hypothesis. Of course, it would be possible to invent ad hoc scenarios to reconcile such a topology with the 2R hypothesis. For example, we might hypothesize a series of events of tandem duplication and of deletion occurring independently of the hypothesized genome duplications. Nonetheless, the widespread occurrence of

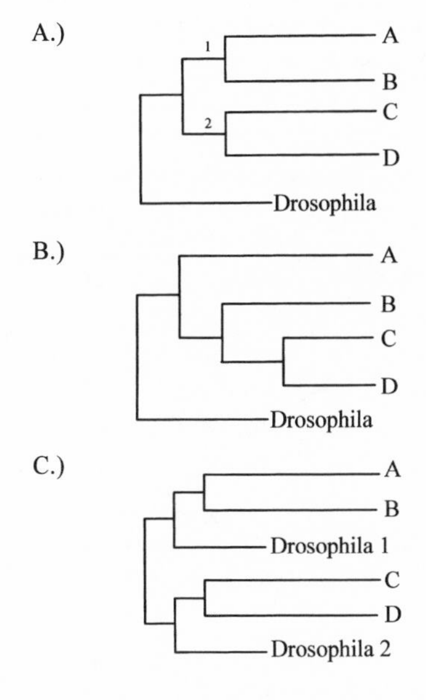

Figure 8.10. Example of possible phylogenies for a gene family having four members (A–D) in vertebrates: **(A)** a topology of the form (AB) (CD), which supports the hypothesis of two rounds of gene duplication early in vertebrate history; **(B)** an example of a phylogeny with topology of the form (A) (BCD); **(C)** a case in which the gene duplication separating the ancestor of A and B from that of C and D occurred prior to the divergence of deuterostomes and protostomes.

topologies of the (A) (BCD) type in gene families having four paralogues in vertebrates would be evidence against the 2R hypothesis.

To test the 2R hypothesis, I reconstructed nine phylogenies of proteins that play important roles in regulating development—which have at least one known homologue in *Drosophila*—and that have four paralogues in vertebrates (Hughes 1999). To provide additional tests of the hypothesis, the results were compared with those of recently published studies of additional gene families having similar properties.

The expectation was that, if the 2R hypothesis is true, in a majority of the families there would be strong support for an (AB) (CD) topology (Figure 8.10A). By contrast, a high proportion of gene families with paralogues that diverged prior to the origin of the vertebrates (Figure 8.10C) or with topologies of the (A) (BCD) type would argue against the 2R hypothesis. From a methodological point of view, it is important to realize that the null hypothesis in such an analysis must be the hypothesis of no effect; in other words, the hypothesis that two rounds of genome duplication did not occur in vertebrate history. Only if the data provide a compelling reason to reject the null hypothesis—i.e., a large proportion of families showing the (AB) (CD) topology—can we reject the null hypothesis and accept the 2R hypothesis.

The results for 13 phylogenies are summarized in Table 8.10. Of these, five showed a topology indicative of gene duplication prior to the origin of vertebrates (CDX, Brachyury, MyoD, NOTCH, and Pax; Table 8.10). In four of these cases, the internal branch of the tree that supported this topology received significant bootstrap support (Table 8.10). An example is the NOTCH family, illustrated in Figure 8.11. The tree was rooted with *Drosophila* crumbs

Table 8.10. Summary of phylogenetic analyses of gene families having four members in vertebrates

Gene	Duplication before vertebrate origin[a]	Topolgy[b]	Source
CDX	+ (47)		Hughes (1999)
dpp	–	(AB) (CD) (58,62)	Hughes (1999)
BMP5-8	–	(A) (BCD) (99)	Hughes (1999)
Elav	–	(A) (BCD) (100)	Hughes (1999)
Egr	–	(A) (BCD) (97)	Hughes (1999)
SP	–	(A) (BCD) (96)	Hughes (1999)
Brachyury	+ (100)		Hughes (1999)
MyoD	+ (96)		Hughes (1999)
NOTCH	+ (100)		Hughes (1999)
FGFR	–	(A) (BCD) (98)	Coulier et al. (1997)
antp	–	(A) (BCD) (<50)	Zhang and Nei (1996)
Hox-linked COL	–	(A) (BCD) (93)	Bailey et al. (1997)
Pax	+ (99)		Balczarek et al. (1997)

a. Percentage of bootstrap support in parentheses. For all genes but *MyoD*, the duplication could be shown to have occurred before the divergence of protostomes and deuterostomes.

b. Topologies are in Figure 8.10. Figure 8.10A represents the (AB) (CD) topology, while Figure 8.10B represents an example of (A) (BCD) topology. Percentage of bootstrap support in parentheses; for (AB) (CD) topology bootstrap percentages for branches 1 and 2 (as in Figure 8.10A) are given.

and vertebrate homologues. In this tree, vertebrate NOTCH4 clustered outside vertebrate NOTCH1-3 and NOTCH from insects; bootstrap support for this clustering pattern was 100%.

Of the remaining eight phylogenies, only one (dpp) showed a topology of the (AB) (CD) form, although bootstrap support for this topology was very weak (Table 8.10). All the others showed topologies of the form (A) (BCD); and in five of these the internal branch supporting this topology received bootstrap support of 96% or better (Table 8.10). The phylogeny of *Drosophila* elav and its vertebrate homologues HuA, –B, –C, and –D provides an example (Figure 8.12). This phylogeny was rooted with sex-lethality of insects (Figure 8.12). *Drosophila* elav clustered outside the vertebrate genes, while among the vertebrate genes HuA clustered outside the other three (Figure 8.12). The branch supporting the position of vertebrate HuA received 100% bootstrap support (Figure 8.12).

The available data from protein phylogenies thus do not support the 2R hypothesis. Other data that might be relevant to this hypothesis include estimates

Figure 8.11. (facing page) Phylogenetic tree of the NOTCH gene family, constructed by the neighbor-joining method on the basis of the proportion of amino acid difference (p). Numbers on the branches are as in Figure 3.2.

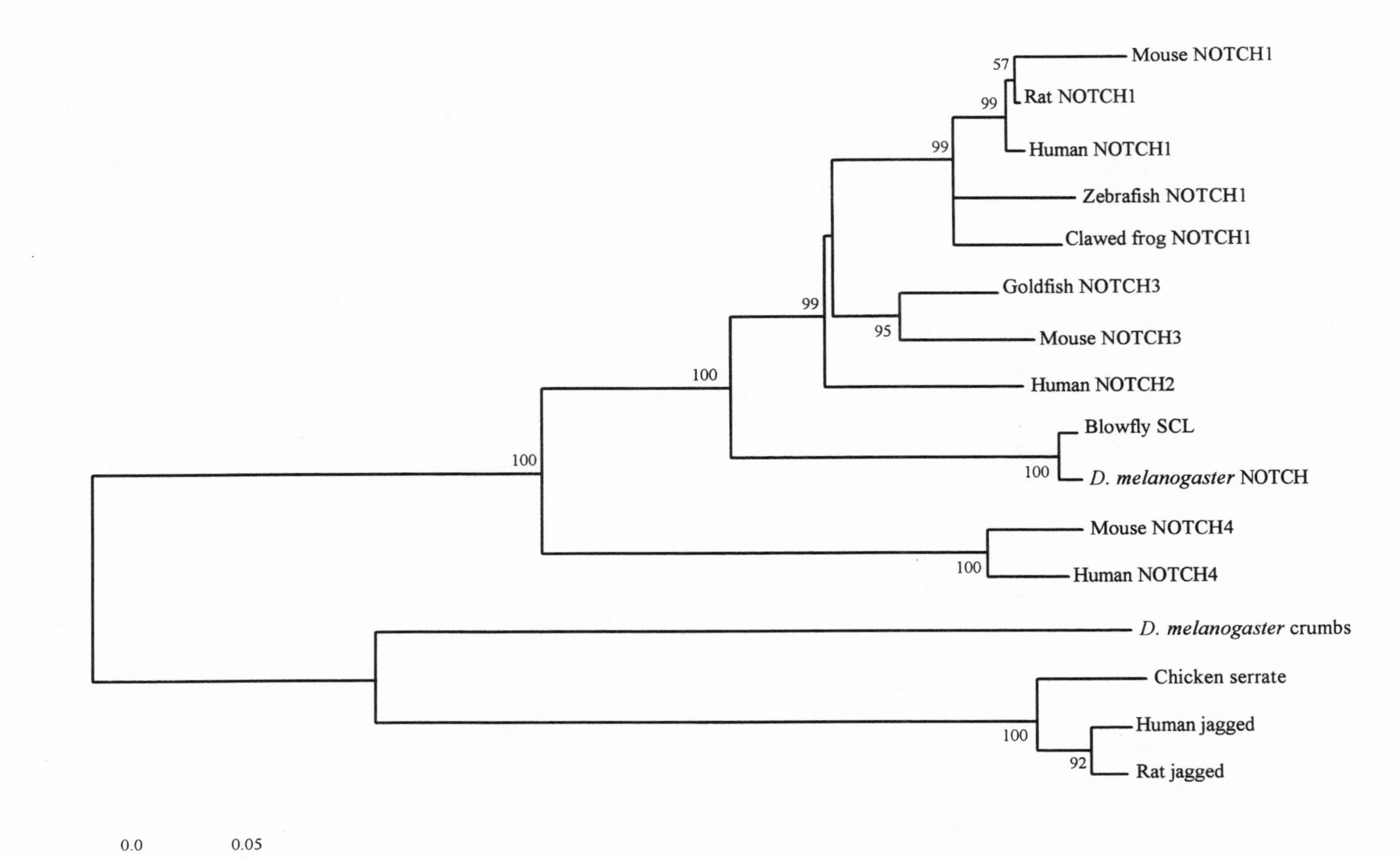
Mouse NOTCH1
Rat NOTCH1
Human NOTCH1
Zebrafish NOTCH1
Clawed frog NOTCH1
Goldfish NOTCH3
Mouse NOTCH3
Human NOTCH2
Blowfly SCL
D. melanogaster NOTCH
Mouse NOTCH4
Human NOTCH4
D. melanogaster crumbs
Chicken serrate
Human jagged
Rat jagged
57
99
99
95
99
100
100
100
100
100
92
0.0
0.05
p

of gene number or of genome size. However, both of these measures are problematic as potential sources of information regarding ancient events of genome duplication. After genome duplication, it is expected that most duplicate genes will eventually be silenced by mutation if they do not evolve new functions (chapter 7, An Alternative Model of Evolution of Novel Proteins). Silenced genes and other noncoding DNA may be lost over time through deletion, a process that may have been accelerated by natural selection in some groups (such as birds).

As the 2R hypothesis is usually stated, it is assumed that one round of genome duplication occurred prior to the divergence of jawed and jawless vertebrates and one round subsequently (Sidow 1996). If this event were still reflected in genome sizes of current-day jawed and jawless vertebrates, we would expect the latter to have genomes about half as large as those of the former. In fact, however, the genome size range of living jawless vertebrates (Agnatha) overlaps those of the classes of jawed vertebrates (Table 8.8), thus providing no obvious support for the 2R hypothesis.

Although the available evidence does not support the 2R hypothesis, the phylogenetic analyses summarized in Table 8.10 are all consistent with the hypothesis that a single genome duplication occurred either at some point in deuterostome history prior to the origin of the vertebrates or within the vertebrate lineage shortly after its origin. However, the available data can be explained easily without hypothesizing any genome-wide duplication event. Known gene phylogenies can be easily explained by independent duplication of individual genes or chromosomal segments, processes well known to occur in eukaryotic organisms.

One reason why the 2R hypothesis had a strong appeal to biologists is that it was formerly believed that the animal phyla all appeared very rapidly in the Cambrian period (the so-called Cambrian explosion). Thus, it was thought that some unusual mechanism must have been responsible for the rapid evolution of some adaptations unique to vertebrates such as the specific immune systems, which includes the immunoglobulins, T cell receptors, and MHC. It was thought that genome duplication might somehow make possible such rapid evolution.

More recently, molecular analyses have suggested that the animal phyla diverged at a much earlier time (Wray et al. 1996). Thus, there is no real need to explain the rapid appearance of vertebrate adaptations. In addition to the specific immune system involving immunoglobulins, T cell receptors, and MHC, the vertebrates possess an innate immune system, which involves a variety of nonspecific defenses. Invertebrate immune systems are also nonspecific, and thus it has been proposed that the vertebrate and invertebrate innate immune systems are descended from a system present in their common ancestor (Janeway

---

Figure 8.12. (facing page) Phylogenetic tree of the elav gene family, constructed by the neighbor-joining method on the basis of the proportion of amino acid difference (p). Numbers on the branches are as in Figure 3.2.

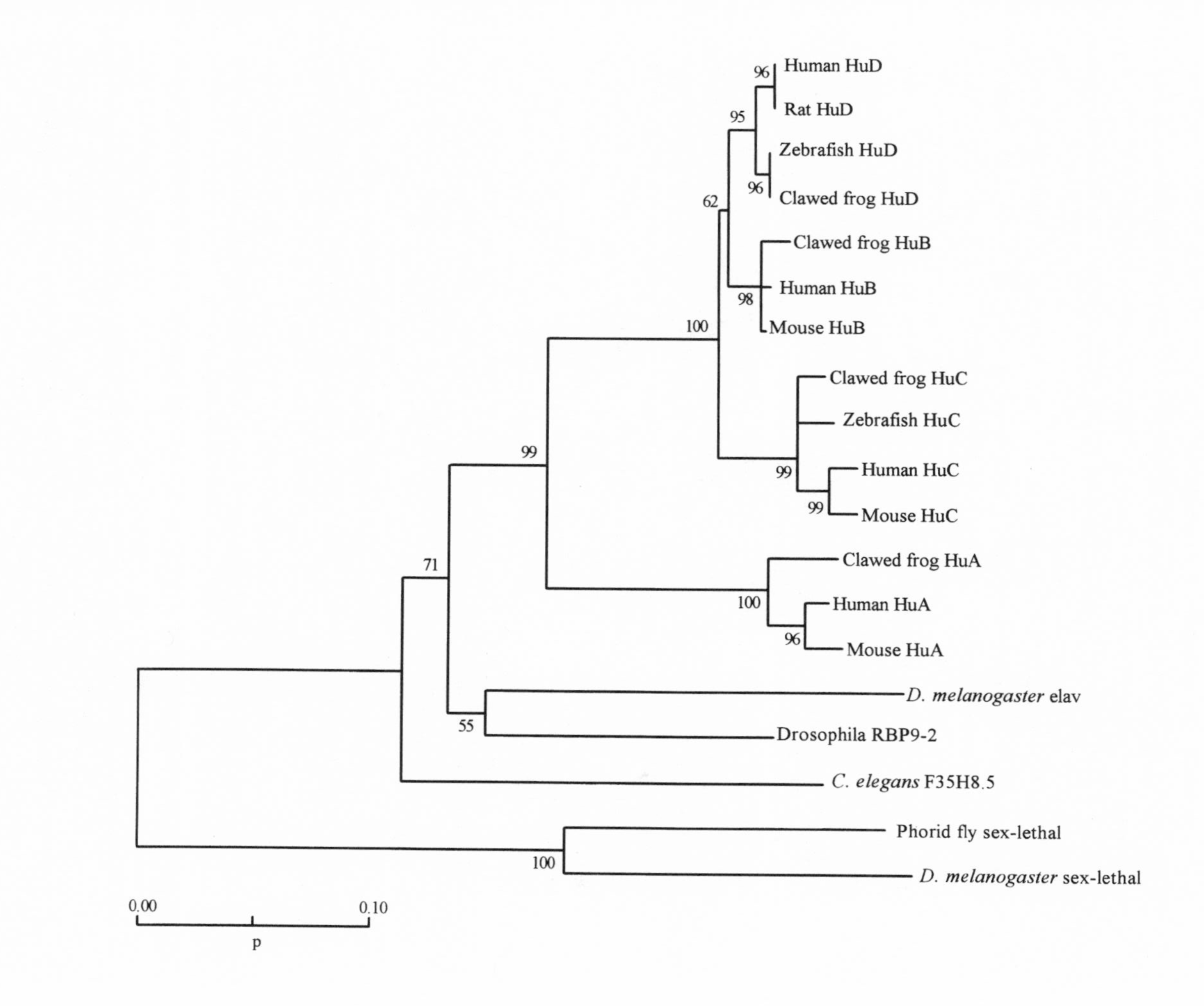

Human HuD
Rat HuD
Zebrafish HuD
Clawed frog HuD
Clawed frog HuB
Human HuB
Mouse HuB
Clawed frog HuC
Zebrafish HuC
Human HuC
Mouse HuC
Clawed frog HuA
Human HuA
Mouse HuA
D. melanogaster elav
Drosophila RBP9-2
C. elegans F35H8.5
Phorid fly sex-lethal
D. melanogaster sex-lethal
96
95
96
62
98
100
99
99
99
71
100
96
55
100
0.00
0.10
p

1989; Hultmark 1993; Fearon 1997). However, phylogenetic analyses of immune system proteins suggest that the vertebrate and invertebrate immune systems evolved independently (Hughes 1998a). This is easily explained if we suppose that the divergence of the major animal lineages occurred in the very distant past.

Finally, it seems very unlikely that genome duplication could give rise to a suite of new adaptations by itself. Adaptation of a protein-coding gene to a new function often requires positive Darwinian selection (chapter 7, An Alternative Model of the Evolution of Novel Proteins). If genome duplication were followed by simultaneous positive selection on a large number of the newly duplicated genes, the result would be a substitutional load that no population could bear.

Adaptive Gene Clustering

It is well known by now that the genomes of organisms frequently contain clusters of linked genes that encode proteins that are functionally related. The operons of bacteria provide the most striking examples. These are clusters of genes involved in a similar function that are regulated together and transcribed in a single messenger RNA. The best known example is the *lac* operon of *E. coli*, which encodes three genes involved in the metabolism of lactose; *lac*Z (encoding β galactosidase), *lac*Y (encoding lactose permease), and *lac*A (encoding thiogalactose acetyltransferase) (Sadler and Tecklenberg 1981). Directly upstream from the *lac*Z gene is the *lacI* gene; *lacI* encodes the lac operon repressor protein, which prevents transcription of the operon in the absence of lactose. These four genes are not homologous; so their presence together is not simply the result of past gene duplication.

Interestingly, the *lac* operon of *Agrobacterium radiobacter* contains a rather different set of genes from that of *E. coli* (Williams et al. 1992). The only homologous genes present in these regions in both species are *lacI* and *lac*Z. There are three other genes in the *A. radiobacter lac* operon that are not shared with *E. coli*. Thus, it appears that most of the genes have been brought together independently in the two species.

In eukaryotes, often there are also clusters of genes whose function is similar or whose products interact functionally. Sometimes, these are related genes, forming a gene family. If so, their linkage can thus be explained as the result of their origin in past gene duplication, and it need not have adaptive significance. However, in some cases there is additional evidence that the linkage of related genes is adaptive. For example, linked loci may be under some degree of joint regulation. In the case of β-globin cluster, a locus control region that is conserved across orders of placental mammals is involved in regulation expression of these genes (Slightom et al. 1997). In addition, gene arrangements may be conserved for very long periods of time, suggesting some selective advantage. The homeobox clusters provide an example; in these, the same gene order is conserved in different animal phyla. Experiments that split one homeobox complex, the bithorax complex in *Drosophila*, had no detectable deleterious effects on development (e.g., Tiong et al. 1987), but the selective advantage of clustering may be too subtle to detect by experiments covering only one generation in the laboratory (Zhang and Nei 1996).

The genomes of vertebrates also include complexes of nonhomologous loci that are functionally related. The MHC gene complex of mammals includes not only the related MHC class I and class II genes but also some unrelated genes whose products are involved in the function of the class I MHC: (1) the genes encoding the LMP2 and LMP7 proteasome subunits, which are involved in generating peptides for presentation by class I molecules; and (2) two subunits of the TAP transporter, which transports peptides across the ER membrane for assembly with the class I molecule (see chapter 4, Explaining MHC Polymorphism).

Another apparent example is the so-called *surfeit* locus. This is the tightest cluster of mammalian genes so far known, consisting of six evolutionarily unrelated tightly linked loci (designated *Surf-1* through *Surf-6*). The function of the gene products is not known, except for *Surf-3*, which encodes the ribosomal protein L7a. But all six genes show a pattern of universal tissue expression characteristic of "housekeeping" genes. The loci and their arrangement are conserved in birds as well as mammals (Colombo et al. 1992). In the pufferfish, *Surf-1*, *Surf-3*, and *Surf-6* are clustered together; *Surf-2* and *Surf-4* are together but not linked to the other surfeit genes; and *Surf-5* is unlinked to any of the surfeit genes (Gilley et al. 1997). In *Drosophila melanogaster* and in the nematode worm *Caenorhabditis elegans*, the known homologues of the *surfeit* genes are not linked at all (Armes and Fried 1995). Thus, the surfeit cluster seems to have been formed in the ancestors of birds and mammals and has since been conserved for at least 300 million years. This conservation suggests that the linkage of these genes may confer some advantage.

When a group of genes are linked together, if their linkage is not adaptive, then it must reflect the result of historical processes. Sometimes a group of genes may be brought together after each is translocated independently from another location. Sometimes groups of related genes are found on two or more chromosomes. The existence of such groups of paralogues has been attributed to ancient block duplication events, either through duplication of a chromosome or a portion of the chromosome or through duplication of the entire genome.

One example involves 15 genes, belonging to 10 gene families (Table 8.11), located on human chromosome 6 in the vicinity of the MHC gene complex (the chromosomal region 6p21.3) There are also members of these 10 families on chromosome 9 (nine of them in the region 9q33–34) (Table 8.11). Of these 10 gene families, six also have members on chromosome 17 of the mouse, which is the MHC chromosome in that species. These six families also have representatives on mouse chromosome 2 (Table 8.11) In addition, four of the gene families with members on human chromosomes 6 and 9 also include members on chromosome 1 (in the regions 1p13 and 1q21–31 (Table 8.11). Two of these four families also have representatives on mouse chromosome 1 (Table 8.11).

Table 8.11. Gene families with members on human chromosomes 6, 9, and 1 and on mouse chromosomes 17, 2, and 1

Family	Human 6p21.3	Mouse 17	Human 9q33–34	Mouse 2	Human 1p13,1q21–25	Mouse 1
Retinoid X receptor (RXR)	RXRB	RXRB	RXRA	RXRA	RXRG	RXRG
α pro-collagen (COL)	COL11A2	COL11A2	COL5A1	COL5A1	—	—
ATP-binding cassette transporter (ABC)	TAP1	TAP1	ABC2	ABC2	—	—
	TAP2	TAP2				
Proteasome component β (PSMB)	LMP2	LMP2	PSM7	PSM7	—	—
	LMP7	LMP7				
Notch (NOTCH)	NOTCH4	NOTCH4	NOTCH1	NOTCH1	NOTCH2	—
Pre-B cell leukemia transcription factor (PBX)	PBX2	—	PBX3	—	PBX1	PBX1
Tenascin (TEN)	TNX	—	TNC	—	TNR	—
C3/C4/C5 complement component (C3/4/5)	C4A	—	C5	—	—	—
	C4B					
Heat-shock protein 70 (HSP70)	HSP70-1	HSP70-1	GRP78	GRP78	—	—
	HSP70-2	HSP70-3				
	HSP70-HOM	HSP70t				
Valyl-tRNA synthetase (VARS)	VARS2	—	VARS1[a]	—	—	—

It is conventional to use lowercase letters to designate mouse genes; that convention has not been followed here in the interest of highlighting orthologous relationships between human and mouse genes.

a. Not mapped to 9q33–34, but only to chromosome 9.

The current chromosomal distribution of these genes has been considered as evidence of past block duplications, perhaps as part of the two rounds of genome duplication that are widely (but, as we have seen, probably incorrectly) believed to have occurred in early in the history of the vertebrates (Kasahara et al. 1996; Katsanis et al. 1996). However, before accepting that a group of genes were duplicated as a block, it is important to reconstruct their phylogenies (Hughes 1998b). I reconstructed the phylogenies of the nine gene families involved in these putative block duplications for which sequences from both human chromosomes 6 and 9 were available (Hughes 1998b). The results showed that, rather than being duplicated as a block, these genes duplicated at different times scattered over at least 1.6 billion years.

By using sequences from different major groups of organisms, I was able to time gene duplication events relative to divergence events of major taxa (kingdoms, phyla, or classes). This approach makes it possible to date gene duplications without assuming a constant rate of molecular evolution ("molecular clock"). Furthermore, such relative dating is independent of estimates of absolute divergence times based on the fossil record.

The results of these analyses are shown in Table 8.12. In one of the gene families (ABC), the duplication of the chromosome 6 and chromosome 9 members took place before eukaryotes diverged from eubacteria (Figure 8.12). In two

Table 8.12. Summary of gene duplications relative to organismal divergences, as indicated by phylogenetic analyses of gene families including paralogues on human chromosomes 6, 9, and 1

Family	Human chromosomes	Before[a]	After[a]
RXR	6–9	Tet-Ost (95) [100]	Deu-Pro (100) [95]
	1–9	Tet-Ost (95) [100]	Deu-Pro (100) [95]
COL	1–9	Mam-Av (100) [100]	—
ABC	6–9	Euk-Eub (95) [93]	—
PSMB	6–9	An-Fn (100) [100]	—
NOTCH	6–9	Deu-Pro (96) [41]	
	1–9	Tet-Ost (53) [60]	—
PBX	6–9	—	Deu-Pro (45) [87]
	1–9	—	Deu-Pro (45) [87]
TEN	6–9	Tet-Ost (99) [82]	—
		Tet-Ost (99) [82]	
C3/4/5	6–9	Gna-Ag (98) [71]	—
HSP70	6–9	An-Fn (100) [99]	—
VARS	6–9	—	Pri-Rod (99) [99]

a. The number in parentheses is the bootstrap percentage for the relevant branch in the NJ tree; the number in square brackets is the bootstrap percentage for the corresponding branch in the MP tree. Organismal divergence events are abbreviated as follows: An-Fn = Animalia-Fungi; Deu-Pro = Deuterostomes-Protostomes; Euk-Eub = Eukaryotes-Eubacteria; Gna-Ag = Gnathostomata-Agnatha; Mam-Av = Mammalia-Aves; Pri-Rod = Primates-Rodentia; Tet-Ost = Tetrapoda-Osteichthyes. From Hughes (1998b).

families (PSMB and HSP70), the duplication took place prior to the divergence of animals and fungi, and in one (NOTCH) before the divergence of protostomes and deuterostomes. The remaining families seem to have duplicated after the origin of vertebrates (Tables 8.12). Assuming a molecular clock, I obtained estimates of divergence times for the paralogues on human chromosomes 6, 9, and 1 (Table 8.13). These range from around 600 million years ago to over 2 billion years ago, and show general agreement with the relative dating based on phylogenetic trees (Table 8.12).

Figures 8.13–8.15 show examples of these phylogenetic analyses. The ATP-binding cassette (or ABC) transmembrane transporters include a wide variety of transmembrane transporters of both eukaryotes and prokaryotes (Higgins 1992). The members of this family on human chromosome 6 are the genes encoding TAP1 and TAP2, the two subunits of the TAP transporter involved in peptide presentations by the class I MHC (chapter 4, Structure and Function of MHC Molecules). The member of this family encoded on human chromosome 9 is called ABC2, and its function is uncertain. Many members of this family have an internally duplicated structure, with homologous N-terminal and C-terminal halves. The phylogenetic analysis (Figure 8.13) was based on the conserved ATP-binding cassette region. In the phylogenetic tree, ABC2 clustered with bacterial sequences such as NodI from *Rhizobium*, while other bacterial sequences clustered with TAP1 and TAP2 (Figure 8.13).

The heat-shock protein 70 (HSP70) family consists of molecular chaperones that assist in protein folding (Schlesinger 1990). Three members of this family are encoded on chromosome 6 (HSP70-1, HSP70-2, and HSP70-HOM), and

Table 8.13. Estimates of divergence times based on the number of amino acid replacements per 100 sites (d_{AA}) calibrated with dates from the fossil record

Family	Human chromosomes	$d_{AA} \pm$ SE	Calibration ($d_{AA} \pm$ SE)	Time estimate ± 99% C.I.
RET	6–1/9	27.2 ± 2.4	Tet-Ost (16.8 ± 1.7)	648 ± 149
	1–9	25.1 ± 2.6	Tet-Ost (16.8 ± 1.7)	596 ± 160
COL	6–9	37.6 ± 3.8	Pri-Rod (3.9 ± 1.6)	974 ± 252
			Mam-Av (7.6 ± 1.4)	1531 ± 396
ABC	6–9	149.2 ± 16.0	Euk-Eub (104.6 ± 7.6)	2140 ± 592
PSMB	6–9	134.1 ± 9.9	An-Fn (59.1 ± 5.7)	2268 ± 430
NOTCH	6–1/9	72.3 ± 5.9	Deu-Pro (38.1 ± 3.8)	1519 ± 318
	1–9	26.8 ± 3.3	Deu-Pro (38.1 ± 3.8)	564 ± 179
PBX	6–1/9	18.7 ± 2.0	Deu-Pro (24.4 ± 2.6)	612 ± 167
	1–9	18.2 ± 2.5	Deu-Pro (24.4 ± 2.6)	599 ± 209
TEN	6–1/9	83.3 ± 4.2	Tet-Ost (47.9 ± 3.2)	696 ± 90
		74.5 ± 4.5	Tet-Ost (47.9 ± 3.2)	623 ± 98
C3/4/5	6–9	131.2 ± 3.9	Gna-Ag (191.9 ± 2.6)	579 ± 44
HSP70	6–9	44.7 ± 2.9	An-Fn (33.4 ± 2.4)	1604 ± 268

Abbreviations for organismal divergence events are as in Table 8.12. Divergence times used in calibration were the following: Euk-Eub, 1500 Mya; An-Fn, 1200 Mya; Deu-Pro, 800 Mya; Tet-Ost, 400 Mya; Pri-Rod, 100 Mya. From Hughes (1998b).

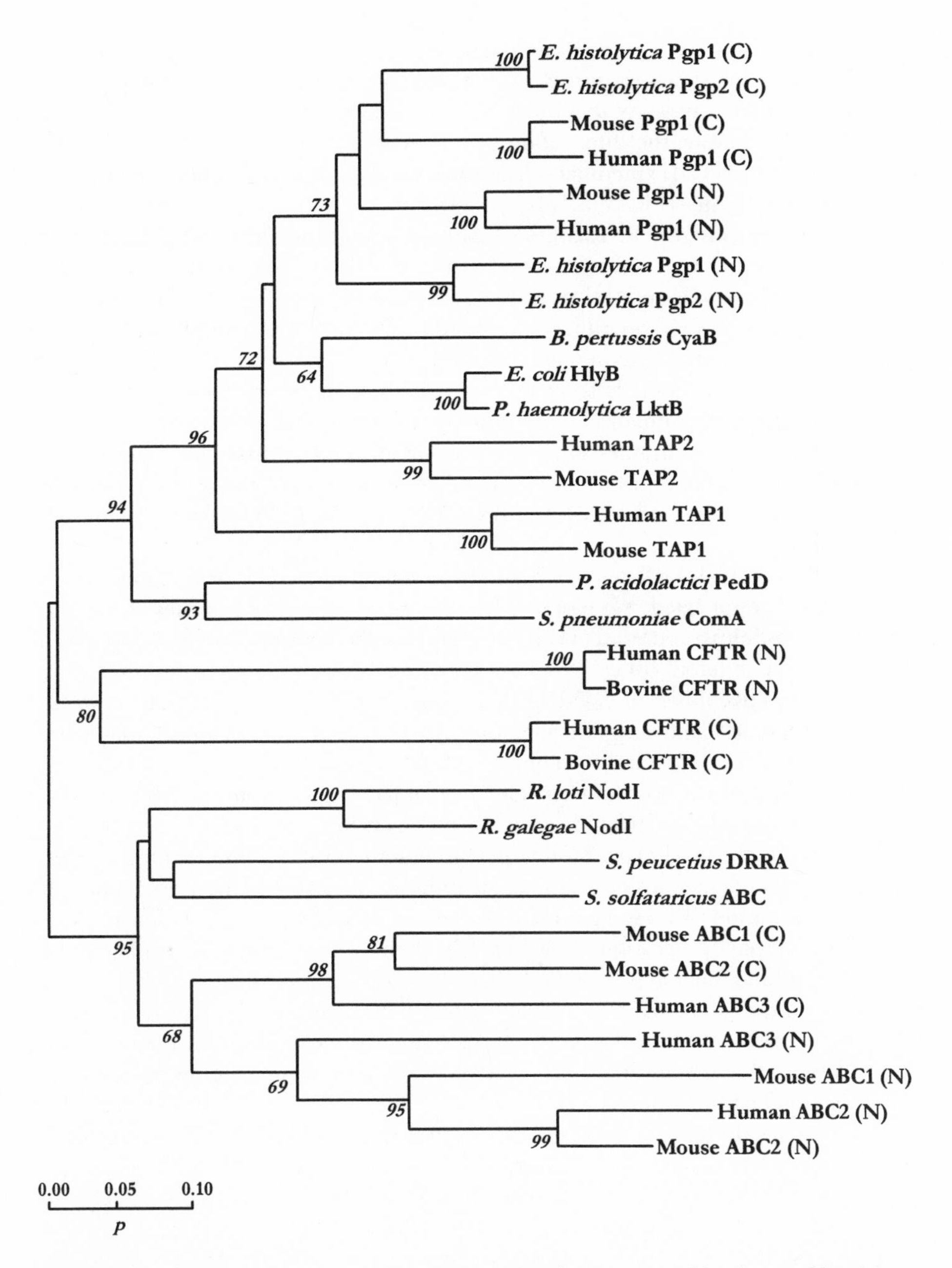

Figure 8.13. Phylogenetic tree of the ABC gene family, constructed by the neighbor-joining method on the basis of the proportion of amino acid difference (p). Numbers on the branches are as in Figure 3.2 From Hughes (1998b).

one (GRP78) on chromosome 9. The fact that there are yeast homologues of both of these shows (Figure 8.14) that they diverged prior to the divergence of animals and fungi. As mentioned in the previous section, in the case of the NOTCH family, the duplication of chromosome 6 (NOTCH4) and chromosome 9 (NOTCH1) members clearly antedated the divergence of deuterostomes and protostomes (Figure 8.12).

By contrast, the RXR family of transcription factors shows gene duplication that occurred much more recently, after the divergence of protostomes and deuterostomes (Figure 8.15). In the phylogenetic tree, RXRA (on human chromosome 9), RXRB (on human chromosome 6), and RXRG (on human chromosome 1) clustered together, apart from their insect homologues (Figure 8.15).

On the basis of these analyses, it can be concluded that the clusters of homologues on human chromosomes 6 and 9 (and the corresponding clusters on mouse chromosomes 17 and 2) did not all result from a block duplication early in vertebrate history or at any other time. Four of the genes on human chromosomes 6 and 9 (RXR, PBX, TEN, and C3/C4/C5 family members) probably did duplicate as a block early in vertebrate history. Shortly after that, RXR, NOTCH, PBX, and TEN family members, now on human chromosomes 9 and 1, duplicated as a block. If the clusters of homologous genes on human chromosomes 6 and 9 did not all duplicate simultaneously but, in fact, duplicated at widely different times, why have they now become clustered together? The obvious possibilities are (1) that their association is the result of chance; or (2) that there is adaptive significance in their clustering; in other words, that is has been favored by natural selection.

There are at least four separate pairs of paralogues (from the ABC, PSMB, NOTCH, and HSP70 families), whose members have been independently translocated to the two linkage groups on chromosomes 6 and 9. Such a large number of independent translocations seems unlikely to have occurred by chance. If the clustering of these genes is adaptive, the basis for this adaptation remains speculative. It is possible, however, that their clustering may be explained by shared functional and structural characteristics.

One characteristic shared by many of the genes on human 6p21.3 is a universal or very broad pattern of expression (Table 8.14). This is true of the class I MHC genes and of the TAP transporter and proteasome components that interact functionally with class I molecules. 6p21.3 also includes genes for essential cellular structural elements, such as histones and β tubulin, and broadly expressed regulators of expression such as PBX2 and ZNF13 (Table 8.14). 9q33–34 also includes a number of broadly expressed genes, most notably the *surfeit* complex. The presence of MHC class I in 6p21.3 and *surfeit* in 9q34 suggests the intriguing hypothesis that such complexes of highly expressed genes may, over evolutionary time, act as strong attractors of other highly expressed genes. It may be advantageous to locate highly expressed genes in regions that are transcriptionally active in most cells. Thus, a translocation that moves a highly expressed gene to one of these regions may be selectively favored.

Another characteristic of many genes in 6p21.3 and 9q33–34 is that they encode unusually long polypeptide chains (Table 8.14). Often, these genes

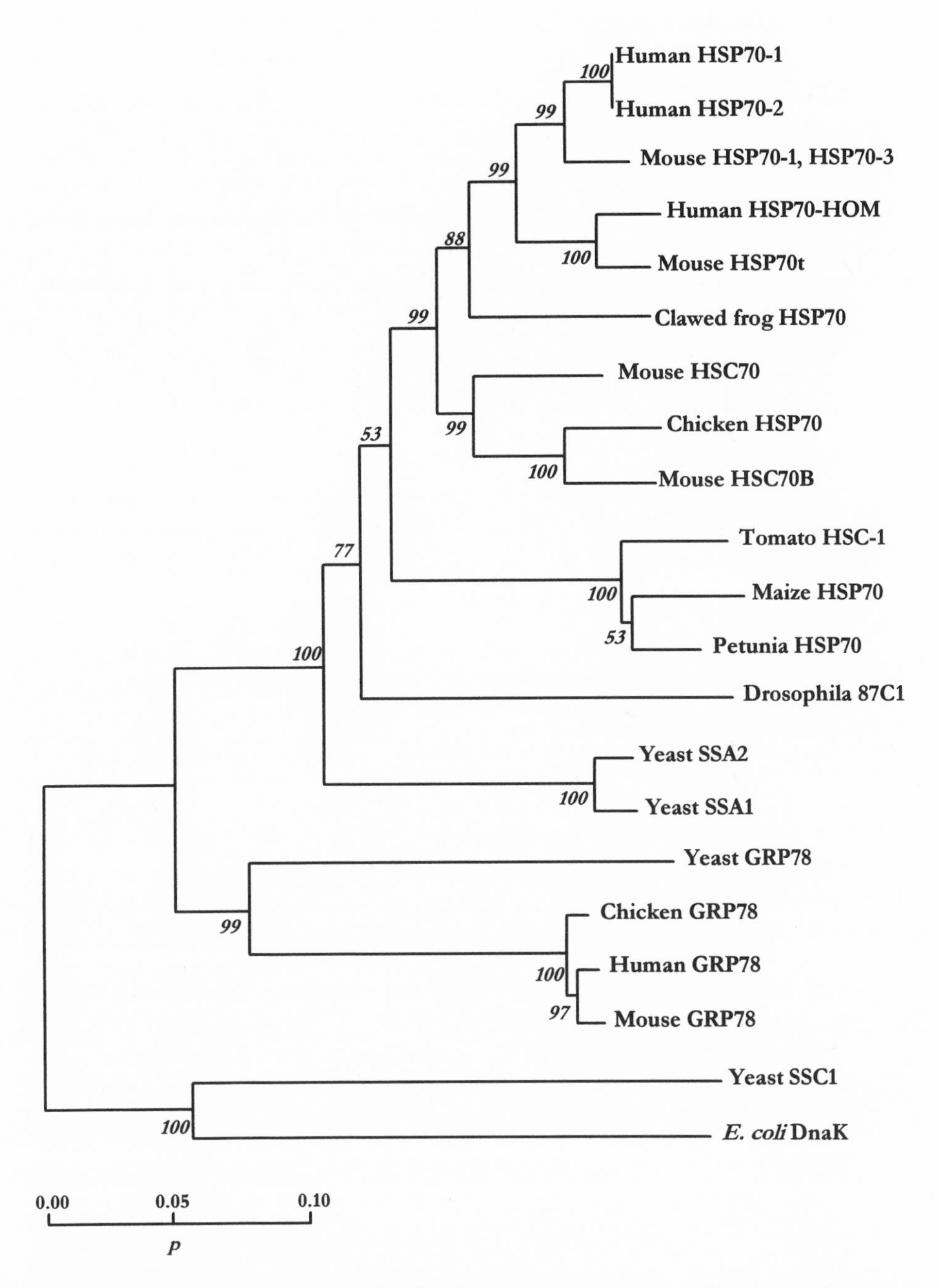

Figure 8.14. Phylogenetic tree of the HSP70 gene family, constructed by the neighbor-joining method on the basis of the proportion of amino acid difference (p). Numbers on the branches are as in Figure 3.2 From Hughes (1998b).

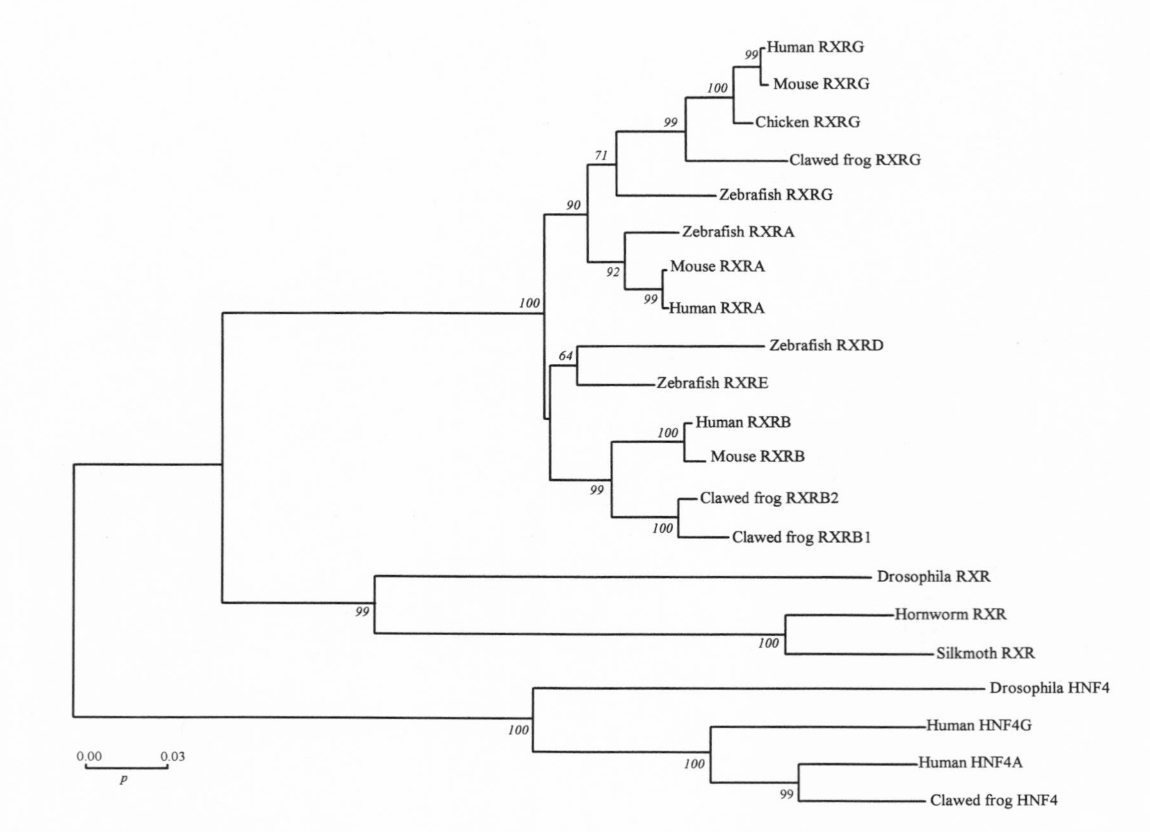

Figure 8.15. Phylogenetic tree of the RXR gene family, constructed by the neighbor-joining method on the basis of the proportion of amino acid difference (p). Numbers on the branches are as in Figure 3.2. From Hughes (1998b).

Table 8.14. Genes with broad to universal expression and genes encoding large (>700 a.a.) proteins located in human 6p21.3 and 9q33–34

	6p21.3	9q33–34
Broad universal expression		
	HLA class Ia, HSP70 homologues, histone H1.D, histone H2A.1, cyclin dependent kinase inhibitor I, serine kinase, Ndr serine/threonine kinase, ZNF173, β tubulin, valine t-RNA synthetase, PBX2, LMP2, LMP7, TAP1, TAP2, RXRB, RD protein	PSMB7, ABC2, PBX3, GRP78, Surfeit gene cluster, gelsolin, RXRA, ribosomal protein L12, CAN
Large proteins[a]		
	C4A (1744), C4B (1699), NOTCH4 (>1095), TENX (3536), valine t-RNA synthetase (1265), COL11A2 (1629–1736), phospholipase D (841) complement factor B (764), complement C2 (753), helicase-like (1245), female sterile homeotic homologue (755)	gelsolin (782), golgin-97 (767), CAN (2090), TENC (2203), C5 (1676), COL5A1 (1839), NOTCH1 (>2444), ABC2 (1472)[b]

[a]The number of amino acid residues is given in parentheses, where known. [b]Based on mouse homologue. From Hughes (1998b).

include large numbers of exons. For example, the C4A gene consists of 41 exons, while COLllA2 has 65 exons. It is possible that it is advantageous to locate such large genes in regions likely to be active transcriptionally, because the process of transcription and splicing of the pre-mRNA is complicated and presumably time consuming. Therefore, clusters of highly expressed genes such as the class I MHC and *surfeit* may serve to attract genes encoding large proteins. If so, the fact that the C2, factor B, and C4 complement components are linked to the MHC in mammals may be due to the fact that they are large proteins rather than (as has sometimes been supposed) the fact that they have an immune system function. (Their immune system function is, in any event, unrelated to that of the MHC).

There is some evidence for the hypothesis that a gene may be translocated into a certain chromosomal region, and then its position there will be conserved because it is adaptively advantageous. As mentioned, the fact that the *surfeit* genes are not linked in invertebrates and only some of them are linked in the pufferfish but their linkage is conserved in birds and mammals suggests that these genes were brought together over the course of evolution and that their current linkage in birds and mammals is selectively advantageous. Another interesting example involves the *DAZ* gene cluster, which encodes proteins involved in spermatogenesis and is located on the Y chromosome in humans. The ancestor of the *DAZ* genes was an autosomal gene that was translocated to the Y chromosome and amplified during primate evolution (Saxena et al. 1996).

9

Conclusions

A distinguished evolutionary biologist of the pre-DNA era once expressed to me the opinion that the availability of molecular sequence data had not made much difference for evolutionary biology: all DNA sequence data had done was to provide further evidence of processes such as genetic drift and natural selection, which we already knew to be at work in populations of organisms. Of course, one easy response on the part of a molecular evolutionist to such a sweeping statement would be to point out that DNA sequencing has brought with it many surprises, facts about genomes of which we were previously completely unaware. For example, the existence of introns in the genes of eukaryotes was unsuspected before DNA sequencing revealed them. But a larger question remains: how much difference does molecular data make for evolutionary theory? Is the Neo-Darwinian theory of evolution developed in the 1920s through 1940s sufficient, perhaps with some minor modifications, to account for what we know about evolution? Or is a new theory of evolution—perhaps a new "New Synthesis"—necessary?

In previous chapters I have discussed evidence for the role of natural selection at the nucleotide sequence level. Natural selection was, of course, central to the theory of evolution developed by Darwin and Wallace. It likewise played a key role in the Neo-Darwinian synthesis. Therefore, it might be expected that what we have learned about natural selection at the molecular level should have important implications for our overall theory of evolution.

In this chapter I try to summarize what molecular data have taught us about evolution, with particular emphasis on adaptive evolution. It will come as no surprise to those who have read the preceding pages that I believe we have learned a great deal that is new. I then briefly consider the implications of what we have learned for the status of evolutionary theory and the directions of future research in this field.

Our Changing Knowledge of Molecular Evolution

One of the first summaries of the findings of molecular evolution was that of Kimura and Ohta (1974), who listed five "principles governing molecular evolution." Nei (1987) produced a similar list, which overlapped with Kimura and Ohta's but added a number of points not covered by them. It may be worthwhile to quote Kimura and Ohta's "principles" in order to discuss them in the light of the much greater amount of molecular data available nearly 25 years later.

> (1) For each protein, the rate of evolution in terms of amino acid substitutions is approximately constant per year per site for various lines, as long as the function and tertiary structure of the molecule remain essentially unaltered. (Kimura and Ohta 1974, p. 427)

This statement relates to the question of the molecular clock, which has generated a vast literature (for reviews, see Gillespie 1991 and Li 1997). Of course, by accelerating the rate of amino acid replacement in a particular lineage, positive selection can lead to violation of the "clock-like" nature of molecular evolution. Because it is predicted that evolutionary rates will be constant "as long as the function of the molecule remain[s] unaltered," looking for changes in evolutionary rate can be a way to look for changes in function that may be indicative of adaptive evolution.

Consider the case of a multigene family encoding a protein with two different domains. Suppose that, in general, one domain (domain 1) is more strongly functionally constrained than the other (domain 2). Even if we have no absolute dates for divergence times of the genes analyzed, we expect the rate of amino acid replacement in domain 2 to be correlated with that in domain 1 (Figure 9.1A). Furthermore, we expect the relationship to be approximately linear, with a slope greater than 1, because domain 2, being less constrained, should evolve at a faster rate than domain 1 (Figure 9.1). However, what if the relationship is not linear (e.g., as in Figure 9.1B)? This suggests that a change in constraint has taken place at some point in evolutionary history, and this, in turn, implies that a change in function has taken place.

An example involves the heat-shock protein 70 (HSP70) gene family. HSP70s are highly conserved chaperone proteins found in all organisms (Figure 8.14). Known members of the family form four subfamilies (A, B, C, and GRP). All four families have members in yeast. The yeast SSC1 is encoded by a nuclear gene, but the protein is present in the mitochondria, and in a phylogenetic analysis yeast SSC1 clustered with HSP70s from bacteria (Figure 8.14). Thus, it seems likely that SSC1 originally was a gene of the bacterial symbiont that gave rise to mitochondria and was subsequently translocated from the mitochondrial to the nuclear genome.

The HSP70 protein forms a homodimer, each subunit of which includes two functionally distinct segments: (1) an ATPase; and (2) a peptide-binding region. The ATPase fragment has two lobes, with a deep cleft at the base of

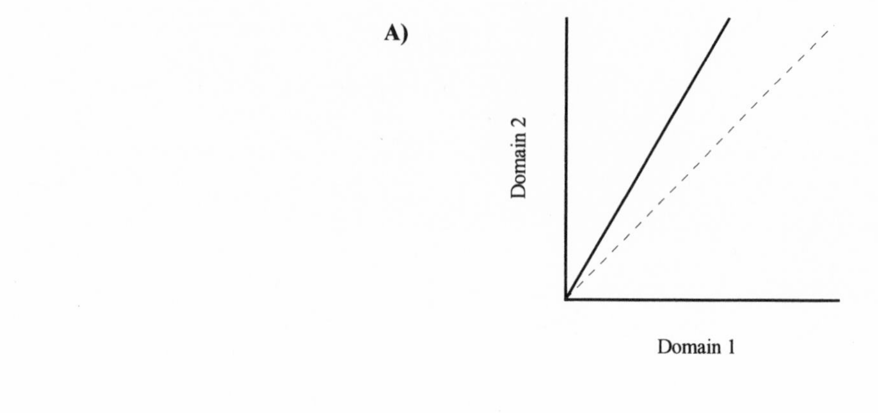

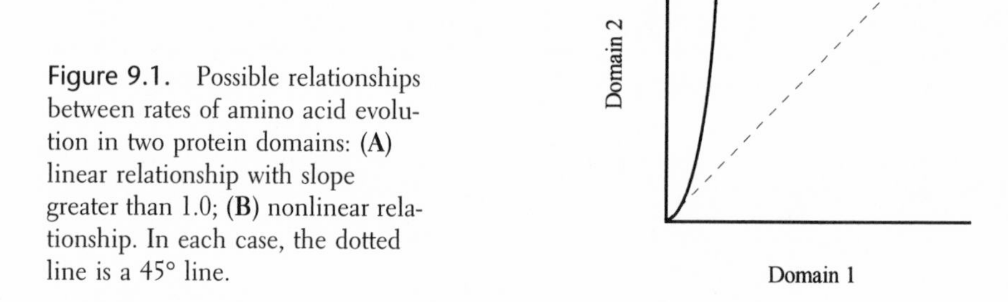

Figure 9.1. Possible relationships between rates of amino acid evolution in two protein domains: (**A**) linear relationship with slope greater than 1.0; (**B**) nonlinear relationship. In each case, the dotted line is a 45° line.

which ATP is bound (Flaherty et al. 1990). The ATPase fragment is composed of four domains (IA, IIA, IB, and IIB); the former two constitute the base of the cleft, while the latter two are believed to be involved in interactions with other proteins. A 21-amino acid segment in IIB has been identified as a calmodulin-binding domain (Stevenson and Calderwood 1990).

When phylogenetic trees were constructed separately for separate domains of HSP70, they revealed striking differences in the relative rate of amino acid replacement in different regions (Figure 9.2). This can be seen in Figure 9.2 by comparing the branches marked 1–2 (which separates subfamilies A and B) and 1–3 (which separates subfamily C from subfamilies A and B). In the IA domain, branch 1–3 is about twice as long as branch 1–2, although the difference is not statistically significant (Table 9.1). Likewise, in the peptide-binding domain, the lengths of branch 1–2 and branch 1–3 are not significantly different (Table 9.1). On the other hand, in domain IIA, branch 1–3 is over five times as long as branch 1–2, while in IIB branch 1–3 is nearly five times as long as branch 1–2, and in IB branch 1–3 is over 16 times as long as branch 1–2 (Table

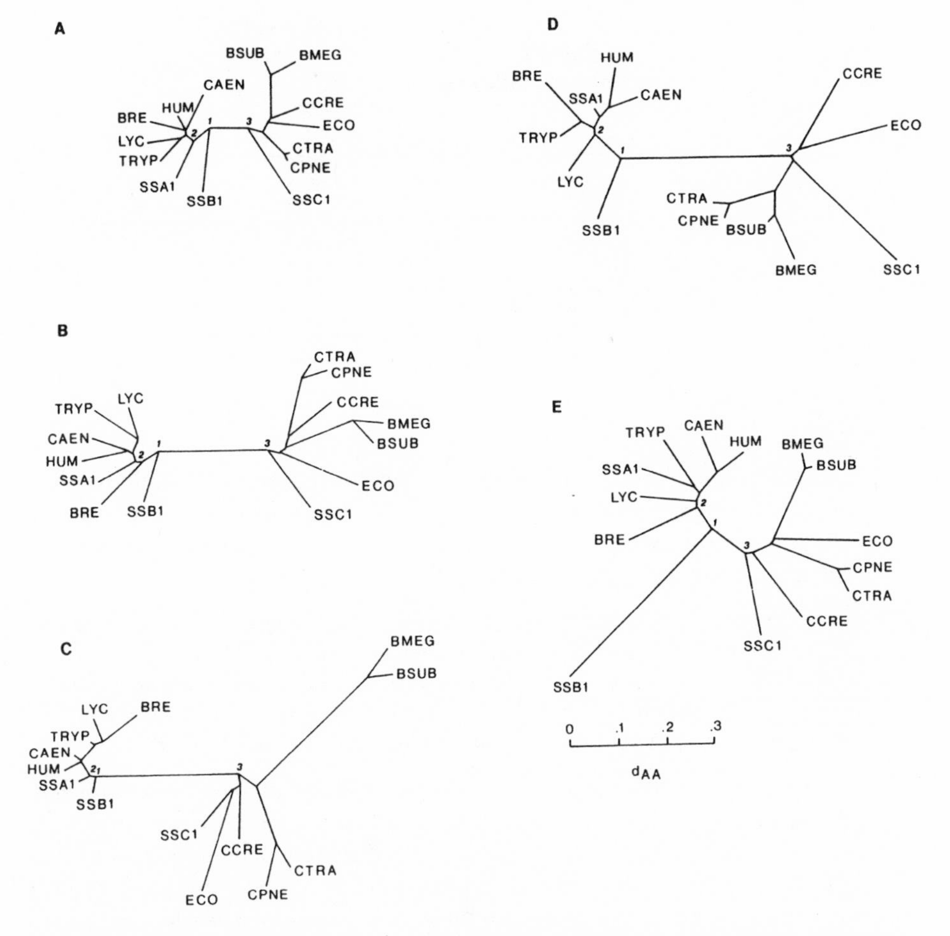

Figure 9.2. Phylogenetic trees for separate domains of HSP70s, constructed by the neighbor-joining method based on the Poisson-corrected estimate of the number of amino acid replacements per site (d_{AA}). **(A)** IA; **(B)** IIA; **(C)** IB; **(D)** IIB; (**E**) peptide-binding domain. Abbreviations for subfamily A are as follows: BRE = *Bremia lactucae;* CAEN = *Caenorhabditis elegans;* HUM = human; LYC = *Lycopersicon esculentum;* SSA1 = yeast SSA1; and TRYP = *Trypanosoma cruzi*. Abbreviations for subfamily C are as follows: BMEG = *Bacillus megaterium;* BSUB = *Bacillus subtilis;* CCRE = *Caulobacter crescentus;* CPNE = *Chlamydia pneumoniae;* CTRA = *C. trachomatis;* ECO = *Escherichia coli;* and SSC1 = yeast SSC1. Yeast SSB1 is the only subfamily B member. From Hughes (1993b).

Table 9.1. Lengths of branches (in terms of amino acid replacements per 100 sites), ±SE, in phylogenies of HSP70 regions (Figure 9.1)

Region	d_{12}	d_{13}
IA	3.5 ± 2.4	8.3 ± 3.3
IIA	3.9 ± 3.1	22.0 ± 5.1*
IB	1.8 ± 3.0	29.5 ± 9.0*
IIB	7.1 ± 3.6	35.2 ± 9.3**
Peptide-binding	5.2 ± 3.0	8.8 ± 3.1

Tests of the hypothesis that a distance equals that in region IA. *$p < .05$; **$p < .01$. Data from Hughes (1993b).

9.1). These differences in relative branch lengths suggest that subfamily C has diverged functionally from subfamilies A and B in domains IIA, IB, and IIB (Hughes 1993b). A similar approach may be useful for detecting functional changes in other multigene families.

> (2) Functionally less important parts of a molecule evolve (in terms of mutant substitutions) faster than more important ones.
>
> (3) Those mutant substitutions that disrupt less the existing structure and function of a molecule (conservative substitutions) occur more frequently in evolution than more disruptive ones. (Kimura and Ohta 1974, p. 428–429)

These two statements express principles that are fundamental to modern molecular biology. They are used as the basis for programs to align molecular sequences and for inferences regarding possibly functionally important parts of molecules. They also provide the underlying logic of the comparison of rates of synonymous and nonsynonymous nucleotide substitution as a test for positive Darwinian selection. When selection favors change at the amino acid level, the usually conservative nature of molecular evolution is reversed. Likewise, they provide a rationale for expecting that only natural selection will lead to a pattern of nonsynonymous substitution, causing change of amino acid properties such as charge to a greater extent than expected under random substitution (Hughes et al. 1990).

> (4) Gene duplication must always precede the emergence of a gene having a new function. (Kimura and Ohta 1974, p. 429)

As we have seen (Chapter 7), this statement is not true in general. The phenomenon of "gene sharing," seen most dramatically in the case of eye lens crystallins (Piatigorsky and Wistow 1991), implies that a single gene can acquire two distinct functions without duplication. Following duplication, each daughter gene can specialize to assume one of these functions. It is likely that this pattern of "gene sharing" preceding gene duplication is a common pattern in the evolution of novel protein function (see Chapter 7), and it may indeed be the most common pattern.

> (5) Selective elimination of definitively deleterious mutants and random fixation of selectively neutral or very slightly deleterious mutants occur far more frequently in evolution than positive Darwinian selection of definitely advantageous mutants. (Kimura and Ohta 1974, p. 430)

The tremendous amount of DNA sequence data that has accumulated in the last 20 years gives overall support to this statement of Kimura and Ohta. At the molecular level, the processes of purifying selection (eliminating deleterious mutation) and chance fixation of neutral mutations clearly predominate. Whether genetic drift also serves to fix "very slightly deleterious mutants" remains controversial. No unambiguous evidence exists on this particular point. Because the alleged effect is a very subtle one, it may indeed be difficult to discriminate between this and alternative hypotheses (but see Ohta 1995).

As reviewed in previous chapters, molecular analyses have also provided strong evidence of positive Darwinian selection. Of course, these cases are still relatively rare. We can still agree with Kimura and Ohta that purifying selection and fixation of neutral mutations occur far more frequently than positive selection. However, I believe that the number of cases of positive selection for which we now have good molecular evidence is considerably greater than would have been predicted by the early neutralists.

Thus, in the time since Kimura and Ohta's (1974) paper, we have seen the triumph of the neutral theory in that its broad predictions have been confirmed by extensive molecular sequence data. At the same time, we have seen increasing evidence of adaptive evolution. Furthermore, as the number of confirmed cases of positive selection increases, we are able to glimpse something of its role in evolution in general. As a result, we may eventually be able to develop strong generalizations about adaptive evolution. Because at the present time the number of cases of positive selection is still small, there is a certain risk in attempting such generalizations. Nonetheless, some patterns seem to be emerging; I will outline these in the next section.

Circumstances of Positive Selection

Known and likely examples of positive selection at the molecular level mostly involve three categories: (1) balancing selection, (2) directional selection between species, and (3) adaptive diversification of genes within multigene families. I will attempt to describe some general features of each of these categories.

Balancing Selection

There are two distinct types of balancing selection observable at the molecular level. The first type might be called transient or opportunistic balancing selection. In this type of selection, it happens that some aspect of the environment gives rise to a form of balancing selection in one species; but this selection is the unique consequence of a situation peculiar to the one species, and thus both the selection and the polymorphism to which it gives rise are transient phenomena over evolutionary time. In the second type, which might be called persistent balancing selection, the factors giving rise to the balancing selection remain constant for long periods of evolutionary time. Thus, such balancing selection can give rise to polymorphisms that are long lasting and predate speciation events. These two types of balancing selection are best viewed as ends of a continuum, rather than as mutually exclusive categories, because there are some known cases that seem to fall somewhere in between these extremes.

The classic example of transient balancing selection is the sickle-cell anemia gene in humans. This polymorphism involves a single amino acid replacement in the β chain of hemoglobin. Without medical intervention, sickle cell homozygotes generally die before reproductive age. However, heterozygotes have improved resistance to *Plasmodium falciparum* malaria. Therefore, overdominant selection has increased the sickle cell allele to a high frequency (about 15%) in areas where malaria is endemic, particularly West Africa. This polymorphism clearly has not been maintained for a very long time. It is unique to humans, being unknown in chimpanzees or other nonhuman primates. How long it has existed in human populations remains uncertain.

It was previously argued that *P. falciparum* has only become a human parasite very recently, perhaps at the time agriculture was introduced into West Africa about 6,000 years ago, and that the sickle-cell mutation first appeared around the same time (Livingstone 1958). Previously, parasitologists assumed that all parasite–host interactions evolve over time in the direction of reduced virulence and ultimately commensalism. On this assumption, it was concluded that *P. falciparum*, being the most virulent of human malaria parasites, must be the most recent. However, the assumption that parasites will inevitably evolve decreased virulence over time is unwarranted. In addition, molecular phylogenies show a close relationship between *P. falciparum* and the chimpanzee malaria *P. reichenowi* (Escalante and Ayala 1994; Hughes and Hughes 1995b). This strongly suggests that *P. falciparum* has coevolved with the hominid lineage since its divergence from the chimpanzee lineage. Nonetheless, it is possible that the environmental changes introduced in West Africa by agriculture did increase the incidence of malaria by providing more habitat for mosquitos. Thus, even if the sickle-cell polymorphism is older than West African agriculture, it may have greatly increased in frequency in the past few thousand years.

The *Adh* polymorphism in *Drosophila melanogaster* (chapter 5, Alcohol Dehydrogenase in *Drosophila melanogaster*) is another case of a transient balanced polymorphism. No similar polymorphism has been found in other well-studied members of this genus, such as *D. pseudoobscura* (Schaeffer and Miller 1992).

The MHC polymorphism is the prototypic example of persistent balancing selection. As discussed previously, the MHC loci are characterized by long-lasting polymorphisms that predate speciation (chapter 4, Trans-species Polymorphism at MHC Loci). Even more remarkedly, the type of selection acting at the MHC loci seems to be the same in all jawed vertebrates (Hughes and Hughes 1995a), and thus presumably the same type of selection has persisted since the origin of vertebrates. The MHC loci themselves have doubtless been turned over by duplication many times in the intervening period, yet the same type of balancing selection has persisted. Thus, for example, mammals and bony fish do not share any orthologous class I or class II MHC loci, yet the pattern of nucleotide substitution indicative of balancing selection is essentially the same in mammals and fish (Hughes and Hughes 1995a). This has apparently happened because the problem of immune defense against parasites has been a persistent one for vertebrates.

The polymorphism at self-incompatibility loci of plants (chapter 5, Self-Incompatibility Genes of Plants) provides another example of persistent balancing selection. In this case, however, the same type of selection has probably not been maintained for as long a time as in the case of the MHC. Different families of flowering plants have totally different systems of self-incompatibility, which have presumably evolved independently.

Many cases of polymorphisms in parasites that arise as a result of selection by the host immune system are probably transient. A possible example would be viral polymorphism selectively maintained within a single patient. As mentioned (chapter 5, Selection on HIV-1), it has been proposed that such selection occurs in the case of HIV-1, although the data on this point are not conclusive at present. Such selection, if it occurs, is presumably a consequence of the individual host's immune system, including the MHC genotype, which, in a highly polymorphic species like the human species, may be shared with few other members of the population.

Balancing selection on the circumsporozoite protein (CSP) of *Plasmodium falciparum* has maintained polymorphism for about 2 million years (chapter 5, The Circumsporozoite Protein of *Plasmodium*). Nonetheless, this is still a relatively transient polymorphism. There is no evidence of similar selection on the corresponding region of the CSP in most other species of the genus *Plasmodium* (Hughes 1991b; Hughes and Hughes 1995b). The one possible exception is the monkey parasite *P. cynomolgi*, whose CSP may be subject to similar selection; but in *P. cynomolgi*, the protein region subject to the strongest selection is not the same as that in *P. falciparum* (Hughes 1991b). Different *Plasmodium* species are, of course, exposed to different host immune systems, and thus it is not really surprising that the type of natural selection acting on their proteins may differ. Thus, it is extremely surprising that MSA-1 polymorphism in *P. falciparum* has been maintained for about 35 million years, which means that it has survived many changes in host MHC molecules (chapter 5, Other *Plasmodium* Surface Proteins). The existence of persistent balancing selection in this case, while other *Plasmodium* surface proteins are subject to relatively transient balancing selection, suggests that the role of MSA-1 in interaction with the immune system may be fundamentally different from that of other malarial antigens. Perhaps region 6 of MSA-1, the region showing ancient polymorphism, interacts with immunoglobulins rather than MHC, because immunoglobulin loci are much more stable over evolutionary time than MHC loci.

Directional Selection between Species

The diversity of species is one of the most striking facts about the organic world. Darwin and Wallace proposed that this diversity is the result of an evolutionary process and, furthermore, that natural selection has played a key role in the process. At the molecular level, it might be supposed that directional selection causing divergence between species would be one of the most commonly observed forms of natural selection. In fact, at the present time, there is much less evidence of such selection than there is either of balancing selection within species or of directional selection between duplicate genes in multigene families (see Chapter 6). The relative lack of examples of directional selection between species may simply result from a lack of study of this question. Alternatively, it may have some implications for our understanding of how adaptive evolution occurs.

The mechanism of speciation remains controversial. One possible mechanism is that, after two populations are isolated by a geographic barrier, chance fixation of neutral alleles in one or both populations results in a barrier to reproduction (Nei et al. 1983). If this model is correct, the initial achievement of reproductive isolation would not depend on positive selection. Of course, either during their initial period of isolation or subsequently, one or both populations might be subject to some directional selection not shared by the other population. Such selection might lead to directional divergence between species at one or more orthologous protein-coding loci.

However, the morphological differences between even closely related species are often quite dramatic; for example, between humans and chimpanzees. It is unlikely that major morphological differences can be caused by differences in the amino acid sequences of proteins. Rather, it has been argued that such differences are likely to be caused by mutations that affect the pattern of gene expression (Nei 1987). These presumably will often involve changes in regulatory region sequences that bind transcription factors, rather than in coding regions. Also, gene duplication and subsequent functional divergence of the duplicated genes may be a factor in morphological change, if patterns of gene duplication in relevant gene families differ between related species.

So far, the study of adaptive evolution of regulatory regions is in its infancy. In the future, comparison of such regions between closely related species may yield important clues regarding adaptive radiation. Because regulatory regions are not coding regions, new strategies of testing for positive selection will be required. In some cases, it may be possible to obtain evidence of convergent evolution in regulatory regions, which in turn, strongly suggests adaptive evolution. The evolution of an AP-1 binding site in squid crystallin genes (Tomarev et al. 1994) may be an example (chapter 7, Bifunctionality Preceding Gene Duplication). Furthermore, although the comparison of synonymous and nonsynonymous rates cannot be used in the case of regulatory regions, it may be possible to test for an enhanced rate of nucleotide substitution in a regulatory region by comparing it with adjacent regions that lack a regulatory function and, thus, are free of selective constraint.

Adaptive Diversification of Duplicate Genes

Extensive study of polymorphism within species has uncovered relatively few convincing examples of balancing selection. Comparison on orthologous genes of related species has revealed still fewer convincing examples of positive Darwinian selection. Comparisons within multigene families seem rather different. Here, analyses have quite frequently revealed evidence of adaptive diversification (chapter 7, Diversifying Selection in Multigene Families). Under the model of evolution of new protein function presented above (chapter 7, An Alternative Model of the Evolution of Novel Proteins), it is expected that when two proteins encoded by related genes have different functions, positive selection will usually have been involved in adapting them to these functions.

It is important to remember that when we compare two related genes that have evolved separate functions, these are hardly a random sample of all duplicated genes. In most cases of gene duplication it is likely that one of the duplicate genes will eventually become a pseudogene, because a mutation eliminating its expression will not be deleterious. In a minority of cases, the two will adapt to separate functions as a result of positive selection. As I have argued previously, there will be a much greater opportunity for such selection if both functions are present in the common ancestor prior to gene duplication. Because of the importance of gene duplication in evolution, it seems likely that directional selection on duplicated genes is the primary type of Darwinian selection operating in nature.

Ohno (1970) emphasized the importance of gene duplication in evolution. However, under his model, the evolution of a new protein function was viewed as the result of a chance process. As we have seen, this model is almost certainly incorrect (Chapter 7). Rather, I would argue that the evolution of a new protein function is the primary focus of positive Darwinian selection and, thus, the main source of new adaptations. Understanding this process should be a primary focus of study of those who are interested in adaptive evolution. Because such selection is unlikely to operate on numerous loci simultaneously, it is improbable that the duplication of entire genomes by polyploidization has played a significant role in adaptive evolution, contrary to the view of Ohno (1970).

Mutation and Selection

Both Nei (1987) and Li (1997) raise the question of the relative importance of mutation and selection in evolution and argue for the greater importance of the former. This may seem a trivial argument because few biologists would contest that adaptive evolution requires both processes: mutation, as the source of new genotypes, and natural selection. However, the emphasis on mutation is understandable as a response to certain ideas that were common among evolutionary biologists prior to the molecular era. For example, it was often said that natural populations of organisms contain enough genetic variation that they can respond to virtually any sort of selection the environment imposes. If true, this would imply that new mutations are not really needed for adaptive evolution; the relevant alleles are already present in advance.

The basis for this idea was, no doubt, the observation that, in artificial selection, most organisms are able to be selected for increased value of some trait of interest. However, it has also been observed that the capacity of an artificially selected population to respond to selection is soon exhausted. Thus, there is enough heritable variation in body size in a species like the fruitfly that we will get a good response if we select for larger fruitflies; but we cannot keep selecting until we get fruitflies as large as houseflies (let alone as large as bumblebees). Clearly, it would seem that a major new mutation, probably one affecting the timing of gene expression in development, would be necessary to get housefly-sized fruitflies.

Furthermore, if we were to take literally the idea that each natural population contains enough genetic variation to respond to any sort of selection, we would be led to the bizarre conclusion that all fixed differences between sibling species existed as polymorphisms in their common ancestor. For example, this would imply that all the morphological and behavioral differences between humans and chimpanzees resulted from fixation of different alleles by different selection pressures acting on two different populations of their ancestral species, and that no new mutations occurred in either lineage. But it seems very doubtful that anyone has ever taken these ideas to such an extreme conclusion.

Thus, Nei (1987) and Li (1997) seem correct in emphasizing the potential importance of major, new mutations ("macromutations") in adaptive evolution. Molecular studies have shown that mutations include not just nucleotide substitutions but such important processes as gene duplication and recombination. Recombination at the molecular level includes not only classic recombination between loci but also the creation of new hybrid loci by gene conversion or exon shuffling (Doolittle 1985; Hughes 1996b). A recombinational event that combines a gene with a new promoter may be a way to dramatically change the pattern of gene expression and, thus, may be important. It is quite plausible that such mechanisms have been important in evolution. However, there remains a need to document cases in which macromutational processes can be linked to adaptive phenotypic changes.

In several of his writings, Kimura (e.g., 1983) suggested that adaptive evolution occurs when a formerly neutral variant becomes selectively favored. He called this the "Dykhuizen-Hartl" effect. Although such events may sometimes take place, there are reasons for believing that their role in adaptive evolution is limited at best. By definition, two selectively equivalent alleles at a protein-coding locus should produce products that are functionally essentially equivalent in the species' usual environment. Such slight differences as might exist between two such allelic products hardly seem sufficient to account for any major adaptive advance.

As mentioned in the previous chapter, minute effect selection models that are derived from quantitative genetics may be applicable to either directional or stabilizing selection on such traits as nucleotide composition, codon usage, or genome size (chapter 8, Natural Selection on Condon Usage). Their application to most phenotypic traits is problematic, however, because these traits are not likely to result from minute additive effects of a large number of loci. At least approximately, however, the type of evolution envisaged by these models may take place when environmental change causes a shift in the optimum value of some phenotypic trait. The changes in bill size of the Galapagos finch tracked by Grant and Grant (1989a) are an apparent example. Larger bill sizes were apparently favored in dry years, and the mean bill size increased in apparent response to such selection. This case and others like it, I would argue, are totally irrelevant to adaptive evolution in the long term. I would predict that major adaptive shifts—such as the evolution of the numerous spectacular bill forms for which the Galapagos finches are famous—involve new macromutations rather than selection on genetic variation already present in the population.

These considerations also lead to another prediction: that, in the course of evolution, natural selection generally precedes the mutations that lead to important adaptive advances. When selection is strong enough, macromutations that produce radical phenotypic changes have a better chance of being fixed than such mutations would ever have if they were selectively neutral. Of course, this is only a prediction which awaits empirical testing.

Is It Time for a New "New Synthesis"?

To return to the question posed at the beginning of the chapter, I think that it is safe to say that molecular data have indeed changed our views of evolution in numerous ways. Considering only adaptive evolution, the topic of this book, the limited amount of data already available has brought to light new processes and phenomena that were previously unknown. For example, balancing selection was known both in theory and by a few examples (such as the sickle-cell gene), but the marked contrast between the two types of balancing selection (persistent and transient) was totally unimagined prior to the availability of DNA sequence data. Several authors, mostly notably Ohno (1970), had proposed an important role for gene duplication in evolution, but many were shackled to an incorrect view that saw the evolution of a new protein function as an essentially random process in which natural selection played no role. Now DNA sequence data have made it clear that gene duplication is indeed important in evolution, but that natural selection plays a key role in the evolution of a new protein function. Indeed, these data suggest that positive Darwinian selection following gene duplication is the main focus of adaptive evolution of protein-coding genes.

It might be asked whether the new theoretical perspectives on evolution that we have obtained from molecular data—including both the neutral theory and new insights into adaptive evolution—constitute a true "scientific revolution" or "paradigm shift" in evolutionary biology in the sense in which these terms were used by Kuhn (1962). Of course, Kuhn's concept of a scientific revolution was descriptive and not normative, and there is no reason to suppose that evolutionary biology, or indeed any science, will always follow his neat schema. The issue is further complicated by the wide dissemination of Kuhn's ideas among scientists. Perhaps paradigm shifts can no longer happen in quite the same way once scientists themselves are aware of the idea of a paradigm shift.

The "Neo-Darwinian Synthesis" or "New Synthesis" formulated by evolutionary biologists from the 1920s to the 1940s (Mayr and Provine 1980) was in some ways a very unusual event in the history of science. The New Synthesis represented the achievement of a consensus among evolutionary biologists, but the process of consensus-making was much more organized and self-conscious than most such processes are likely to be. One reason for the self-conscious posture of the Neo-Darwinists was the powerful position in both society and in science itself of antievolutionary views. This was particularly true in the United States, where the successful prosecution in 1925 of Tennessee high school teacher John Scopes for breaking a law against the teaching of evolution was followed by a period of about three decades in which evolutionary ideas were largely absent from public education. Even at the university level, quite a few state-supported universities as recently as the 1960s listed courses in evolution under the innocuous title of "theoretical biology" to avoid the wrath of fundamentalist state legislators. If we keep in mind the prevailing climate of near-persecution, some of the less pleasant aspects of the Neo-Darwinian synthesis are understandable if not exactly justifiable; for example, the tendency of the Neo-Darwinians to come down very hard on dissenters within the fold and to enforce an "orthodoxy" that would seem antithetical to the spirit of scientific inquiry.

So if the question is whether we need a new synthesis in the spirit of that previous synthesis, my answer would be an emphatic, "No." Rather, I would hope that evolutionary biology has matured as a science in a way that makes the self-conscious closing of ranks of the old New Synthesis unthinkable today. As for Kuhn's metaphor of a scientific revolution, my own taste in metaphors runs more to that of Otto Neurath, who likened science to a boat that we continually repair while remaining afloat in it (Quine 1960).

I am also perfectly comfortable with the retention of the term "Neo-Darwinian" for the general approach taken by the majority of researchers currently working in evolutionary biology, including myself. Neo-Darwinian primarily refers to the synthesis of Mendelism and Darwinism. If, as has been argued (Orel 1996), modern molecular biology represents the flowering of Mendelism, molecular evolutionary biology represents a far more profound synthesis of Mendelism and Darwinism than was possible in the 1920s–1940s. This usage differs from that of Nei (1987), who confines Neo-Darwinism to the dominant view among evolutionary biologists during the period of the New Synthesis, and contrasts his own position with that view. Nei (1987) also apparently considers Neo-Darwinism a kind of aberration, and emphasizes the similarity of his own position with that of Morgan in the era preceding the New Synthesis. Certainly, Morgan emphasized the importance of mutation, but it would be anachronistic to give him too much credit for anticipating modern ideas. In keeping with my Neurathian view of the scientific enterprise, I prefer to emphasize a theme of continuity with change as evolutionary biology passes from the era of classic Neo-Darwinism into the molecular era.

Future Directions

Throughout the tremendous flowering of the biological sciences that has taken place in the second half of this century, most experimental biologists have been content to claim an allegence to evolution as a general explanatory principle in biology; but, if asked, I suspect that most would admit that evolution has little to do with their day-to-day work. There has been a sharp separation between evolutionary biologists—often field or museum researchers viewed as hopelessly old-fashioned by their peers in molecular biology—and experimental biological or biomedical scientists. The advent of molecular sequence data has already changed this situation to some extent. As mentioned here, experimental biologists implicitly make use of insights derived from the neutral theory in framing hypotheses about which regions of an amino acid or DNA sequence might be functionally important. And phylogenetic trees are sprouting up with increasing frequency even in biochemical journals. (It is another issue, of course, whether or not the theoretical bases of the tree-making methods employed is widely understood.)

The study of adaptive evolution may still seem esoteric to molecular biologists, but I would argue that it can have important practical consequences. For example, it may be very important in our attempt to cure AIDS to determine whether polymorphism selectively maintained by the host immune system plays a key role in the pathogenicity of the HIV-1 virus (chapter 5, Selection on HIV-1). Similarly, in the case of other human pathogens, the evolutionary dynamics of the host–parasite interaction may yield important clues to their pathogenicity and may suggest new strategies of prevention.

Furthermore, just as evidence that a particular protein domain is subject to strong purifying selection can be useful in identifying it as functionally important, evidence of positive selection on a domain can suggest hypotheses regarding function. Imagine if, before we knew anything about the function of MHC molecules, we had known that natural selection acts to diversify the amino acid sequence in a certain region of the molecule. Such a finding would suggest that the relevant region must be important for these molecules' function and might, in turn, lead to a quick discovery that the region in question binds peptides. Of course, the history of our knowledge of the MHC took a different path, but in the future there may be cases where we discover evidence of positive selection before we know a molecule's function.

The *Lyb-2* locus of the mouse provides an example. This polymorphic locus encodes a cell-surface protein that may play a role in communication between B cells and T cells (Van de Velde et al. 1991). Estimation of rates of synonymous and nonsynonymous substitution among mouse *Lyb-2* alleles (Table 9.2) showed that positive selection has acted to diversify the distal extracellular portion of the molecule (Hughes 1993c). Although the function of *Lyb-2* is poorly understood, the fact that natural selection is acting on the molecule in this region implies that diversity there confers a selective advantage. This, in turn, may suggest future experiments to elucidate the molecule's function.

In the future, the concept of positive Darwinian selection will, I hope, become part of the conceptual framework of every working molecular biologist, just as the concept of purifying selection has now become (even if the term "purifying selection" is itself rarely used). Likewise, the methods used to test for positive selection will become important tools in the arsenal of any one involved in sequence analysis. Although cases of neutral evolution greatly outnumber

Table 9.2. Mean number of synonymous (d_S) and nonsynonymous (d_N) nucleotide substitutions per 100 sites (±SE) among alleles at the *Lyb*-2 locus of the mouse

Domain	No. codons	d_S	d_N
Cytoplasmic	95	1.8 ± 1.3	0.6 ± 0.4
Trans-membrane	21	3.8 ± 3.9	1.5 ± 1.5
Proximal extracellular	154	2.7 ± 1.3	2.5 ± 0.7
Distal extracellular	77	1.4 ± 1.4	8.6 ± 1.8*
Entire sequence	348	2.2 ± 0.8	3.2 ± 0.5

*d_S significantly different from d_N ($p < .05$). From Hughes (1993c).

those of positive selection, because of the importance of the latter in adaptive evolution, it will be important to be able to recognize them and to discriminate statistically between the two.

While helping to heal the long-standing split between evolutionary and experimental biology, the advent of the molecular era has served to accentuate fault lines within evolutionary biology itself. Provine (1991, p. 212) has lamented that "evolutionary biology has already been bifurcated into molecular and phenotypic by Motoo Kimura's neutral theory of molecular evolution." This, I think, is giving the neutral theory far too much credit. The bifurcation between phenotype and genotype is part of the nature of biological reality itself. It is true that many evolutionary biologists in the premolecular era convinced themselves that they could understand what was going on at the genotypic level merely by studying phenotypes, but this belief was largely delusional. It is time that evolutionary biologists face the reality of the gap between genotype and phenotype, because merely denying its existence will not make it go away.

Actually, there are two gaps between the genotype (i.e., the DNA sequence) and the kind of gross morphological phenotypes (like bill shape or feather color of birds) studied by traditional morphological evolutionists. The first is the gap between DNA sequence and phenotypes at the molecular level, including the protein level, that underly phenotypes at the morphological or whole-organism level. Then there is a second gap between molecular phenotypes and their whole-organism effects.

I believe that molecular evolutionary studies of adaptive evolution have begun to bridge the first gap and have at least shown signs of bridging the second gap. For example, the MHC provides an example of the connection between positively selected nucleotide substitutions and the peptide-binding properties of important immune system recognition proteins (Chapter 4). These peptide-binding properties can, in turn, be related at least in some cases to phenotypes at a whole-organism level, such as the ability to resist a specific parasite.

A second type of molecular phenotype involves properties of the genome itself or "nucleotypes" (Chapter 8). At the present time, complete genomes of a number of organisms have been sequenced, and that number will no doubt increase soon. The resulting data will give us a greatly enhanced ability to study the evolution of genomes and to assess the relative contributions of chance processes and natural selection to the evolution of genome properties. In the last chapter (chapter 8, Adaptive Gene Clustering), I showed the inadequacy of some hypotheses that attempted to explain current linkage relationships in the mammalian genome purely on the basis of presumably selectively neutral duplication events. As in this example, it seems likely that current linkage relationships have arisen as a result of both neutral and selective processes—that is, both genetic events (such as duplication and translocation that have been fixed by chance) and other such events that proved to be selectively advantageous.

Clearly, bridging the second gap remains a major challenge for molecular evolutionary biology. One type of study that will help bridge this gap is detailed evolutionary study of key proteins in complex physiological systems such as the vertebrate immune system, the vertebrate eye, and the vertebrate nervous system. In addition, it will be important to unravel the molecular basis of major morphological changes between closely related species, such as upright posture in humans in contrast to our closest relatives, the chimpanzees. As mentioned, study of the adaptive evolution of regulatory sequences may play a key role.

Ecologists and ethologists have drawn attention to some specialized types of natural selection, and it would contribute greatly to bridging the gap between molecular and phenotypic evolution if we were able to examine these at the molecular level. One example is kin selection, which is natural selection acting through genes shared with nondescendant relatives rather than through the individual's own reproductive success (Hamilton 1963, 1964). Kin selection has been hypothesized as the mechanism by which "altruistic" behaviors have evolved, particularly those of the sterile castes of social insects, which work to raise the offspring of relatives but do not usually reproduce themselves. Evidence of kin selection at the molecular level might be obtained by studying the molecular evolution of melittin, the active ingredient of honey bee venom. In stinging a vertebrate intruder in the hive, a honeybee dies—the ultimate act of "altruism" in defense of the young being raised by the colony. By sequencing homologues of the melittin gene from related subsocial and solitary bee species, it should be possible to reconstruct the evolutionary steps whereby the honeybee's ancestors evolved a protein toxic to vertebrates.

Another type of selection that has attracted a great deal of attention at the phenotypic level but so far is little understood at the molecular level is sexual selection; that is, natural selection arising from a differential ability of individuals to obtain mates (Darwin 1871). One idea that seems to closely relate to studies of molecular variation is the idea that, in animals, females choose their mates on the basis of "good genes." This idea is controversial in the behavioral literature. In some animal species, males provide parental care or resources needed by the females to reproduce; or they defend such resources from other males, allowing access to females in exchange for mating. In these cases, it seems likely that females choose mates primarily on the basis of the quality of the parental care or resource and only secondarily (if at all) on the basis of the male's genetic quality. In other cases, however, males provide no parental care or resources.

In lekking species, males perform elaborate courtship displays, and mating is highly nonrandom. At least in these species, it has been proposed that females use information in males' displays to assess their genetic quality as a way of enhancing the genetic quality of their offspring.

The idea of genetic quality seems problematic from what we know about evolution at the molecular level. Theoretically, if there were a locus undergoing directional selection, females might choose mates bearing the favored allele. But the neutral theory suggests that directional selection is relatively rare, and molecular studies seem to back up this prediction. It is hard to imagine how a propensity for females to choose genetically favored mates could evolve and be maintained if there rarely was an opportunity to use it. By contrast, Hamilton and Zuk (1982) hypothesized that females might choose males bearing currently favored alleles in the case of frequency-dependent selection driven by parasites. The problem with this model is that in the best known case where a polymorphism is maintained by parasites—the MHC—the selection is probably overdominant, rather than frequency dependent (chapter 4, Trans-species Polymorphism at MHC Loci).

Still, molecular data show that by far the predominant form of natural selection at the molecular level is purifying selection—that is, the elimination of deleterious mutations. Might female choice of mates serve the role of avoiding deleterious mutations? It has been noted that male courtship displays of many species serve as an "honest advertisement" (i.e., provide accurate information) regarding the male's overall level of health (Bradbury and Vehrencamp 1998). Furthermore, in many animal species, the female's response to male courtship often includes nonreceptivity behaviors that discourage all but the most vigorous males (Thornhill and Alcock 1983). Given the persistence of deleterious mutation, there seems an obvious benefit, from the female's point of view, to choice behaviors that prevent mating with males bearing slightly deleterious alleles. Such female choice would accelerate the effect of purifying selection in eliminating deleterious mutations. This hypothesis could be tested by comparing levels of nonsynonymous nucleotide polymorphism in two species with similar effective population sizes but different mating systems, such that female choice of male "quality" is thought to play a major role in one but not the other.

I am confident that the simple methods of molecular data analysis described in this book—including reconstruction of phylogenetic trees, estimation of patterns of nucleotide substitution in functionally distinct gene regions, and reconstruction of ancestral amino acid sequences—have the power to unravel for us many of the mysteries of past adaptive evolution. At the same time, however, I think it is important for evolutionary biologists to be aware of the distinction between our ability to explain some phenomenon *in principle* and our actual explanation of it. For instance, Darwinists have long held that the processes of mutation and natural selection can, in principle, explain the evolution of complex structures such as the vertebrate eye. However, we should be clear that we have not yet reached the point where we are actually able to explain all the steps in the evolution of the vertebrate eye. Molecular studies of eye development have made tremendous progress toward that goal within the past few decades, but we are still a long way from a complete account. Similarly, while kin selection can, in principle, account for the evolution of insect social behavior, we have as yet no direct evidence of how it acted to give rise to the complex adaptations of social insects.

I think that evolutionary biologists need to be honest about what has really been demonstrated and what remains simply an attractive hypothesis. We must avoid mistaking "adaptive stories" (Gould and Lewontin 1979) for proven explanations. But we should realize that there can be nonadaptive stories as well, and that these can represent just as potent a lure for the unwary as can adaptive stories. For example, I discussed in the previous chapter (chapter 8, Genome Duplication in the Vertebrates) the hypothesis that the vertebrate genome has undergone two rounds of duplication—a story that has proven extraordinarily attractive to biologists even though there is no unambiguous evidence for it. Whether a proposed evolutionary scenario involves adaptive evolution or simply chance genetic events, it is important to subject it to rigorous testing before we consider it established.

Finally, it is important to be humble about what we can and cannot know. In evolutionary biology we are always attempting to reconstruct events in the past, and our ability to do so will always remain dependent on the quality of the evidence that has survived. In many cases, crucial evidence may simply be lacking. Often, apparently adaptive changes in protein sequences have taken place so long ago that we cannot reliably reconstruct the path of evolution—let alone prove that natural selection, rather than chance fixation of neutral mutations, actually played a role. We must realize that the molecular techniques now available to us have opened a fascinating but limited window on the mechanisms by which over millions of years of life as we know it has evolved. Let us be grateful for that window, while accepting that there will always be much that is mysterious about the history of life on earth.

REFERENCES

Ackerman SJ, Gleich GJ, Loegering DA, Richardson BA, Butterworth AE (1985) Comparative toxicity of purified human eosinophil granule cationic proteins for schistosomula of *Schistosoma mansoni*. Am J Trop Med Hyg 34: 735–745.

Akashi H (1994) Synonymous codon usage in *Drosophila melanogaster*: natural selection and translational accuracy. Genetics 136: 927–935.

Alcami A, Smith GL (1992) A soluble receptor for interleukin-1β encoded by vaccinia virus: a novel mechanism of virus modulation of the host response to infection. Cell 71: 153–167.

Alexander RD (1979) *Darwinism and Human Affairs*. University of Washington Press, Seattle.

Alvarez LW (1983) Experimental evidence that an asteroid impact led to the extinction of many species 65 million years ago. Proc Natl Acad Sci USA 80: 627–642.

Andersson L, Sigurdardottir S, Borsch C, Gustafsson K (1991) Evolution of MHC polymorphism: extensive sharing of polymorphic sequence motifs between human and bovine DRB alleles. Immunogenetics 33: 235–243.

Antohi S, Brumfield V (1984) Polycation–cell surface interactions and plasma membrane compartments in mammals: interference of oligocation with polycation condensation. Z Naturforsch 396: 767.

Arden B, Klein J (1982). Biochemical comparison of major histocompatibility complex molecules from different subspecies of *Mus musculus*: evidence for trans-species evolution of alleles. Proc Natl Acad Sci USA 79: 2342–2346.

Armes N, Fried M (1995) The genomic organization of the region containing the *Drosophila melanogaster* rpL7a (Surf-3) gene differs from those of mammalian and avian surfeit loci. Mol Cell Biol 15: 2367–2373.

Arnold E, Arnold GF (1991) Human immunodeficiency virus structure: implications for antiviral design. Adv Viral Res 39: 1–87.

Arnot DE (1989) Malaria and the major histocompatibility complex. Parasitol Today 5: 138–142.

Ayala FJ, Valentine JW (1979) Genetic variability in the pelagic environment: a paradox? Evology 60: 24–29.

Bailey DW, Kohn HI (1965) Inherited histocompatibility changes in progeny of irradiated and unirradiated inbred mice. Genet Res 6: 330–340.

Bailey GS, Poulter RTM, Stockwell AP (1978) Gene duplication in tetraploid fish: model for gene silencing at unlinked duplicate loci. Proc Natl Acad Sci USA 75: 5575–5579.

Bailey WJ, Kim J, Wagner GP, Ruddle FH (1997) Phylogenetic reconstruction of the vertebrate Hox cluster duplication. Mol Biol Evol 14: 843–853.

Bajwa W, Torchia TE, Hopper JE (1988) Yeast regulatory gene GAL3: carbon regulation; UAS_{Gal} elements in common with *GAL1*, *GAL2*, *GAL7*, *GAL10*, *GAL80*, and *MEL1*; encoded protein strikingly similar to yeast and *Escherichia coli* galactokinases. Mol Cell Biol 8: 3439–3447.

Balczarek KA, Lai Z-C, Kumar S (1997) Evolution and functional diversification of the paired box (*Pax*) DNA-binding domains. Mol Biol Evol 14: 829–842.

Begun DJ, Aquadro CF (1991) Molecular population genetics of the distal portion of the X chromosome in *Drosophila*: evidence for genetic hitchhiking of the *yellow-achaete* region. Genetics 129: 1147–1158.

Begun DJ, Aquadro CF (1992) Levels of naturally occurring DNA polymorphism correlate with recombination rates in *D. melanogaster*. Nature 356: 519–520.

Begun DJ, Aquadro CF (1995) Evolution at the tip and base of the X chromosome in an African population of *Drosophila melanogaster*. Mol Biol Evol 12: 382–390.

Belich MP, Glynne RJ, Senger G, Sheer D, Trowsdale J (1994) Proteasome components with reciprocal expression to that of the MHC-encoded LMP proteins. Curr Biol 4: 769–776.

Bender BS, Crogham T, Zhang L, Small PA Jr (1992) Transgenic mice lacking class I major histocompatibility complex restricted T cells have delayed viral clearance and increased mortality after influenza virus challenge. J Exp Med 175: 1143–1145.

Bennett MD (1971) The duration of meiosis. Proc R Soc Lond B 178: 277–299.

Bergstrom TF, Josefsson A, Erlich HA, Gyllensten UB (1997) Analysis of intron sequences at the class II HLA-DRB1 locus: implications for the age of allelic diversity. Hereditas 127: 1–5.

Berke G (1994) The binding and lysis of target cells by cytotoxic lymphocytes: molecular and cellular aspects. Annu Rev Immunol 12: 735–773.

Bernardi G (1993) The vertebrate genome: isochores and evolution. Mol Biol Evol 10: 186–204.

Bernardi G. Olofsson B, Filipski J, Zerial M. Salinas J, Cuny G, Meunier-Rotival M, Rodier F (1985) The mosaic genome of warm-blooded vertebrates. Science 228: 953–958.

Berry A, Kreitman M (1993) Molecular analysis of an allozyme cline: alcohol dehydrogenase in *Drosophila melanogaster* on the east coast of North America. Genetics 134: 869–893.

Berry AJ, Ajioka JW, Kreitman M (1991) Lack of polymorphism on the *Drosophila* fourth chromosome resulting from selection. Genomics 129: 1111–1117.

Bertoletti A, Sette A, Chisari F, Penna A, Levrero M, De Carli M, Flaccadori F, Ferrari C (1994) Natural variants of cytotoxic epitopes are T-cell receptor antagonists for antiviral cytotoxic T cells. Nature 369: 407–410.

Bhat PJ, Oh D, Hopper JE (1990) Analysis of the *GAL3* signal transduction pathway activating Gal4 protein-dependent transcription in *Saccharomyces cerevisiae*. Genetics 125: 281–291.

Bingulac-Popovic J, Figueroa F, Sato A, Talbot WS, Johnson SL, Gates M, Postlethwait

JH, Klein J (1997) Mapping of *Mhc* class I and class II regions to different linkage groups in the zebrafish, *Danio rerio*. Immunogenetics 46: 129–134.

Bird AP (1986) CpG-rich islands and the function of DNA methylation. Nature 321: 209–213.

Bisbee CA, Bakes MA, Wilson AC, Irandokht H-A, Fischberg M (1977) Albumin phylogeny for clawed frogs (*Xenopus*). Science 195: 785–787.

Bjorkman PJ, Saper MA, Samraoui B, Bennet WS, Strominger JL, Wiley DC (1987a) Structure of the human class I histocompatibility antigen, HLA-A2. Nature 329: 506–512.

Bjorkman PJ, Saper MA, Samraoui B, Bennet WS, Strominger JL, Wiley DC (1987b) The foreign antigen binding site and T cell recognition regions of class I histocompatibility antigens. Nature 329: 512–518.

Blackman MJ, Heidrich H-G, Donachie S, McBride JS, Holder AA (1990) A single fragment of a malaria merozoite surface protein remains on the parasite during red cell invasion and is the target of invasion-inhibiting antibodies. J Exp Med 172: 372–382.

Bock WJ (1980) The definition and recognition of biological adaptation. Am Zool 20: 217–227.

Bodmer JG, Marsh SG, Albert ED, Bodmer WF, Bontrop RE, Charron D, Dupont B, Erlich HA, Fauchet R, Mach B, Mayr WR, Parham P, Sasazuki T, Schreuder GM, Strominger JL, Svejgaard A, Terasaki PI (1997) Nomenclature for factors of the HLA system, 1996. Eur J Immunogenet 24: 105–151.

Bradbury JW, Vehrencamp SL (1998) *Principles of Animal Communication*. Sinauer, Sunderland, MA.

Brandon RN (1978) Adaptation and evolutionary theory. Stud Hist Philos Sci 9: 181–206.

Braud V, Jones EY, McMichael A (1997) The human major histocompatibility complex class Ib molecule binds signal sequence-derived peptides with primary anchor residues at positions 2 and 9. Eur J Immunol 27: 1164–1169.

Briscoe DA, Robertson A, Malpica J-M (1975) Dominance at *Adh* locus in response of adult *Drosophila melanogaster* to environmental alcohol. Nature 255: 148–149.

Brown JH, Jardetsky T, Saper MA, Samraoui B, Bjorkman PJ, Wiley DC (1988) A hypothetical model of the foreign antigen binding site of class II histocompatibility molecules. Nature 332: 845–850.

Brown JH, Jardetsky T, Saper MA, Samraoui B, Bjorkman PJ, Wiley DC (1993) Three-dimensional structure of the human class II histocompatibility antigen HLA-DR1. Nature 364: 33–39.

Buck L, Axel R (1991) A novel multigene family may encode odorant receptors: a molecular basis for odor recognition. Cell 65: 175–187.

Bulmer M (1991) The selection-mutation-drift theory of synonymous codon usage. Genetics 129: 897–907.

Burton DW, Bickham JW, Genoways HH (1989) Flow-cytometric analyses of nuclear DNA content in four families of Neotropical bats. Evolution 43: 756–765.

Cain AJ (1951) So-called non-adaptive or neutral characters in evolution. Nature 166: 424.

Callard R, Gearing A (1994) *The Cytokine Factsbook*. Academic Press, London.

Carman WF, Boner W, Fattovich G, Colman K, Dornan ES, Thursz M, Hadziyannis S (1997) Hepatitis B virus core protein mutations are concentrated in progressive disease and in T helper cell epitopes during clinical remission. J Infect Diseases 175: 1093–1100.

Cavalier-Smith T (1985) Cell volume and the evolution of eukaryote genome size. Pp.

105–184 in Cavalier-Smith T, ed. *The Evolution of Genome Size*. Wiley, New York.
Cavalli-Sforza LL, Feldman MW (1981) *Cultural Transmission: A Quantitative Approach*. Princeton University Press, Princeton, NJ.
Cereb N, Hughes AL, Yang SY (1997) Locus-specific conservation of the *HLA* class I introns by intralocus recombination. Immunogenetics 47: 30–36
Chambers GK (1988) The *Drosophila* alcohol dehydrogenase gene-enzyme system. Adv Genet 25: 39–107.
Charlesworth B (1992) New gene sweep clean. Nature 356: 475–476.
Charlesworth B, Morgan MT, Charlesworth D (1993) The effect of deleterious mutations on neutral molecular variation. Genetics 134: 1289–1303.
Chen R, Greer A, Dean AM (1995) A highly active decarboxylating dehydrogenase with rationally inverted coenzyme specificity. Proc Natl Acad Sci USA 92: 11666–11670.
Choudhary M, Laurie CC (1991) Use of *in vitro* mutagenesis to analyze the molecular basis of the difference in *Adh* expression associated with the allozyme polymorphism of *Drosophila melanogaster*. Genetics 129: 481–488.
Clark AG, Kao T-H (1991) Excess nonsymonymous substitution at shared polymorphic sites among self-incompatibility alleles of Solanaceae. Proc Natl Acad Sci USA 88: 9823–9827.
Clarke AG (1971) The effects of maternal pre-immunization on pregnancy in the mouse. J Reprod Fertil 24: 369–375.
Clarke B, Kirby DRS (1966) Maintenance of histocompatibility complex polymorphisms. Nature 211: 999–1000.
Cochrane AH, Santoro F, Nussenzweig V, Gwadz RW, Nussenzweig RS (1982) Monoclonal antibodies identify the protective antigens of sporozoites of *Plasmodium knowlesi*. Proc Natl Acad Sci USA 79: 5651–5655.
Collette Y (1997) Towards a consensus for a role of Nef in both viral replication and immunomodulation? Res Virol 148: 22–30.
Colombo, P, Yon J, Garson K, Fried M (1992) Conservation of the organization of five tightly clustered genes over 600 million years of divergent evolution. Proc Natl Acad Sci USA 89: 6358–6362.
Coulier F, Pontarotti P, Roubin R, Hartung H, Goldfarb M, Birnbaum D (1997) Of worms and men: an evolutionary perspective on the fibroblast growth factor (FGF) and FGF receptor families. J Mol Evol 44: 43–56.
Cracraft J (1982) Phylogenetic relationships and monophyly of loons, grebes, and hesperornithiform birds, with comments on the early history of birds. Syst Zool 31: 35–56.
Crisanti A, Muller H-M, Hilbich C, Sinigaglia F, Matile H, McKay M, Scaife J, Beyreuther K, Bujard H (1988) Epitopes recognized by human T cells map within the conserved part of the GP190 of *P. falciparum*. Science 240: 1324–1326.
Cuny G, Soriano P, Macaya G, Bernardi G (1981) The major components of the mouse and human genomes: preparation, basic properties, and compositional heterogeneity. Eur J. Biochem 111: 227–233.
Damian RT (1964) Molecular mimicry: antigen sharing by parasite and host and its consequences. Am Nat 98: 129–149.
Darwin C (1871) *The Descent of Man*. Murray, London.
David JR, Capy P (1988) Genetic variation of *Drosophila melanogaster* natural populations. Trends Genet 4: 106–111.
Dean AM, Golding B (1997) Protein engineering reveals ancient adaptive replacements in isocitrate dehydrogenase. Proc Natl Acad Sci USA 94: 3104–3109.
DeCloux A, Woods AS, Cotter RJ, Soloski ML, Forman J (1997) Dominance of a single peptide bound to the class Ib molecule $Qa1^b$. J Immunol 158: 6089–6098.

Diaz M, Flajnik MF (1998) Evolution of somatic hypermutation and gene conversion in adaptive immunity. Immunol Rev 162: 13–24.

Dobson DE, Praeger EM, Wilson AC (1984) Stomach lysozymes of ruminants: I. Distribution and catalytic properties. J Biol Chem 259: 11607–11616.

Dobzhansky T (1956) What is an adaptive trait? Am Nat 90: 337–347.

Dobzhansky T (1968a) On some fundamental concepts of Darwinian biology. Evol Biol 2: 1–34.

Dobzhansky T (1968b) Adaptedness and fitness. Pp. 109–121 in Lewontin RC, ed. *Population Biology and Evolution*. Syracuse University Press, Syracuse, NY.

Doherty PC, Zinkernagel RM (1975) Enhanced immunologic surveillance in mice heterozygous at the *H-2* gene complex. Nature 256: 50–52.

Domachowske JB, Rosenberg HF (1997) Eosinophils inhibit retroviral transduction of human target cells by a ribonuclease-dependent mechanism. J Leukoc Biol 62: 363–368.

D'Onofrio G, Bernardi G (1992) A universal compositional correlation among codon positions. Gene 110: 81–88.

Doolittle WF (1985) The genealogy of some recently evolved vertebrate proteins. Trends Biochem Sci 10: 233–237.

Doolittle RF (1994) Convergent evolution: the need to be explicit. Trends Biochem Sci 19: 15–18.

Driscoll J, Finley D (1992). A controlled breakdown: antigen processing and the turnover of viral proteins. Cell 68: 823–825.

Driscoll M, Dean E, Reilly E, Bergholz E, Chalfie M (1989) Genetic and molecular analysis of a *Caenorhabditis elegans* β-tubulin that coveys benzimidazole sensitivity. J Cell Biol 109: 2993–3003.

Dunbar RIM (1982) Adaptation, fitness, and the evolutionary tautology. Pp. 9–28 in King's College Sociobiology Group, eds., *Current Problems in Sociobiology*. Cambridge University Press, Cambridge.

Dykhuizen D, Hartl DL (1980) Selective neutrality of 6PGD allozymes in *E. coli* and the effects of genetic background. Genetics 96: 801–817.

Egid K, Brown JL (1989) The major histocompatibility complex and female mating preferences in mice. Anim Behav 38: 548–549.

Eklund A, Egid K, Brown JL (1991) The major histocompatibility complex and mating preferences of male mice. Anim Behav 42: 693–694.

Elard L. Comes AM, Humbert JF (1996) Sequences of β-tubulin cDNA from benzimidazole-susceptible and -resistant strains of *Teladorsagia circumcincta*, a nematode parasite of small ruminants. Mol Biochem Parasitol 79: 249–253.

Endler JA (1986) *Natural Selection in the Wild*. Princeton University Press, Princeton, NJ.

Endo T, Ikeo K, Gojobori T (1996) Large-scale search for genes on which positive selection may operate. Mol Biol Evol 13: 685–690.

Escalante AA, Ayala FJ (1994) Phylogeny of the malarial genus *Plasmodium* derived from rRNA gene sequences. Proc Natl Acad Sci USA 91: 11373–11377.

Fearon DT (1997) Seeking wisdom in innate immunity. Nature 388: 323–324.

Feduccia A (1996) *The Origin and Evolution of Birds*. Yale University Press, New Haven, CT.

Felsenstein J (1985) Confidence limits on phylogenies: an approach using the bootstrap. Evolution 39: 783–791.

Filipski J (1987) Correlation between molecular clock ticking, codon usage, fidelity of

DNA repair, chromosome banding and chromatin compactness in germline cells. FEBS Lett 217: 184–186.

Fisher AG, Ensoli B, Looney D, Rose A, Gallo RC, Saag MS, Shaw GM, Hahn BH, Wong-Staal F (1988) Biologically diverse molecular variants within a single HIV-1 isolate. Nature 344: 440–444.

Flaherty KM, DeLuca-Flaherty C, McKay DB (1990) Three-dimensional structure of the ATPase fragment of a 70k heat shock cognate protein. Nature 346: 623–628.

Flynn JL, Goldstein MM, Triebold KJ, Koller B, Bloom BR (1992) Major histocompatibility complex class I-restricted T cells are required for resistance to *Mycobacterium tuberculosis* infection. Proc Natl Acad Sci USA 89: 12013–12017.

Forde DE (1934) *Habitat, Economy, and Society*. Methuen, London.

Franklin FCH, Lawrence MJ, Franklin-Tong VE (1995) Cell and molecular biology of self-incompatibility in flowering plants. Int Rev Cytol 158: 1–64.

Fu Y-X, Li W-H (1993) Statistical tests of neutrality of mutations. Genetics 133: 693–709.

Fujimoto K, Okino N, Kawabata S-I, Iwanga S, Ohnishi E (1995) Nucleotide sequence of the cDNA encoding the proenzyme of phenol oxidase A_1 of *Drosophila melanogaster*. Proc Natl Acad Sci USA 92: 7769–7773.

Galinski MR, Arnot DE, Cochrane AH, Barnwell JW, Nussenzweig RS, Enea V (1987) The circumsporozoite gene of the *Plasmodium cynomolgi* complex. Cell 48: 311–319.

Ganz T, Selsted ME, Lehrer RI (1989) Defensins. Eur J Haematol 44: 1–8.

Garboczi DN, Ghosh P. Utz U, Fan QR, Biddison WE, Wiley DC (1996) Structure of the complex between human T-cell receptor, viral peptide, and HLA-A2. Nature 384: 131–141.

Garrett TP, Saper MA, Bjorkman PJ, Strominger JL, Wiley DC (1989) Specifying pockets for the side chains of peptide antigens in HLA-Aw68. Nature 342: 692–696.

Gaykema WPJ, Hol WGJ, Vereijken JM, Soter NM, Bak, HJ, Beintema JJ (1984) 3.2 A structure of the copper-containing, oxygen-carrying protein *Panulirus interruptus* hemocyanin. Nature 309: 23–29.

Germain RN (1994) MHC-dependent antigen processing and peptide presentation: providing ligands for T-lymphocyte activation. Cell 76: 287–299.

Gilbert AN, Yamazaki K, Beauchamp GK, Thomas L (1986) Olfactory discrimination of mouse strains (*Mus musculus*) and major histocompatibility complex types by humans (*Homo sapiens*). J Comp Psychol 100: 262–265.

Gilbert SC, Plebanski M, Gupta S, Morris J, Cox M, Aidoo M, Kwiatkowski D, Greenwood BM, Whittle HC, Hill AVS (1998) Association of malaria parasite population structure, HLA, and immunological antagonism. Science 279: 1173–1177.

Gillespie JH (1991) *The Causes of Molecular Evolution*. Oxford University Press, New York.

Gilley J, Armes N, Fried M (1997) *Fugu* genome is not a good mammalian model. Nature 385: 305–306.

Golding GB, Dean AM (1998) The structural basis of molecular adaptation. Mol Biol Evol 15: 355–369.

Good MF, Pombo D, Quakyi IA, Riley EM, Houghten RA, Menon A, Allings DW, Berzofsky JA, Miller LH (1988). Human T-cell recognition of the circumsporozoite protein of *Plasmodium falciparum*: immunodominant T-cell domains map to the polymorphic regions of the molecule. Proc Natl Acad Sci USA 85: 1199–1203.

Goodman M, Moore GW, Matsuda G (1975) Darwinian evolution in the genealogy of haemoglobin. Nature 253: 603–608.

Gould SJ (1983) The hardening of the modern synthesis. Pp. 71–93 in Grene M, ed., *Dimensions of Darwinism*. Cambridge University Press, Cambridge.

Gould SJ, Lewontin RC (1979) The spandrels of San Marco and the Panglossian paradigm: a critique of the adaptationist programme. Proc R Soc Lond B 205: 581–598.

Goulder PJR, Phillips RE, Colbert RA, McAdam S, Ogg G, Nowak MA, Giangrande P, Luzzi G, Morgan B, Edwards A, McMichael AJ, Rowland-Jones S (1997) Late escape from an immunodominant cytotoxic T-lymphocyte response associated with progression to AIDS. Nat Med 3: 212–217.

Graf JD, Kobel HR (1991) Genetics of *Xenopus laevis*. Pp. 19–34 in Kay BK, Peng HB, eds., Xenopus laevis: *Practical Uses in Cell and Molecular Biology*. Academic Press, New York.

Grant BR, Grant PR (1989a) Natural selection in a population of Darwin's finches. Am Nat 133: 377–393.

Grant BR, Grant PR (1989b) *Evolutionary Dynamics of a Natural Population*. University of Chicago Press, Chicago.

Grantham R, Gautier C, Gouy M, Mercier R, Pave A (1980) Codon catalog usage and the genome hypothesis. Nucleic Acids Res 8: r49–r62.

Graur D, Li W-H (1991) Neutral mutation hypothesis test. Nature 354: 114–115.

Guo H-C, Jardetzky TS, Garrett TP, Lane WS, Strominger JS, Wiley DC (1992). Different length peptides bind to HLA-Aw68 similarly at their ends but bulge out in the middle. Nature 360: 364–366.

Gyllensten UB, Erlich HA (1989) Ancient roots for polymorphism at the HLA-DQA locus in primates. Proc Natl Acad Sci USA 86: 9986–9990.

Gyllensten UB, Lashkari D, Erlich HA (1990) Allelic diversification at the class II DQB locus of the mammalian major histocompatibility complex. Proc Natl Acad Sci USA 87: 1835–1839.

Gyllensten UB, Sundvall M, Erlich HA (1991) Allelic diversity is generated by intraexon sequence exchange at the *DRB1* locus of primates. Proc Natl Acad Sci USA 88: 3686–3690.

Hahn B, Shaw GM, Taylor ME, Redfield RR, Markham PD, Salahuddin SZ, Wong-Staal F, Gallo RC, Parks ES, Parks DW (1986) Genetic variation in HTLV-III/LAV over time in patients with AIDS or at risk for AIDS. Science 232: 1548–1553.

Haldane JBS (1957) The cost of natural selection. J Genet 55: 511–524.

Hall M, Scott T, Sugumaran M, Soderhall K, Law JH (1995) Proenzyme of *Manduca sexta* phenol oxidase: purification, activation, substrate specificity of the active enzyme, and molecular cloning. Proc Natl Acad Sci USA 92: 7764–7768.

Hamann KJ, Barker RL, Ten RM, Gleich GJ (1991) The molecular biology of eosinophil granule proteins. Int Arch Allergy Appl Immunol 94: 202–209.

Hamilton WD (1963) The evolution of altruistic behavior. Am Nat 97: 354–356.

Hamilton WD (1964) The genetical evolution of social behavior I, II. J Theor Biol 7: 1–16, 17–52.

Hamilton WD, Zuk M (1982) Heritable true fitness and bright birds: A role for parasites? Science 218: 384–387.

Hara TJ (1975) Olfaction in fish. Prog Neurobiol 5: 271–335.

Harvey PH, Pagel MD (1991). *The Comparative Method in Evolutionary Biology*. Oxford University Press, Oxford.

Hedges SB, Parker PH, Sibley CG, Kumar S (1996) Continental breakup and the ordinal diversification of birds and mammals. Nature 381: 226–229.

Hedrick PW, Black FL (1997) HLA and mate selection: no evidence in South Amerindians. Am J Hum Genet 61: 505–511.

Hedrick PW, Thomson G (1983) Evidence for balancing selection at HLA. Genetics 104: 449–456.

Hennig W (1950) *Grundzuge einer Theorie der phylogenetischen Systematik*. Deutscher Zentralverlag, Berlin.

Hennig W (1965) Phylogenetic systematics. Annu Rev Entomol 10: 97–116.

Higgins CF (1992) ABC transporters: from microorganisms to man. Annu Rev Cell Biol 8: 67–113.

Hill AVS, Allsopp CEM, Kwiatkowski D, Anstey NM, Twumasi, P, Rowe PA, Bennett S, Brewster D, McMichael AJ, Greenwood BM (1991) Common West African HLA antigens are associated with protection from severe malaria. Nature 352: 595–600.

Hill AVS, Elvin J, Willis AC, Aidoo M, Allsopp CEM, Gotch FM, Gao XM, Takiguchi M, Greenwood BM, Townsend ARM, McMichael AJ, Whittle HC (1992) Molecular analysis of the association of HLA-B53 and resistance to severe malaria. Nature 360: 434–439.

Hill RE, Hastie ND (1987) Accelerated evolution in the reactive centre regions of serine protease inhibitors. Nature 326: 96–99.

Holder AA, Riley EM (1996) Human immune response to MSP-1. Parasitol Today 12: 173–174.

Holder AA, Sandhu JS, Hillman Y, Davey LS, Nicholls SC, Cooper H, Lockyer MJ (1987) Processing of the precursor to the major merozoite surface antigens of *Plasmodium falciparum*. Parasitology 94: 199–208.

Holland K, Spindler K, Horodyski F, Grabau E, Nichol S, Van de Pol S (1982) Rapid evolution of RNA genomes. Science 215: 1577–1585.

Holland PWH, Garcia-Fernandez J, Williams NA, Sidow A (1994) Gene duplications and the origins of vertebrate development. Development Suppl: 125–133.

Holmes EC, Zhang LQ, Simmonds P, Ludlam CA, Leigh Brown AJ (1992) Convergent and divergent sequence evolution in the surface envelope glycoprotein of human immunodeficiency virus type 1 within a single infected patient. Proc Natl Acad Sci USA 89: 4835–4839.

Howard JC (1987) MHC organization in the rat: evolutionary considerations. Pp. 397–427 in Kelso G, Schulze DH, ed., *Evolution and Vertebrate Immunity*. University of Texas Press, Austin.

Hudson RR (1990) Gene genealogies and the coalescent process. Oxford Surv Evol Biol 7: 1–44.

Hughes AL (1988) *Evolution and Human Kinship*. Oxford University Press, New York.

Hughes AL (1989) Interaction between strains in the social relations of inbred mice. Behav Genet 19: 685–700.

Hughes AL (1991a) Evolutionary origin and diversification of the mamalian CD1 antigen genes. Mol Biol Evol 8: 185–201.

Hughes AL (1991b) Independent gene duplications, not concerted evolution, explain relationships among class I MHC genes of murine rodents. Immunogenetics 33: 367–373.

Hughes AL (1991c) Circumsporozoite protein genes of malaria parasites (*Plasmodium spp.*): evidence for positive selection on immunogenic regions. Genetics 127: 345–353.

Hughes AL (1992a) Positive selection and interallelic recombination at the merozoite surface antigen-1 (MSA-1) locus of *Plasmodium falciparum*. Mol Biol Evol 9: 381–393.

Hughes AL (1992b) Coevolution of the vertebrate integrin α and β chains. Mol Biol Evol 9: 216–234.

Hughes AL (1993a) Positive selection in a multi-gene family: catfish olfactory receptors. Trends Ecol Evol 8: 273–274.

Hughes AL (1993b) Nonlinear relationships among evolutionary rates identify regions of functional divergence in heat-shock protein 70 genes. Mol Biol Evol 10: 243–255.
Hughes AL (1993c) Evidence of positive selection at the *Lyb-2* locus of the mouse. Immunogenetics 38: 54–56.
Hughes AL (1994a) The evolution of functionally novel proteins after gene duplication. Proc R Soc Lond B 256: 119–124.
Hughes AL (1994b) Evolution of cystein proteinases in eukaryotes. Mol Phylo Evol 3: 310–321.
Hughes AL (1995a) Origin and evolution of HLA class I pseudogenes. Mol Biol Evol 12: 247–258.
Hughes AL (1995b) The evolution of the type I interferon gene family in mammals. J Mol Evol 41: 539–548.
Hughes AL (1996a) Evolution of the HLA complex. Pp. 73–92 in Jackson MS, Strachan T, Dover G. eds., *Human Genome Evolution*. Bios, Oxford.
Hughes AL (1996b) Gene duplication and recombination in the evolution of mammalian Fc receptors. J Mol Evol 43: 4–10.
Hughes AL (1997) Rapid evolution of immunoglobulin superfamily C2 domains expressed in immune system cells. Mol Biol Evol 14: 1–5.
Hughes AL (1998a) Protein phylogenies provide evidence of a radical discontinuity between arthropod and vertebrate immune systems. Immunogenetics 47: 283–296.
Hughes AL (1998b) Phylogenetic tests of the hypothesis of block duplication of homologous genes on human chromosomes 6, 9, and 1. Mol Biol Evol 15: 854–870.
Hughes AL (1999) Phylogenies of developmentally important proteins do not support the hypothesis of two rounds of genome duplication early in vertebrate history. J Mol Evol 48: 565–576.
Hughes MK, Hughes AL (1993a) Evolution of duplicate genes in a tetraploid animal, *Xenopus laevis*. Mol Biol Evol 10: 1360–1369.
Hughes AL, Hughes MK (1993b) Adaptive evolution in the rat olfactory receptor gene family. J Mol Evol 36: 249–254.
Hughes AL, Hughes MK (1995a) Natural selection on the peptide-binding regions of major histocompatibility complex molecules. Immunogenetics 42: 233–243.
Hughes MK, Hughes AL (1995b) Natural selection on *Plasmodium* surface proteins. Mol Biochem Parasitol 71: 99–113.
Hughes AL, Hughes MK (1995c) Small genomes for better fliers. Nature 377: 391.
Hughes AL, Nei M (1988) Pattern of nucleotide substitution at MHC class I loci reveals overdominant selection. Nature 335: 167–170.
Hughes AL, Nei M (1989a) Nucleotide substitution at major histocompatibility complex class II loci: evidence for overdominant selection. Proc Natl Acad Sci USA 86: 958–962.
Hughes AL, Nei M (1989b) Evolution of the major histocompatibility complex: independent origin of nonclassical class I genes in different groups of mammals. Mol Biol Evol 6: 559–579.
Hughes AL, Nei M (1990) Evolutionary relationships of class II MHC genes in mammals. Mol Biol Evol 7: 491–514.
Hughes AL, Nei M (1992) Maintenance of MHC polymorphism. Nature 355: 402–403.
Hughes AL, Verra F (1998) Ancient polymorphism and the hypothesis of a recent bottleneck in the malaria parasite *Plasmodium falciparum*. Genetics 150: 511–513.
Hughes AL, Yeager M (1997a) Comparative evolutionary rates of introns and exons. J. Mol Evol 45: 125–130.

Hughes AL, Yeager M (1997b) Coordinated amino acid changes in the evolution of mammalian defensins. J Mol Evol 44: 675–682.
Hughes AL, Yeager M (1998a) Erratum: Comparative evolutionary rates of introns and exons. J Mol Evol 54: 497.
Hughes AL, Yeager M (1998b) Natural selection and the evolutionary history of major histocompatibility complex loci. Front Biosci 3: 509–516.
Hughes AL, Yeager M (1998c) Natural selection at major histocompatibility complex loci of vertebrates. Annu Rev Genet 32: 415–435.
Hughes AL, Ota T, Nei M (1990) Positive Darwinian selection promotes charge profile diversity in the antigen binding cleft of class I MHC molecules. Mol Biol Evol 7: 515–524.
Hughes AL, Hughes MK, Watkins DI (1993) Contrasting roles of interallelic recombination at the HLA-A and HLA-B loci. Genetics 133: 669–680.
Hughes AL, Hughes MK, Howell CY, Nei M (1994) Natural selection at the class II major histocompatibility complex loci of mammals. Philos Trans R Soc Lond B 345: 359–367.
Hultmark D (1993) Immune reactions in *Drosophila* and other insects: a model for innate immunity. Trends Genet 9: 178–183.
Iismaa TP, Shine J (1992) G protein-coupled receptors. Curr Opin Cell Biol 4: 195–202.
Ikemura T (1981) Correlation between the abundance of *Escherichia coli* transfer RNAs and the occurrence of the respective codons in its proteins genes: a proposal for a synonymous codon choice that is optimal for the *E. coli* translational system. J Mol Biol 151: 389–409.
Ikemura T (1985) Codon usage and tRNA content in unicellular and multicellular organisms. Mol Biol Evol 2: 1–13.
Ioerger TR, Clark AG, Kao T-H (1990) Polymorphism at the self-incompatibility locus in Solanaceae predates speciation. Proc Natl Acad Sci USA 87: 9732–9735.
Ikuta A, Szeto S, Yoshida A (1986) Three human alcohol dehydrogenase subunits: cDNA structure and molecular and evolutionary divergence. Proc Natl Acad Sci USA 83: 634–638.
James DA (1965) Effects of antigenic dissimilarity between mother and foetus on placental size in mice. Nature 205: 613–614.
James DA (1967) Some effects of immunological factors on gestation in mice. J Reprod Fertil 14: 265–275.
Jamroz RC, Beintema JJ, Stam WT, Bradfield JY (1996) Aromatic hexamerin subunit from adult female cock roaches (*Blaberus discoidalis*): molecular cloning, suppression by juvenile hormone, and evolutionary perspectives. J Insect Physiol 42: 114–124.
Janeway CA (1989) Approaching the asymptote? Evolution and revolution in immunology. Cold Spring Harbor Symp Quant Biol 54: 1–13.
Jensen RA (1976) Enzyme recruitment in the evolution of new function. Annu Rev Microbiol 30: 409–425.
Jensen RA, Byng GS (1981) The partitioning of biochemical pathways with isozyme systems. Isozymes 5: 143–174.
Johnston M (1987) A model fungal gene regulatory mechanism: The *GAL* genes of *Saccharomyces cerevisiae*. Microbiol Rev 51: 458–476.
Jolles P, Jolles J (1984) What's new in lysozyme research? Mol Cell Biochem 63: 165–189.
Jones DT, Taylor WR, Thornton JM (1992) The rapid generation of mutation data matrices from protein sequences. Comput Appl Biosci 8: 275–282.
Jones G, Manczak M, Horn M (1993) Hormonal regulation and properties of a new

group of basic hemolymph proteins expressed during insect metamorphosis. J Biol Chem 268: 1284–1291.

Jongwutiwes S, Tanabe K, Hughes MK, Kanbara H, Hughes AL (1994) Allelic variation in the circumsporozoite protein of *Plasmodium falciparum* from Thai field isolates. Am J Trop Med Hygiene 51: 659–667.

Jukes TH, Cantor CR (1969) Evolution of protein molecules. Pp. 21–132 in HN Munro, ed., *Mammalian Protein Metabolism*. Academic Press, New York.

Kabat EA, Wu TT, Perry HM, Gottesman KS, Foeller C (1991) *Sequences of Proteins of Immunological Interest*. U.S. Department of Health and Human Services, Government Printing Office, Washington, DC.

Kagan BL, Selsted ME, Ganz T, Lehrer RI (1990) Antimicrobial defensin peptides form voltage-dependent ion-permeable channels in planar lipid bilayer membrane. Proc Natl Acad Sci USA 87: 210–214.

Kao T-H, McCubbin AG (1996) How flowering plants discriminate between self and non-self pollen to prevent inbreeding. Proc Natl Acad Sci USA 93: 12059–12065.

Kaplan NL, Hudson RR, Langley CH (1989) The "hitchhiking effect" revisited. Genetics 123: 887–899.

Kasahara M, Hayashi M, Tanaka K, Inoku H, Sugaya K, Ikemura T, Ishibashi T (1996) Chromosomal localization of the proteasome Z subunit gene reveals an ancient chromosomal duplication involving the major histocompatibility complex. Proc Natl Acad Sci USA 93: 9096–9101.

Katsanis N, Fitzgibbon J, Fisher EMC (1996) Paralogy mapping: identification of a region in the human MHC triplicated onto human chromosomes 1 and 9 allows the prediction and isolation of novel *PBX* and *NOTCH* loci. Genomics 35: 101–108.

Kawabata T, Yasuhara Y, Ochiai M, Matsuura S, Ashida M (1995) Molecular cloning of insect prophenol oxidase: a copper-containing protein homologous to arthropod hemocyanin. Proc Natl Acad Sci USA 92: 7774–7778.

Kimura M (1955) Solution of a process of random genetic drift with a continuous model. Proc Natl Acad Sci USA 41: 144–150.

Kimura M (1964) Diffusion models in population genetics. J Appl Prob 1: 177–232.

Kimura M (1968) Evolutionary rate at the molecular level. Nature 217: 624–626.

Kimura M (1977) Preponderance of synonymous changes as evidence for the neutral theory of molecular evolution. Nature 267: 275–276.

Kimura M (1980) A simple method for estimating evolutionary rates of base substitutions through comparative studies of nucleotide sequences. J. Mol Evol 16: 111–120.

Kimura M (1981) Possibility of extensive neutral evolution under stabilizing selection with special reference to nonrandom usage of synonymous codons. Proc Natl Acad Sci USA 78: 5773–5777.

Kimura M (1983) *The Neutral Theory of Molecular Evolution*. Cambridge University Press, Cambridge.

Kimura M (1987) Molecular evolutionary clock and the neutral theory. J Mol Evol 26: 24–33.

Kimura M, Crow JF (1978) Effect of overall phenotypic selection on genetic change at individual loci. Proc Natl Acad Sci USA 75: 6168–6171.

Kimura M, Ohta T (1974) On some principles governing molecular evolution. Proc Natl Acad Sci USA 71: 2848–2852.

King JL, Jukes TH (1969) Non-Darwinian evolution. Science 164: 788–798.

Kingman JFC (1982) On the genealogy of large populations. J. Appl Prob 19A: 27–43.

Klein J (1986) *Natural History of the Major Histocompatibility Complex*. Wiley, New York.

Klein J, Figueroa F (1986) Evolution of the major histocompatibility complex. CRC Crit Rev Immunol 6: 295–386.

Klenerman P, Rowland-Jones S, McAdam S, Edwards J, Daenke S, Lalloo D, Koppe B, Rosenberg W, Boyd D, Edwards A, Glangrande P, Phillips RE, McMichael AJ (1994) Cytotoxic T-cell activity antagonized by naturally occurring HIV-1 Gag variants. Nature 369: 403–407.

Kliman RM, Hey J (1994) The effects of mutation and natural selection on codon bias in the genes of *Drosophila*. Genetics 137: 1049–1056.

Kobel HR, Du Pasquier L (1986) Genetics of polyploid *Xenopus*. Trends Genet 2: 310–315.

Kocher TD, Wilson AC (1991) Sequence evolution of mitochondrial DNA in humans and chimpanzees: control region and a protein-coding region. Pp. 391–413 in Osawa S, Honjo T, eds., *Evolution of Life: Fossils, Molecules, Culture*. Springer, Tokyo.

Koenig S, Conley AJ, Brewah YA, Jones GM, Leath S, Boots LJ, Davey V, Pantaleo G, Demarest JF, Carter C, Wannebo C, Yannelli JR, Rosenburg SA, Lane HC (1995) Transfer of HIV-1-specific cytotoxic T lymphocytes to an AIDS patient leads to selection for mutant HIV variants and subsequent disease progression. Nat Med 1: 330–336.

Kornegay JR, Schilling JW, Wilson AC (1994) Molecular adaptation of a leaf-eating bird: stomach lysozyme of the hoatzin. Mol Biol Evol 11: 921–928.

Kraut J (1977) Serine proteases: structure and mechanism of catalysis. Annu Rev Biochem 46: 331–358.

Kreitman M (1983) Nucleotide polymorphism at the alcohol dehydrogenase locus in *Drosophila melanogaster*. Nature 304: 412–417.

Kreitman M, Hudson RR (1991) Inferring the evolutionary histories of the *Adh* and *Adh-dup* loci in *Drosophila melanogaster* from patterns of polymorphism and divergence. Genetics 127: 565–582.

Kropshofer H, Hammerling GJ, Vogt AB (1997) How HLA-DM edits the MHC class II peptide repertoire: survival of the fittest? Immunol Today 18: 77–82.

Kuhn T (1962) *The Structure of Scientific Revolutions*. University of Chicago Press, Chicago.

Kumar S, Hedges SB (1998) A molecular timescale for vertebrate evolution. Nature 392: 917–920.

Kumar S, Tamura K, Nei M (1993) *MEGA: Molecular Evolutionary Genetic Analysis*, version 1.0. Pennsylvania State University, University Park.

Kusaba M, Nishio T, Satta Y, Hinata K, Ockedon D (1997) Striking sequence similarity in inter- and intraspecific comparisons of class I *SLG* alleles from *Brassica oleracea* and *Brassica campestris*: implications for the evolution and recognition mechanism. Proc Natl Acad Sci USA 94: 7673–7678.

Lacey E (1990) Mode of action of benzimidazoles. Parasitol Today 6: 112–115.

Langley CH, MacDonald J, Miyashiyta N, Aguade M (1993) Lack of correlation between interspecific divergence and intraspecific polymorphism at the suppressor of forked region in *Drosophila melanogaster* and *Drosophila simulans*. Proc Natl Acad Sci USA 90: 1800–1803.

Larson KA, Olson EV, Madden BJ, Gleich GJ, Lee NA, Lee JJ (1996) Two highly homologous ribonuclease genes expressed in mouse eosinophils identify a larger subgroup of the mammalian ribonuclease superfamily. Proc Natl Acad Sci USA 93: 12370–12375.

Laurie CC, Bridgham JT, Choudhary M (1991) Associations between DNA sequence

variation and variation in expression of the *Adh* gene in natural populations of *Drosophila melanogaster*. Genetics 129: 489–499.
Lawlor DA, Ward FF, Ennis PD, Jackson AP, Parham P (1988) HLA-A, –B polymorphisms predate the divergence of humans and chimpanzees. Nature 335: 268–271.
Lawlor DA, Zemmour J, Ennis PD, Parham P (1990) Evolution of class I MHC genes abd proteins: from natural selection to thymic selection. Annu Rev Immunol 8: 23–64.
Leahy DJ, Hendrickson WA, Aukhil I, Erickson HP (1992) Structure of a fibronectin type III domain from tenascin phased by MAD analysis of the selenomethionyl protein. Science 258: 987–991.
Lee TDG (1991) Helminthotoxic responses of intestinal eosinophils to *Trichinella spiralis* newborn larvae. Infect Immun 59: 4405–4411.
Levinson G, Hughes AL, Letvin N (1992) Sequence and diversity of rhesus monkey T-cell receptor β chain genes. Immunogenetics 35: 75–88.
Lewis D (1949) Incompatibility in flowering plants. Biol Rev Camberidhe Philos Soc 24: 472–496.
Lewis D (1960) Genetic control of specificity and activity of the S antigen in plants. Proc R Soc Lond B 151: 468–477.
Lewontin RC (1974) *The Genetic Basis of Evolutionary Change*. Columbia University Press, New York.
Lewontin RC, Hubby JL (1966) A molecular approach to the study of genic heterozygosity in natural populations. II. Amount of variation and degree of heterozygosity in natural populations of *Drosophila pseudoobscura*. Genetics 54: 595–609.
Li W-H (1980) Rate of gene silencing at duplicate loci: a theoretical study and interpretation of data from tetraploid fishes. Genetics 95: 237–258.
Li W-H (1982) Evolutionary change of duplicate genes. Isozymes 6: 55–92.
Li W-H (1983) Evolution of duplicate genes and pseudogenes. Pp. 14–37 in Nei M, Koehn RK, ed., *Evolution of Genes and Proteins*. Sinauer, Sunderland, MA.
Li W-H (1987) Models of nearly neutral mutations with particular implications for nonrandom usage of synonymous codons. J Mol Evol 24: 337–345.
Li W-H (1993) Unbiased estimation of the rates of synonymous and nonsynonymous substitution. J Mol Evol 36: 96–99.
Li W-H (1997) *Molecular Evolution*. Sinauer, Sunderland, MA.
Li W-H, Graur D (1991) *Fundamentals of Molecular Evolution*. Sinauer, Sunderland, MA.
Li W-H, Wu, C-I, Luo C-I (1985) A new method for estimating synonymous and nonsynonymous rates of nucleotide substitution considering the relative likelihood of nucleotide and codon change. Mol Biol Evol 2: 150–174.
Li W-H, Tanimura M, Sharp PM (1988) Rates and dates of divergence between AIDS virus nucleotide sequences. Mol Biol Evol 5: 313–320.
Li X, Zelenka PS, Piatigorsky J (1993) Differential expression of the two δ-crystallin genes in lens and nonlens tissues: shift favoring δ2 expression from embryonic adult chickens. Dev Dynam 196: 114–123.
Li X, Wistow GJ, Piatigorsky J (1995) Linkage and expression of the argininosuccinate lyase/δ-crystallin genes of the duck: insertion of a CR1 element in the intergenic spacer. Biochim Biophys Acta 1261: 25–34.
Lidicker WZ Jr, Patton JL (1987) Patterns of dispersal and genetic structure in populations of small rodents. Pp. 144–161 in Chepko-Sade BD, Halpin ZT, eds., *Mammalian Dispersal Patterns*. University of Chicago Press, Chicago.

Livingstone FB (1958) Anthropological implications of sickle cell gene distribution in West Africa. Am Anthropol 60: 531–561.

Lockwood SF, Derr JN (1992) Intra- and interspecific genome-size variation in the Salmonidae. Cytogenet Cell Genet 59: 303–306.

Long M, Langley CH (1993) Natural selection and the origin of *jingwei*, a chimeric processed functional gene in *Drosophila*. Science 260: 91–95.

Lopez de Castro JA, Strominger JL, Strong DM, Orr HT (1982) Structure of crossreactive human histocompatibility antigens HLA-A28 and HLA-A2: possible implications for the generation of HLA polymorphism. Proc Natl Acad Sci USA 79: 3813–3817.

Lumsden DJ, Wilson EO (1981) *Genes, Mind, and Culture*. Harvard University Press, Cambridge, MA.

Lyon JA, Geller RH, Haynes JD, Chulay JD, Weber JL (1986) Epitope map and processing scheme for the 195,000-dalton surface glycoprotein of *Plasmodium falciparum* merozoites deduced from cloned overlapping segments of the gene. Proc Natl Acad Sci USA 83: 2989–2993.

McAdam ST, Boyson JE, Liu X, Garber TL, Hughes AL, Bontrop RE, Watkins DI (1994) A uniquely high level of recombination at the *HLA-B* locus. Proc Natl Acad Sci USA 91: 5893–5897.

McAdam ST, Boyson JE, Liu X, Garber TL, Hughes AL, Bontrop RE, Watkins DI (1995) Chimpanzee MHC class I alleles are related to only one of the six families of human A locus alleles. J Immunol 154: 6421–6429.

McBride JS, Heidrich H-G (1987) Fragments of the polymorphic M_r 185000 glycoprotein from the surface of isolated *Plasmodium falciparum* merozoites form an antigenic complex. Mol Biochem Parasitol 23: 71–84.

McDonald JH, Kreitman M (1991) Adaptive protein evolution at the *Adh* locus in *Drosophila*. Nature 351: 652–654.

McFadden G, ed. (1995) *Viroceptors, Virokines and Related Immune Modulators Encoded by DNA Viruses*. Landes, Austin, TX.

McMichael AJ, Phillips RE (1997) Escape of human immunodeficiency virus from immune control. Annu Rev Immunol 15: 271–296.

Maddison WP, Maddison DR (1992) *MacClade: Analysis of Phylogeny and Character Evolution*. Sinauer, Sunderland, MA.

Manning CJ, Potts WK, Wakeland EK, Dewsbury DA (1992) What's wrong with MHC mate choice experiments? Pp. 229–235 in Doty RL, Muller-Schwarze D, eds., *Chemical Signals in Vertebrates* 6. Plenum, New York.

Martin-Campos JM, Cameron JM, Miyashita N, Aguade M (1992) Intraspecific and interspecific variation in the *y-ac-sc* region of *Drosophila simulans* and *Drosophila melanogaster*. Genetics 130: 805–816.

Maruyama T, Nei M (1981) Genetic variability maintained by mutation and overdominant selection in finite populations. Genetics 98: 441–459.

Maxam AM, Gilbert W (1977) A new method for sequencing DNA. Proc Natl Acad Sci USA 74: 560–564.

May RM, Anderson RM (1990) Parasite–host coevolution. Parasitology 100: S89–S101.

Mayer WE, Jonker D, Klein D, Ivanyi P, van Seventer G, Klein J (1988) Nucleotide sequence of chimpanzee MHC class I alleles: evidence for transspecies mode of evolution. EMBO J 7: 2765–2774.

Maynard Smith J, Haigh J (1974) The hitch-hiking effect of a favorable gene. Genet Res 23: 23–35.

Mayr E (1981) Biological classification: Toward a synthesis of opposing methodologies. Science 214: 5109–5116.

Mayr E (1942) *Systematics and the Origin of Species*. Columbia University Press, New York.
Mayr E (1982a) Adaptation and selection. Biol Zentral 101: 161–174.
Mayr E (1982b) *The Growth of Biological Thought*. Belknap, Cambridge, MA.
Mayr E, Provine W, eds. (1980) *The Evolutionary Synthesis*. Harvard University Press, Cambridge, MA.
Medawar PB (1951) Problems of adaptation. New Biol 11: 10–26.
Meyer J, Walker-Jonah A, Hollenberg CP (1991) Galactokinase encoded by *GAL1* is a bifunctional protein required for induction of the *GAL* genes in *Kluyveromyces lactis* and is able to suppress the *gal3* phenotype in *Saccharomyces cerevisiae*. Mol Cell Biol 11: 5454–5461.
Michaelson D, Rayner J, Conto M, Ganz T (1992) Cationic defensins arise fro charge-neutralized propeptides: a mechanism for avoiding leukocyte autocytotoxicity? J Leukoc Biol 51: 632–639.
Michener CD, Sokal RR (1957) A quantitative approach to a problem in classification. Evolution 11: 130–162.
Miklos GLG, Rubin GM (1996) The role of the genome project in determining gene function. Cell 86: 521–529.
Milkman R (1972) How much room is left for NonDarwinian evolution? Pp. 217–219 in Smith HH, ed., *Evolution of Genetic Systems*. Gordon and Breach, New York.
Milkman R (1978) Selection differentials and selection coefficients. Genetics 88: 391–403.
Milkman R (1982) Toward a unified selection theory. Pp. 105–118 in Milkman R, ed., *Perspectives on Evolution*. Sinauer, Sunderland, MA.
Minkoff EC (1965) The effect on classification of slight alterations in numerical technique. Syst Zool 14: 196–213.
Mitton JB (1997) *Selection in Natural Populations*. Oxford University Press, Oxford.
Mitton JB, Koehn RK (1975) Genetic organization and adaptive response of allozymes to ecological variables in *Fundulus heteroclitus*. Genetics 79: 97–111.
Miyata T, Miyazawa S, Yasunaga T (1979) Two types of amino acid substitution in protein evolution. J Mol Evol 12: 219–236.
Modrow S, Hahn BH, Shaw GM, Gallo RC, Wong-Staal F, Wolf H (1987) Computer-assisted analysis of envelope protein sequences of seven human immunodeficiency virus isolates: prediction of antigenic epitopes in conserved and variable regions. J Virol 61: 570–578.
Mollon J (1991) Hue and the hetahelicals. Nature 351: 696–697.
Monaco JJ (1992) A molecular model of MHC class-I-restricted antigen processing. Immunol Today 13: 173–178.
Monos DS, Tekolf WA, Shaw S, Cooper HL (1984) Comparison of structural and functional variation in class II HLA molecules: the role of charged amino acid substitutions. J Immunol 132: 1379–1385.
Moriyama EN, Powell JR (1997) Codon usage bias and tRNA abundance in *Drosophila*. J Mol Evol 45: 514–523.
Mortimer W (1972) *Organize!: My Life as a Union Man*. Beacon, Boston.
Muller HJ (1949) The Darwinian and modern conceptions of natural selection. Proc Am Philos Soc 93: 459–470.
Murphy PM (1993) Molecular mimicry and the generation of host defense protein diversity. Cell 72: 823–826.
Nakamura Y, Gojobori T, Ikemura T (1998) Codon usage tabulated from the international DNA sequence databases. Nucleic Acids Res 26: 334.

Nathenson SG, Geleibter J, Pfaffenbach GM, Zeff RA (1986) Murine major histocompatibility complex class-I mutants: molecular analysis and structure-function implications. Annu Rev Immunol 4: 471–502.

Nei M (1987) *Molecular Evolutionary Genetics*. Columbia University Press, New York.

Nei M, Gojobori T (1986) Simple methods for estimating the numbers of synonymous and nonsynonymous nucleotide substitutions. Mol Biol Evol 3: 418–426.

Nei M, Hughes AL (1992) Balanced polymorphism and evolution by the birth-and-death process. Pp. 27–38 in Tuji K, Aizawa M, Sasazuki T, eds., *Proceedings of the 11th Histocompatibility Workshop and Conference*. Oxford University Press, Oxford.

Nei M, Jin L (1989) Variances of the average numbers of nucleotide substitutions within and between populations. Mol Biol Evol 6: 290–300.

Nei M, Li W-H (1980) Nonrandom association between electromorphs and inversion chromosomes in finite populations. Genet Res 35: 65–83.

Nei M, Maruyama T, Wu C-I (1983) Models of evolution of reproductive isolation. Genetics 103: 557–579.

Nei M, Gu X, Sitnikova T (1997) Evolution by the birth-and-death process in multigene families of the vertebrate immune system. Proc Natl Acad Sci USA 94: 7799–7806.

Ngai J, Dowling MM, Buck L, Axel R, Chess A (1993). The family of genes encoding odorant receptors in the channel catfish. Cell 72: 657–666.

Nickerson JM, Wawrousek EF, Hawkins JW, Wakil AS, Wistow GJ, Thomas G, Norman BL, Piatigorsky J (1985) The complete sequence of the chicken δ1 crystallin gene in its 5′ flanking region. J Biol Chem 260: 9100–9106.

Nickerson JM, Wawrousek EF, Borras T, Hawkins JW, Norman BL, Filpula DR, Nagle JW, Ally AH, Piatogorsky J (1986) Sequence of the chicken δ2 crystallin gene and its intergenic spacer. J Biol Chem 261: 552–557.

Nowak MA, May RM, Phillips RE, Rowland-Jones S, Lalloo DG, McAdam S, Klenerman P, Koppe B, Sigmund K, Bangham CRM, McMichael AJ (1995) Antigenic oscillations and shifting immunodominance in HIV-1 infections. Nature 375: 606–611.

Oakeshott JG, Gibson JB, Anderson PR, Knibb WR, Anderson DG, Chambers GK (1982) Alcohol dehydrogenase and glycerol-3-phosphate dehydrogenase clines in *Drosophila melanogaster* on different continents. Evolution 36: 86–96.

Ober C, Weitkamp LR, Cox N, Dytch H, Kostyu D, Elias S (1997) HLA and mate choice in humans. Am J Hum Genet 61: 497–504.

Ohno S (1970) *Evolution by Gene Duplication*. Springer-Verlag, New York.

Ohno S (1973) Ancient linkage groups and frozen accidents. Nature 244: 259–262.

Ohno S, Wolf U, Atkin NB (1968) Evolution from fish to mammals by gene duplication. Hereditas 59: 169–187.

Ohta T (1982) Allelic and nonallelic homology of a supergene family. Proc Natl Acad Sci USA 79: 3251–3254.

Ohta T (1991) Evolution of the multigene family: a case of dynamically evolving genes at major histocompatibility complex. Pp. 145–159 in Osawa S, Honjo Y, eds., *Evolution of Life: Fossils, Molecules, and Culture*. Springer-Verlag, Tokyo.

Ohta T (1995) Synonymous and nonsynonymous substitutions in mammalian genes and the nearly neutral theory. J Mol Evol 40: 56–63.

Orel V (1996) *Gregor Mendel: The First Geneticist*. Oxford University Press, Oxford.

Orgel LE (1977) Gene-duplication and the origin of proteins with novel functions. J Theor Biol 67: 773.

Orlowski M (1990) The multicatalytic proteinase complex: A major extralysosomal proteolytic system. Biochemistry 29: 10289–10297.

Orzack SH, Sober E (1994a) Optimality models and the test of adaptationism. Am Nat 143: 361–380.

Orzack SH, Sober E (1994b) How (not) to test an optimality model. Trends Ecol Evol 9: 265–267.

Osawa S, Jukes TH, Watanabe K, Muto A (1992) Recent evidence for the evolution of the genetic code. Microbiol Rev 56: 229–264.

Ota T, Nei M (1994a) Variance and covariances of the numbers of synonymous and nonsynonymous substitutions per site. Mol Biol Evol 11: 613–619.

Ota T, Nei M (1994b) Estimation of the number of amino acid substitutions per site when the substitution rate varies among sites. J Mol Evol 38: 642–643.

Ota T, Nei M (1994c) Divergent evolution and evolution by the birth-and-death process in the immunoglobulin V_H gene family. Mol Biol Evol 11: 469–482.

Ota T, Nei M (1995) Evolution of immunoglobulin VH pseudogenes in chickens. Mol Biol Evol 12: 94–102.

Pagel M, Johnstone RA (1992) Variation across species in the size of the nuclear genome supports the junk-DNA explanation for the C-value paradox. Proc R Soc Lond B 249: 119–124.

Palumbi SR (1994) Genetic divergence, reproductive isolation and marine speciation. Annu Rev Ecol Syst 25: 547–572.

Palumbi SR, Metz EC (1991) Strong reproductive isolation between closely related tropical sea urchins (genus Echinometra). Mol Biol Evol 8: 227–239.

Pamer EG, Wang C-R, Flaherty L, Fischer-Lindahl K, Bevan MJ (1992) H-2M3 presents a Listeria monocytogenes peptide to cytotoxic T lymphocytes. Cell 70: 215–223.

Parker DS, Wawrousek EF, Piatigorsky J (1988) Expression of the δ-crystallin genes in the embryonic chicken lens. Dev Biol 126: 375–381.

Peters PJ, Neefjes JJ, Oorschot V, Ploegh HL, Geuze HJ (1991) Segregation of MHC class II molecules from MHC class I molecules in the Golgi complex for transport to lysosomal compartments. Nature 349: 669–676.

Phillips RE, Rowland-Jones S, Nixon DF, Gotch FM, Edwards JP, Ogunlesi AO, Elvin JG, Rothbard JA, Bangham CRM, Rizza CE, McMichael AJ (1991) Human immunodeficiency virus genetic variation that can escape cytotoxic T cell recognition. Nature 354: 453–459.

Piatigorsky J, Horwitz J (1996) Characterization and enzyme activity of argininosuccinate lyase/δ-crystallin of the embryonic duck lens. Biochim Biophys Acta 1295: 158–164.

Piatigorsky J, Wistow G (1991) The recruitment of crystallins: new functions precede gene duplication. Science 252: 1078–1079.

Potts WK, Manning CJ, Wakeland EK (1991) Mating patterns in seminatural populations of mice influenced ny MHC genotype. Nature 352: 619–621.

Price DA, Goulder PJR, Klenerman P, Sewell AK, Easterbook PJ, Troop M, Bangham CRM, Phillips RE (1997) Positive selection of HIV-1 cytotoxic T lymphocyte escape variants during primary infection. Proc Natl Acad Sci USA 94: 1890–1895.

Provine W (1991) Mechanisms of speciation: a review. Pp. 145–159 in Osawa S, Honjo Y, eds., *Evolution of Life: Fossils, Molecules, and Culture*. Springer-Verlag, Tokyo.

Quakyi IA, Currier J, Fell A, Taylor DW, Roberts T, Houghten RA, England RD, Berzofsky JA, Miller LH, Good MF (1994) Analysis of human T cell clones specific for conserved peptide sequences within malaria proteins: paucity of clones responsive to intact parasites. J Immunol 153: 2082–2092.

Quine WVO (1960) *Word and Object*. MIT Press, Cambridge, MA.

Rada C, Lorenzi R, Powis SJ, van den Bogaerde J, Parham P, Howard JC (1990) Concerted evolution of class I genes in the major histocompatibility complex of murine rodents. Proc Natl Acad Sci USA 87: 2167–2171.

Rammensee H-G, Friede T, Stevanovic S (1995) MHC ligands and peptide motifs: first listing. Immunogenetics 41: 178–228.

Reeder JC, Brown GV (1996) Antigenic variation and immune evasion of *Plasmodium falciparum* malaria. Immunol Cell Biol 74: 546–554.

Rehermann B, Pasquinelli C, Mosier SM, Chisari FV (1995) Hepatitis B virus (HBV) sequence variation in cytotoxic T lymphocyte epitopes is not common in patients with chronic HBV infection. J Clin Invest 96: 1527–1534.

Rich SM, Licht MC, Hudson RR, Ayala FJ (1998) Malaria's Eve: evidence of a recent population bottleneck throughout the world populations of *Plasmodium falciparum*. Proc Natl Acad Sci USA 95: 4425–4430.

Richman AD, Uyenoyama MK, Kohn JR (1996) S-allele diversity in a natural population of *Physalis crassifolia* (ground cherry) assessed by RT-PCR. Heredity 76: 497–505.

Rivett AJ (1993) Proteasomes: multicatalytic proteinase complexes. Biochem J 291: 1–10.

Rosenberg HF, Dyer KD, Tiffany HL, Gonzalez M (1995) Rapid evolution of a unique family of primate ribonuclease genes. Nat Genet 10: 219–223.

Rothenfluh HS, Blanden RV, Steele EJ (1995) Evolution of V genes: DNA sequence structure of functional germline genes and pseudogenes. Immunogenetics 42: 159–171.

Rusche JR, Javaherian K, McDanal C, Petro J, Lynn DL, Grimaila R, Langlois A, Gallo RC, Arthur LO, Fischinger PJ, Bolognesi DP, Putney SD, Matthews TJ (1988) Antibodies that inhibit fusion of human immunodeficiency virus-infected cells bind a 24-amino acid sequence of the viral envelope, gp120. Proc Natl Acad Sci USA 85: 3198–3202.

Rzepcyk CM, Ramasamy R, Mutch DA, Ho P C-L, Battistutta D, Anderson KL, Parkinson D, Doran TJ, Honeyman M (1989) Analysis of human T cell response to *Plasmodium falciparum* merozoite surface antigens. Eur J Immunol 19: 1797–1802.

Rzhetsky A, Nei M (1992a) A simple method for estimating and testing minimum-evolution trees. Mol Biol Evol 9: 945–967.

Rzhetsky A, Nei M (1992b) Statistical properties of the ordinary least-squares, generalized least-squares, and minimum-evolution methods of phylogenetic inference. J Mol Evol 35: 367–375.

Sabatini LM, Ota T, Azen EA (1993) Nucleotide sequence analysis of the human salivary protein genes *HIS1* and *HIS2* and evoloution of the *STATH/HIS* gene family. Mol Biol Evol 10: 497–511.

Sadler JR, Tecklenburg M (1981) Cloning and characterization of the natural lactose operon. Gene 13: 13–23.

Saitou N, Imanishi M (1989) Relative efficiencies of the Fitch-Margoliash, maximum parsimony, maximum likelihood, minimum evolution, and neighbor-joining methods of phylogenetic tree construction in obtaining the correct tree. Mol Biol Evol 6: 514–524.

Saitou N, Nei M (1987) The neighbor-joining method: a new method for reconstructing phylogenetic trees. Mol Biol Evol 4: 406–425.

Saper MA, Bjorkman PJ, Wiley DC (1991) Refined structure of the human histocompatibility antigen HLA-A2 at 2.6 A resolution. J Mol Biol 219: 277–319.

Saxena R, Brown LG, Hawkins T, Alagappan RK, Skaletsky H, Reeve MP, Reijo R, Rozen S, Dinulos MB, Disteche CM, Page DC (1996) The *DAZ* gene cluster on the human Y chromosome arose from an autosomal gene that was transposed, repeatedly amplified and pruned. Nat Genet 14: 292–299.

Schaeffer SW, Miller EL (1992) Molecular population genetics of an electrophoretically monomorphic protein in the alcohol dehydrogenase region of *Drosophila pseudoobscura*. Genetics 132: 163–198.

Schlesinger MJ (1990) Heat shock proteins. J Biol Chem 265: 12111–12114.

Schmidt-Nielsen K (1979) *Animal Physiology*, 2nd ed. Cambridge University Press, Cambridge.

Seibert SA, Howell CY, Hughes MK, Hughes AL (1995) Natural selection on the *gag*, *pol*, and *env* genes of human immunodeficiency virus 1 (HIV-1). Mol Biol Evol 12: 803–813.

Sessions SK, Larson A (1987) Developmental correlates of genome size in plethodontid salamanders and their implications for genome evolution. Evolution 41: 1239–1251.

Sharp PM (1991) Determinants of DNA sequence divergence between *Escherichia coli* and *Salmonella typhimurium*: codon usage, map position, and concerted evolution. J Mol Evol 33: 23–33.

Sharp PM, Li W-H (1986) An evolutionary perspective on synonymous codon usage in unicellular organisms. J Mol Evol 24: 28–38.

Sharp PM, Tuohy TMF, Mosurski KR (1986) Codon usage in yeast: cluster analysis clearly differentiates highly and lowly expressed genes. Nucleic Acids Res 14: 5125–5143.

Shaw A, McRee DE, Vacquier VD, Stout CD (1993) The crystal structure of abalone sperm lysin. Science 262: 1864–1867.

Shaw A, Fortes PGA, Stout CD, Vacquier VD (1995) Crystal structure and subunit dynamics of the abalone sperm lysin dimer: egg envelopes dissociate dimers, the monomer is the active species. J Cell Biol 130: 1117–1125.

She JX, Boehme SA, Wang TW, Bonhomme F, Wakeland EK (1991) Amplification of major histocompatibility complex class II gene diversity by intraexonic recombination. Proc Natl Acad Sci USA 88: 453–457.

Sherwood SW, Patton JL (1982) Genome evolution in pocket gophers (genus *Thomomys*). Chromosoma 85: 163–179.

Shields DC, Sharp PM, Higgins DG, Wright F (1988) "Silent" sites in *Drosophila* genes are not neutral: evidence of selection among synonymous codons. Mol Biol Evol 5: 704–716.

Sibley CG, Ahlquist J (1990) *Phylogeny and Classification of Birds*. Yale University Press, New Haven.

Sidow A (1996) Gen(ome)e duplications in the evolution of early vertebrates. Curr Opin Genet Dev 6: 715–722.

Silver ML, Guo H-C, Strominger JL, Wiley DC (1992) Atomic structure of a human MHC molecule presenting an influenza virus peptide. Nature 360: 367–369.

Simmonds P, Balfe P, Ludlam CA, Bishop JO, Leigh Brown AJ (1990) Analysis of sequence diversity in hypervariable regions of the external glycoprotein of human immunodeficiency virus type 1. J Virol 64: 5840–5850.

Sitnikova T, Nei M (1998) Evolution of immunoglobulin kappa chain variable region genes in vertebrates. Mol Biol Evol 15: 50–60.

Slifman NR, Loegering DA, McKean DJ, Gleich GJ (1986) Ribonuclease activity associated with human eosinophil-derived neurotoxin and eosinophil cationic protein. J Immunol 137: 2913–2917.

Slightom JL, Bock JH, Tagle DA, Gumucio DL, Goodman M, Stojanovic N, Jackson JD, Miller W, Hardison R (1997) The complete sequences of the galago and rabbit b-globin locus control region: extended sequence and functional conservation outside the cores of DNase hypersensitive sites. Genomics 39: 90–94.

Smith AT (1993) The natural history of inbreeding and outbreeding in small mammals.

Pp. 329–351 in Thornhill NW, ed., *The Natural History of Inbreeding and Outbreeding*. University of Chicago Press, Chicago.

Sneath PHA (1961) Recent developments in theoretical and quantitative taxonomy. Syst Zool 10: 118–139.

Sneath PHA, Sokal RR (1973) *Numerical Taxonomy*. W. H. Freeman, San Francisco.

Sokal RR, Rohlf FJ (1981) *Biometry*, 2nd ed. W. H. Freeman, San Francisco.

Soker S, Takashima S, Miao HQ, Neufeld G, Klagsbrun M (1998) Neuropilin-1 is expressed by endothelial and tumor cells as an isoform-specific receptor for vascular endothelial growth factor. Cell 92: 735–745.

Stephan W, Wiehe THE, Lenz WW (1992) The effect of strongly selected substitutions on neutral polymorphism: analytical results based on diffusion theory. Theor Pop Biol 41: 237–254.

Stephens JC (1985) Statistical methods of DNA sequence analysis: Detection of intragenic recombination or gene conversion. Mol Biol Evol 2: 539–556.

Stern JT Jr (1970) The meaning of "adaptation" and its relation to the phenomenon of natural selection. Evol Biol 4: 39–66.

Stevenson MA, Calderwood SK (1990) Members of the 70-kilodalton heat shock protein family contain a highly conserved calmodulin-binding domain. Mol Cell Biol 10: 1234–1238.

Stewart C-B, Schilling JW, Wilson AC (1987) Adaptive evolution in the stomach lysozymes of foregut fermenters. Nature 330: 401–404.

Strobeck C (1983) Expected linkage disequilibrium for a neutral locus linked to a chromosomal arrangement. Genetics 103: 545–555.

Swofford DL (1993) *PAUP: Phylogenetic Analysis Using Parsimony* Illinois Natural History Survey, Champaign.

Szarski H (1976) Cell size and nuclear DNA content in vertebrates. Int Rev Cytol 44: 93–209.

Szarski H (1983) Cell size and the concept of wasteful and frugal evolutionary strategies. J Theor Biol 105: 201–209.

Tai P-C, Banik D, Lin G-I, Pai S, Pai K, Lin M-H, Yuoh G, Che S, Hsu SH, Chen T-C, Kuo T-T, Lee C-S, Yang C-S, Shih C (1997) Novel and frequent mutations of hepatitis B virus coincide with a major histocompatibility complex class I-restricted T-cell epitope of the surface antigen. J Virol 71: 4852–4856.

Tajima F (1983) Evolutionary relationship of DNA sequences in finite populations. Genetics 105: 437–460.

Takahata N, Maruyama T (1979) Polymorphism and loss of duplicate gene expression: a theoretical study with application to tetraploid fish. Proc Natl Acad Sci USA 76: 4521–4525.

Takahata N, Nei M (1985) Gene genealogy and variance of interpopulational nucleotide differences. Genetics 110: 325–344.

Takahata N, Nei M (1990) Allelic genealogy under overdominant and frequency-dependent selection and polymorphism of major histocompatibility complex loci. Genetics 124: 967–978.

Tanabe K, Mackay M, Goman M, Scaife JB (1987) Allelic dimorphism in a surface antigen of the malaria parasite *Plasmodium falciparum*. J Mol Biol 195: 273–287.

Tanaka T, Nei M (1989) Positive Darwinian selection observed at the variable-region genes of immunoglobulin. Mol Biol Evol 6: 447–459.

Taylor M, Feyerisen R (1996) Molecular biology and evolution of resistance to toxicants. Mol Biol Evol 13: 719–734.

Thomas L (1974) Biological signals for self-identification. Pp. 239–247 in Brent L, Holborrow J, eds., *Progress in Immunology II*. North-Holland, Amsterdam.

Thornhill R, Alcock J (1983) *The Evolution of Insect Mating Systems* Harvard University Press, Cambridge, MA.

Thursz MR, Kwiatkowski D, Allsopp CEM, Greenwood BM, Thomas HC, Hill AVS (1995) Association between an MHC class II allele and clearance of hepatitis B virus in the Gambia. N Engl J Med 332: 1065–1069.

Thursz MR, Thomas HC, Greenwood BM, Hill AVS (1997) Heterozygote advantage for HLA class-II type in hepatitis B virus. Nat Genet 17: 11–12.

Tibayrenc M, Kjellberg F, Arnaud J, Oury B, Breniere SF, Darde M-L, Ayala FJ (1991) Are eukaryoyic microorganisms clonal or sexual? A population genetics vantage. Proc Natl Acad Sci USA 88: 5129–5133.

Tiersch TR, Mumme RL (1993) An evaluation of the use of flow cytometry to identify sex in the Florida Scrub Jay. J Field Ornithol 64: 18–26.

Tiersch TR, Wachtel SS (1991) On the evolution of genome size of birds. J. Hered 82: 363–368.

Tiersch TR, Simco BA, Davis KB, Chandler RW, Wachtel SS, Carmichael GJ (1990) Stability of genome size among stocks of the channel catfish. Aquaculture 87: 15–22.

Tiong SY, Whittle JR, Gribbin MC (1987) Chromosomal continuity in the abdominal region of the bithorax complex of *Drosophila* is not essential for its contribution to metameric identify. Development 101: 135–142.

Tomarev SI, Duncan MK, Roth HJ, Cvekl A, Piatigorsky J (1994) Convergent evolution of crystallin gene regulation in squid and chicken: the AP-1/ARE connection. J Mol Evol 39: 134–143.

Tonegawa S (1983) Somatic generation of antibody diversity. Nature 302: 575–581.

Trivers RL (1972) Parental investment and sexual selection. Pp. 136–179 in Campbell B, ed., *Sexual Selection and the Descent of Man: 1871–1971*. Aldine, Chicago.

Vacquier VD, Swanson WJ, Lee Y-H (1997) Positive Darwinian selection on two homologous fertilization proteins: what is the selective pressure driving their divergence? J Mol Evol 44: S15–S22.

Van de Velde H, von Hoegen I, Luo W, Parnes JR, Thielmans K (1991) The B-cell surface protein CD72/Lyb-2 is the ligand for CD5. Nature 351: 662–665.

Van Tyne J, Berger AJ (1959). *Fundamentals of Ornithology*. Wiley, New York.

Voit R, Feldmaier-Fuchs G (1990) Arthropod hemocyanins: molecular cloning and sequencing of cDNAs encoding the tarantula hemocyanin submits a and e. J Biol Chem 265: 19447–19452.

Wachtel SS, Tiersch TR (1993) Variations in genome mass. Comp Biochem Physiol 104B: 207–213.

Walsh JB (1995) How often do duplicated genes evolve new functions? Genetics 139: 421–428.

Wang JH, Yan YW, Garrett TP, Liu JH, Rodgers DW, Garlick RL, Tarr GE, Husain Y, Reinherz EL, Harrison SC (1990) Atomic structure of a fragment of human CD4 containing two immunoglobulin-like domains. Nature 348: 411–418.

Watkins DI, Chen ZW, Hughes AL, Evans MG, Tedder TF, Letvin NL (1990) Evolution of the MHC class I genes of a New World primate from ancestral homologues of human non-classical genes. Nature 346: 60–63.

Watkins DI, Garber TL, Chen ZW, Toukatly G, Hughes AL, Letvin NL (1991) Unusually limited nucleotide sequence variation of the expressed major histocompatibility complex class I genes of a New World primate species (*Saguinus oedipus*). Immunogenetics 33: 79–89.

Watkins DI, McAdam SN, Liu X, Strang CR, Milford EL, Levine CG, Garber TL, Dogon AL, Lord CI, Ghim SH, Troup GM, Hughes AL, Letvin NL (1992) New recombinant *HLA-B* alleles in a tribe of South American Amerindians indicate rapid evolution of MHC class I loci. Nature 357: 329–333.

Watt WB (1992) Eggs, enzymes, and evolution—natural genetic variants change insect fecundity. Proc Natl Acad Sci USA 89: 10608–10612.

Weber JL (1988) Molecular biology of malaria parasites. Exp Parasitol 66: 143–170.

Wegman TG (1984) Foetal protection against abortion: is it immunosuppression or immunostimulation? Ann Immunol 135B: 307–312.

Weiner A, Erickson AL, Kansopon J, Crawford K, Muchmore E, Hughes AL, Houghton M, Walker CM (1995) Persistent hepatitis C virus infection in a chimpanzee is associated with emergence of a cytotoxic T lymphocyte escape variant. Proc Natl Acad Sci USA 92: 2755–2759.

White KL, Snyder HL, Krzych V (1996) MHC class I-dependent presentation of exoerythrocytic antigens to CD8+ lymphocytes is required for protective immunity against *Plasmodium berghei*. J Immunol 156: 3374–3381.

Whittam TS, Nei M (1991) Neutral mutation hypothesis test. Nature 354: 115–116.

Wiley DC, Wilson IA, Skehel JJ (1981) Structural identification of the antibody-binding sites of Hong Kong influenza haemagglutinin and their involvement in antigenic variation. Nature 289: 373–378.

Williams GC (1966) *Adaptation and Natural Selection*. Princeton University Press, Princeton, NJ.

Williams GC (1985) A defense of reductionism in evolutionary biology. Oxford Surveys Evol Biol 2: 1–27.

Williams SG, Greenwood AJ, Jones CW (1992) Molecular analysis of the *lac* operon encoding the binding-protein-dependent lactose transport system and β-galactosidase in *Agrobacterium radiobacter*. Mol Microbiol 6: 1755–1768.

Willott E, Wang X-Y, Wells MA (1989) cDNA and gene sequence of *Manduca sexta* arylphorin, an aromatic amino acid-rich larval serum protein. J Biol Chem 26: 19052–19059.

Wistow GJ, Piatigorsky J (1990) Gene conversion and splice-site slippage in the argininosuccinate lyases/δ-crystallins of the duck lens: members of an enzyme superfamily. Gene 96: 263–270.

Wolfe KH, Sharp PM (1993) Mammalian gene evolution: nucleotide sequence divergence between mouse and rat. J Mol Evol 37: 441–456.

Wolfe KH, Sharp PM, Li W-H (1989) Mutation rates differ among regions of the mammalian genome. Nature 337: 283–285.

Wray GA, Levinton JS, Shapiro LH (1996) Molecular evidence for deep Precambian divergences among metazoan phyla. Science 274: 568–573.

Wright S (1931) Evolution in Mendelian populations. Genetics 16: 97–159.

Wright S (1939) The distribution of self-sterility alleles in populations. Genetics 24: 538–552.

Wright S (1949) Adaptation and selection. Pp. 365–389 in Jepsen GL, Simpson GG, Mayr E, eds., *Genetics, Paleontology, and Evolution*. Princeton University Press, Princeton, NJ.

Wyles JS, Kunkel JG, Wilson AC (1983) Birds, behavior, and anatomical evolution. Proc Natl Acad Sci USA 80: 4394–4397.

Yamaguchi M, Yamazaki K, Boyse EA (1978) Mating preference tests with recombinant congenic strain BALB.HTG. Immunogenetics 6: 261–264.

Yamazaki K, Boyse EA, Mike V, Thaler HT, Mathieson BJ, Abbott J, Boyse J, Zayas ZA,

Thomas L (1976) Control of mating preferences in mice by genes in the major histocompatibility complex. J Exp Med 144: 1324–1335.
Yamazaki K, Yamaguchi M, Andrews PW, Peake B, Boyse EA (1978) Mating preferences of F2 segregants of crosses between MHC-congenic mouse strains. Immunogenetics 6: 253–259.
Yamazaki K, Yamaguchi M, Baranoski L, Bard J, Boyse EA, Thomas L (1979) Recognition among mice: evidence from the use of a Y-maze differentially scented by congenic mice of different major histocompatibility types. J Exp Med 150: 755–760.
Yamazaki K, Beauchamp GK, Bard J, Boyse EA (1990) Single MHC gene mutations alter urine odour constitution in mice. Pp. 255–259 in MacDonald DW, Muler-Schwarze D, Natynczuk SE, eds., *Chemical Signals in Vertebrates* 5. Oxford University Press, Oxford.
Yang Z, Kumar S, Nei M (1995) A new method of inference of ancestral nucleotide and amino acid sequences. Genetics 141: 1641–1650.
Yano K-I, Fukasawa T (1997) Galactose-dependent reversible interaction of Gal3p with Gal80p in the induction pathway of Gal4p-activated genes of *Saccharomyces cerevisiae*. Proc Natl Acad Sci USA 94: 1721–1726.
Yeager M, Hughes AL (1996) Interallelic recombination has not played a major role in the history of the *HLA-C* locus. Immunogenetics 44: 128–133.
Yeager M, Kumar S, Hughes AL (1997) Sequence convergence in the peptide-binding region of primate and rodent MHC class Ib molecules. Mol Biol Evol 14: 1035–1041.
Yokoyama S, Hetherington LE (1982) The expected number of self-incompatibility alleles in finite plant populations. Heredity 48: 299–303.
Yoshida R, Nussenzweig RS, Potocnjar P, Nussenzweig V, Aikawa M (1980) Hybridoma produces protective antibodies directed against the sporozoite stages of malaria prasite. Science 207: 71–73.
Zangenberg G, Muang M-M, Arnheim N, Erlich H (1995) New HLA-DPB1 alleles generated ny interallelic gene conversion detected by analysis of sperm. Nat Genet 10: 407–414.
Zenke FT, Engels R, Vollenbroich V, Meyer J, Hollenberg CP, Breunig (1996) Activation of Gal4p by galactose-dependent interaction of galactokinase and Gal80p. Science 272: 1662–1665.
Zevering Y, Khamboonruang C, Good MF (1994) Effect of polymorphism of sporozoite antigens on T-cell activation. Res Immunol 145: 469–476.
Zhang J, Kumar S (1997) Detection of convergent and parallel evolution at the amino acid sequence level. Mol Biol Evol 14: 527–536.
Zhang J, Nei M (1996) Evolution of antennapedia-class homeobox genes. Genetics 142: 295–303.
Zhang J, Kumar S, Nei M (1997) Small-sample tests of episodic adaptive evolution: a case study of primate lysozymes. Mol Biol Evol 14: 1335–1338.
Zhang J, Rosenberg HF, Nei M (1998) Positive Darwinian selection after gene duplication in primate ribonuclease genes. Proc Natl Acad Sci USA 95: 3708–3713.
Zinkernagel RM, Doherty PC (1974) Immunological surveillance against altered self components by sensitized T lymphocytes in lymphocytic choriomeningitis. Nature 251: 547–548.

INDEX